Springer Series in Reliability Engineering

Series Editor

Professor Hoang Pham
Department of Industrial Engineering
Rutgers
The State University of New Jersey
96 Frelinghuysen Road
Piscataway, NJ 08854-8018
USA

Other titles in this series

Hongzhou Wang and Hoang Pham

Reliability and Optimal Maintenance

With 27 Figures

 Springer

Hongzhou Wang, PhD
Lucent Technologies
Whippany, New Jersey
USA

Hoang Pham, PhD
Department of Industrial Engineering
Rutgers
The State University of New Jersey
Piscataway, New Jersey
USA

British Library Cataloguing in Publication Data
A catalogue record for this book is available from the British Library

Library of Congress Control Number: 2006926891

Springer Series in Reliability Engineering series ISSN 1614-7839
ISBN-10: 1-84628-324-8   e-ISBN 1-84628-325-6   Printed on acid-free paper
ISBN-13: 978-1-84628-324-6

© Springer-Verlag London Limited 2006

Printed in Germany

9 8 7 6 5 4 3 2 1

Springer Science+Business Media
springer.com

To Connie and Grace.

– Hongzhou Wang

To Michelle, Hoang Jr., and David.

– Hoang Pham

Preface

This book aims to present a state-of-the-art survey of theories and methods of reliability, maintenance, and warranty with emphasis on multi-unit systems, and to reflect current hot topics: imperfect maintenance, economic dependence, opportunistic maintenance, quasi-renewal processes, warranty with maintenance and economic dependency, and software testing and maintenance. This book is distinct from others because it consists mainly of research work published on technical journals and conferences in recent years by us and our co-authors.

Maintenance involves preventive and unplanned actions carried out to retain a system at or restore it to an acceptable operating condition. Optimal maintenance policies aim to provide optimum system reliability and safety performance at the lowest possible maintenance costs. Proper maintenance techniques have been emphasized in recent years due to increased safety and reliability requirements of systems, increased complexity, and rising costs of material and labor. For some systems, such as aircraft, submarines, and nuclear power stations, it is extremely important to avoid failure during actual operation because it is dangerous and disastrous.

Special features of our book include: (a) Imperfect maintenance. Imperfect maintenance has being receiving a great deal of attention in reliability and maintenance literature. In fact, its study indicates a significant breakthrough in reliability and maintenance theory; (b) Quasi-renewal processes. Quasi-renewal processes, including renewal processes as a special case, have been proven to be an effective tool to model hardware imperfect maintenance and software reliability growth; (c) Economic dependence and opportunistic maintenance. For maintenance of multi-component systems, economic dependency is one of the major concerns. Due to it, preventive maintenance (PM) on non-failed but deteriorated components may be carried out when corrective maintenance (CM) activities are performed since PM with CM can be executed without substantial additional expense. Accordingly, 'opportunistic' maintenance is resulted in; (d) Correlated failure and repair; (e) Emphasis on multicomponent systems; (f) Combined criteria on maintenance optimization. Usual criteria on maintenance optimization are based on maintenance cost measures only. In this book, optimization criteria are based on both cost and reliability indices; (g) Multiple degraded systems and inspection-maintenance; (h) Software reliability and maintenance models based on quasi-renewal processes; (i) Warranty cost models with imperfect maintenace, dependence, and emphasis on multi-component systems; (j) Monte Carlo reliability simulation techniques and issues.

This book is a valuable resource for understanding the latest developments in reliability, maintenance, inspection, warranty, software reliability models, and Monte Carlo reliability simulation. Postgraduates, researchers, and practitioners in reliability engineering, maintenance engineering, operations research, industrial engineering, systems engineering, management science, mechanical engineering, and statistics will find this book a state-of-the-art survey of the field. This book can serve as a textbook for graduate students, and a reference book for researchers and practioners.

Chapter 1 provides an introduction to recent hot topics in reliability and maintenance engineering, and sketch the framework of this book. Chapter 2 surveys imperfect repair and dependence research in detail to summarize significant approaches to model imperfect maintenance and dependence while Chapter 3 overviews various maintenance policies in the literature and practice, such as age-dependent PM policy, and repair limit policy.

In Chapter 4, we introduce a new modeling tool for imperfect maintenance: a quasi-renewal process which includes the ordinary renewal process as a special case and model imperfect maintenance of one-unit systems using the quasi-renewal process. Eleven imperfect maintenance models based on the quasi-renewal process are discussed in this chapter.

In practice, many systems are series systems or can be simplified into series systems. Chapter 5 investigates reliability and maintenance costs of a series system with n components subject to imperfect repair, and correlated failure and repair, and looks into its optimal maintenance. Some important properties of reliability and maintenance cost of the series system are discussed. Imperfect repairs are modeled through increasing and decreasing quasi-renewal processes.

Opportunistic maintenance of a system with $(n+1)$ subsystems and economic dependence among them is discussed in Chapter 6, in which whenever a subsystem fails its repair is combined with PM of the functioning one having increasing failure rate if it reaches some age. Two different imperfect modeling methods are used in this chapter.

In Chapter 7 we look into preparedness maintenance policies for a system with $n+1$ subsystems, economic dependency and imperfect maintenance. In this preparedness maintenance policy, the system is placed in storage and is called on to perform a given task only if a specific but unpredictable emergency occurs. Some maintenance actions resulting in optimal system "preparedness for field use" may be taken while the system is in storage.

Chapter 8 presents three new opportunistic maintenance models for a k-out-of-n:G system with economic dependency and imperfect maintenance. The results, including at least 13 existing maintenance models as special cases, generalize and unify some previous work.

Chapter 9 studies multi-state degraded systems subject to multiple competing failure processes including two independent degradation processes and random shocks. We first model system reliability and then discuss optimal condition-based inspection-maintenance. A quasi-renewal process is employed to establish the inter-inspection sequence.

In current highly competitive markets, warranty policies become more and more complex. Chapter 10 discusses warranty cost models of repairable complex systems from the manufacturers' point of view under three types of existing warranty policies: free repair warranty (FRPW), free replacement warranty (FRW), and pro-rata warranty (PRW), and two new warranty policies: renewable full service warranty (RFSW) and repair-limit risk-free warranty (RLRFW). Imperfect or minimal repair is considered. Monte Carlo simulation techniques and a new modeling tool – *a truncated quasi-renewal process* – will be used. The focus is on multi-component systems.

Chapter 11 models software reliability and testing costs using the quasi-renewal process, and discusses optimal software testing and release policies. Several software reliability and cost models are presented. Optimum testing policies incorporating both reliability and cost measures are investigated.

To obtain the optimal maintenance policy for a complex system, it is necessary to determine system availability or MTBF. However, there are some difficulties in evaluating complex large-scale system reliability and availability given confidence levels using classical statistics. In Chapter 12 Monte Carlo reliability, availability and MTBF simulation techniques will be examined together with variance reduction methods, simulation errors, *etc.*

We would like to express our appreciation to our wives, Xuehong Connie Wang and Michelle Pham, and to our families for their patience, understanding, and assistance during the preparation of this book. The constructive comments, encouragement, and support of our colleagues are very much appreciated. We are indebted to Kate Brown, Anthony Doyle and the Springer staff for their editorial work.

Hongzhou Wang
Bell Laboratories
Whippany, New Jersey

Hoang Pham
Rutgers University
Piscataway, New Jersey

November 2005

Contents

1

Introduction

Maintenance involves preventive (planned) and unplanned actions carried out to retain a system at or restore it to an acceptable operating condition. Optimal maintenance policies aim to provide optimum system reliability and safety performance at the lowest possible maintenance costs. Proper maintenance techniques have been emphasized due to increased safety and reliability requirements of systems, increased complexity, and rising costs of material and labor (Sheriff and Smith 1981). For some systems, such as aircraft, submarines, military systems, aerospace systems, it is extremely important to avoid failure during actual operation because it is dangerous and disastrous. One important research area in reliability engineering is the study of various maintenance policies in order to improve system reliability, to prevent the occurrence of system failure, and to reduce maintenance costs (Pham and Wang 1996).

In the past several decades, maintenance, replacement and inspection problems have been extensively studied. The literature in this area is vast, making it impossible to give a short overview of the subject. There are to date many models for reliability, maintenance, replacement and inspection, and recent research has attempted to unify some of them. McCall (1963), Barlow and Proschan (1965), Pieskalla and Voelker (1976), Osaki and Nakagawa (1976), Sherif and Smith (1981), Jardine and Buzacott (1985), Valdez-Flores and Feldman (1989), Cho and Parlar (1991), Jensen (1995), Dekker (1996), Pham and Wang (1996), Dekker *et al.* (1997), and Wang (2002) survey and summarize the research and practice in this area in different ways. As modern systems grow in complexity, so do reliability and maintenance challenges. This book aims to present recent research on theories and methods in reliability and optimal maintenance in a realistic way, and the focus is on the emerging areas: imperfect maintenance, dependence, correlates failure and repair. A general introduction to each of them will be given in the following subsections.

Most equipment offers some level of warranty to gain some advantages in the highly competitive markets. The warranty assures the buyer that a faulty item will either be repaired or replaced at no cost or at reduced cost. Some buyers may infer that a product with a relatively long warranty period is a more reliable and longer-lasting product than one with a shorter warranty period. Warranty cost could be a significant percentage of overall product cost. So far, many warranty cost models and polices have been proposed and practiced as summarized in Brennan (1994), Blischke and Murthy (1994, 1996), Sahin and Polatogu (1998), and Bai and Pham (2006b). A simple but relatively complete taxonomy of warranty policies can be found in Blischke and Murthy (1993), and Bai and Pham (2006b). Today maintenance may be considered during warranty, and both consumers and manufacturers may benefit from proper maintenance during warranty. In addition, economic dependence may exist in some multi-component systems and maintenance may be imperfect. These issues in warranty research will be further stated in Section 1.3.

Many modern systems contain both hardware and software, and software problems may account for half of system failures for some systems such as telecommunication systems. Software reliability, testing and maintenance have received much attention in recent years, and some significant work is summarized in Musa *et al.* (1987), Lyu (1996), and Pham (2000). In addition to hardware reliability and maintenance, reliability and optimal testing and debugging for software will also be discussed in this book.

To obtain the optimal maintenance policy for a complex system, we may first need to determine system reliability, availability, or MTBF. However, there are some difficulties in evaluating complex large-scale system reliability and availability using classical statistics: for example, the system reliability structure may be complicated, or subsystems may follow different failure distributions. In this book, Monte Carlo reliability, availability and MTBF simulation algorithms will be discussed together with variance reduction methods, simulation errors, *etc.*

1.1 Imperfect Maintenance

Maintenance can be classified by two major categories: corrective and preventive. Corrective maintenance (CM) is the maintenance that occurs when the system fails. Some researchers refer to CM as repair and we will use them alternatively throughout this study. According to MIL-STD-721B, CM means all actions performed as a result of failure, to restore an item to a specified condition. Obviously, CM is performed at unpredictable time points because an item's failure time is not known. CM is typically carried out in three steps: (1) Diagnosis of the problem, (2) Repair and/or replacement of faulty component(s), and (3) Verification of the repair action. Preventive maintenance (PM) is the maintenance that occurs when the system is operating. According to MIL-STD-721B, PM means all actions performed in an attempt to retain an item in specified condition by providing systematic inspection, detection, and prevention of incipient failures. Maintenance can also be classified according to the *degree* to which the operating condition of an item is restored by maintenance in the following way:

a) **Perfect repair** or **perfect maintenance**: maintenance actions which restore a system operating condition to 'as good as new'. That is, upon a perfect maintenance, a system has the same lifetime distribution and failure rate function as a new one. Complete overhaul of an engine with a broken connecting rod is an example of perfect repair. Generally, replacement of a failed system by a new one is a perfect repair.

b) **Minimal repair**[1] or minimal maintenance: maintenance actions which restore a system to the same failure rate as it had when it failed. Minimal repair was first studied by Barlow and Hunter (1960). The system operating state after the minimal repair is often called 'as bad as old' in the literature. Changing a flat tire on a car is an example of minimal repair because the overall failure rate of the car is essentially unchanged. The mathematical definition of minimal repair is given in the Appendix to Chapter 4.

c) **Imperfect repair** or **imperfect maintenance**: maintenance actions which make a system not 'as good as new' but younger. Usually, it is assumed that imperfect maintenance restores the system operating state to somewhere between 'as good as new' and 'as bad as old'. Clearly, imperfect repair (maintenance) is a general repair (maintenance) which can include two extreme cases: minimal and perfect repair (maintenance). Engine tune-up is an example of imperfect maintenance. Chapter 2 will discuss imperfect maintenance in detail.

d) **Worse repair** or **worse maintenance**: maintenance actions which undeliberately make the system failure rate or actual age increase but the system does not break down. Thus, upon worse repair a system's operating condition becomes worse than that just prior to its failure.

e) **Worst repair** or **worst maintenance**: maintenance actions which undeliberately make the system fail or break down.

According to the above classification, we can say that a PM is a minimal, perfect, imperfect, worst or worse one. Similarly, a CM could be a minimal, perfect, imperfect, worst or worse CM. We will refer to imperfect CM and PM as imperfect maintenance later. The type and degree of maintenance used in practice depend on types of systems, their costs as well as reliability and safety requirements.

In the previous literature, most studies assume that the system after CM or PM is 'as good as new' (perfect maintenance) or 'as bad as old' (minimal maintenance). In practice, the perfect maintenance assumption may be plausible for systems with one component which is structurally simple. On the other hand, the minimal repair assumption seems reasonable for failure behavior of systems when one of its many, non-dominating components is replaced by a new one (Kijima 1989). However, many maintenance activities may not result in these two extreme situations but in a complicated intermediate one. For example, an engine may not be 'as good as new' or 'as bad as old' after tune-up, a type of PM. It usually becomes "younger" than at the time just prior to PM and enters some state between 'as good as new' and 'as bad as old'. Therefore, perfect maintenance and minimal maintenance are not practical in many actual instances and realistic imperfect maintenance should be modeled. In recent years, imperfect CM and PM have received more attention in reliability and maintenance literature. In fact, we can say

---

[1] Here we refer to physical minimal repair instead of statistical (black box) minimal repair; see Natvig (1990).

that imperfect maintenance study indicates a significant *breakthrough* in reliability and maintenance theory.

Helvik (1980) believes that imperfectness of maintenance is related to the skill of the maintenance personnel, the quality of the maintenance procedure, and the maintainability of the system. Obviously, maintenance expenditure and reliability requirements also have important effects on imperfectness of maintenance.

Brown and Proschan (1982) state some possible causes for imperfect, worse or worst maintenance due to the maintenance performer:

- Repairing the wrong part
- Only partially repairing the faulty part
- Repairing (partially or completely) the faulty part but damaging adjacent parts
- Incorrectly assessing the condition of the unit inspected
- Performing the maintenance action not when called for but at his / her convenience (the timing for maintenance is off the schedule)

Nakagawa (1987) suggests three reasons causing worse or worst maintenance:

- Hidden faults and failures which are not detected during maintenance
- Human errors such as wrong adjustments and further damage done during maintenance
- Replacement with faulty parts.

According to Brown and Proschan (1982), maintenance policies based on planned inspections are "periodic inspection", and "inspection interval dependent on age". By periodic inspections, a failed unit is identified (*e.g.*, spare battery, a fire detection device, *etc.*), or it is determined whether the unit is functioning or not. With aging of the unit, the inspection interval may be shortened. These inspection methods are subject to imperfect maintenance caused by randomness in the actual time of inspection in spite of the schedule, imperfect inspection, and cost structure. Therefore, realistic and valid maintenance models must incorporate random features of the inspection policy.

So far, only a small portion of literature concerning the stochastic behavior of repairable systems and their maintenance has considered imperfect maintenance, and most work on imperfect maintenance has been limited to the one-unit system. Kay (1976), and Chan and Downs (1978) have studied the worst PM; Nakagawa (1987) has investigated worst and worse CM and PM. In this book, imperfect maintenance will be one of the major concerns, especially for multi-unit systems.

1.2 Dependence

In recent years, there has existed an increasing interest in multicomponent maintenance models. Schouten (1996) states a good reason: the fact that the vast majority of the maintenance models were concerned with a single piece of equipment operating at a fixed environment was considered as an intrinsic barrier for applications.

Maintenance of a multicomponent system differs from that of a single-unit system because there exists dependence in multicomponent systems. One kind of dependence is the *economic dependence*. For example, due to economic dependence, PM to non-failed subsystems can be performed at a reduced additional cost while failed subsystems are being repaired. Another kind of dependence is *failure dependence*, or *correlated failures*. For example, the failure of one subsystem may affect one or more of the other functioning subsystems, and times to failures of different units are then statistically dependent (Nakagawa and Murthy 1993).

Economic dependency is common in most continuous operating systems. Examples of such systems include aircraft, ship, power plants, telecommunication systems, chemical processing facilities, and mass production lines. For this type of system, the cost of system unavailability (one-time shut-down) may be much higher than component maintenance costs. Therefore, there is often great potential cost savings by implementing an opportunistic maintenance policy (Huang and Okogbaa 1996).

Obviously, the joint maintenance of two or more subsystems tends to spend less cost and less time (economic dependency), and the failures of different subsystems in multicomponent system may not be independent (failure dependency). Thus, each subsystem may not be considered as a single-unit system individually and to apply the existing optimum maintenance models of a single-unit system to each of such subsystems may not lead to a global optimal maintenance policy for the system as a whole.

Imperfect maintenance also exists in the repairable multicomponent system. For example, an aircraft consists of many subsystems which are repairable. If one of its subsystems fails, it can be repaired by replacing some of its parts. Clearly, reliability measures of the repaired subsystem are improved after repair but it might not be as good as new (imperfect CM), and consequently the entire system will no longer function as well as a new one. On the other hand, some subsystems in this example may become worse or break down after repair, and accordingly the whole aircraft system may work worse or break down, that results in worse or worst maintenance.

Realistic imperfect maintenance associated with individual subsystems and, accordingly, systems should be modeled. According to Valdez-Flores and Feldman (1989), "Systems used in the production of goods and delivery of services constitute the vast majority of most industry's capital. These systems are subject to deterioration with usage and age. System deterioration is often reflected in higher production costs and lower product quality. To keep production costs down while maintaining good quality, PM is often performed on such deteriorating systems". Obviously, this kind of system is often composed of many subsystems whose maintenance is often imperfect or sometimes even worse. It is necessary to point out that considering the entire system as a single unit and applying a minimal repair model to it may not be plausible for large-scale systems, such as the above two systems. Such maintenance modeling may also be too rough for complex systems since economic and failure dependencies may exist. Besides, individual maintenance procedures are often scheduled for individual subsystems.

In practice, some subsystems are inspected and tested separately, and their reliability performances are also evaluated individually. Especially, lifetime distributions of all new subsystems may be known through reliability tests and statistical inference before they are put into field use for some systems. Thus, we can evaluate reliability measures and system maintenance costs for the whole system based on failure information, maintenance costs, and maintenance degrees of all subsystems. Therefore, we may say that a realistic method is to treat a system as one with many subsystems which are subject to imperfect maintenance respectively, economic dependence, and failure dependence; to model imperfect maintenance of the system through modeling imperfect maintenance of all subsystems and at the same time to model economic and failure dependence in the system in order to obtain global optimum maintenance policies for the system.

In summary, it would be realistic to consider both imperfect maintenance and dependence among subsystems when studying reliability measurements, maintenance costs and optimum PM policy of multicomponent systems. Economic dependence, in addition to imperfect maintenance, will be another major factor of interest in this book.

In the maintenance literature, a basic assumption tends to be the independence of the time to failure and time to repair. In practice, the repair time of a unit may depend on its time to failure, that is, time to failure and time to repair are not independent. Goel (1989) states that it is a common experience of system engineers that, in most systems, an early failure leads to a short repair time and vice versa. He uses the bivariate exponential distribution to model this class of dependencies. Later we will refer to this dependence as *correlated failures and repairs, i.e.,* time to failure and time to repair are correlated.

This class of dependencies differs from the failure dependence mentioned in Section 1.2: correlated failures and repairs indicate the dependence between the time to failure and time to repair of a unit but the failure dependence or the correlated failures indicates the dependence between the time to failure of one subsystem and that of other subsystems. Correlated failure and repair may exist in either a single-unit system or a multicomponent system but failure dependence can only exist in a multicomponent system.

It is worthwhile to mention that "dependence" will be a general term in this book and can mean every kind of dependency: economic dependence, failure dependence, or dependence of failure and repair time, *i.e.,* correlated failures and repairs.

1.3 Warranty, Dependence, Imperfect Maintenance

A traditional way to model warranty cost of multi-component systems is the black-box approach that does not utilize the information of system structure. In fact, system architecture information is very important in modeling warranty cost since economic dependence may be present. Chapter 10 will discuss the importance of system structure information for four types of systems: series, parallel, series-parallel and parallel-series.

The majority of warranty cost models for repairable products assume perfect repair or minimal repair. As pointed out in Section 1.1, imperfect repair is more realistic. Chapter 10 will discuss warranty cost modeling given repair is imperfect.

Maintenance may be incorporated in the warranty period. For example, if a warranted product failed, the failed component(s) or subsystem(s) that cause the system failure will be replaced; in addition, a PM action could be carried out to reduce the chance of future failure. Both consumers and manufacturers will benefit from this policy. Therefore, warranty policies with integrated PM actions may become more and more attractive. Note that PM could also be imperfect in practice if it is performed.

One of the primary questions to be answered in warranty analysis is how much a warranty program will cost. Due to the random nature of warranty cost, most warranty cost models would prefer to use the expected warranty cost (EWC) as the answer. In contrast to the EWC, the expected value of discounted warranty cost (DWC), which incorporates the value of time, may provide a better cost measure for warranties. This is because in general warranty cost can be treated as a random cash flow in the future. Warranty issuers do not have to spend all the money at the stage of the warranty planning. Instead, they can allocate it over the life cycle of warranted products. Another reason that one should consider the value of time is that for the purpose of determining warranty reserve, a fund can be set up specifically to meet future warranty claims. This book will mainly use DWC as warranty cost measures.

1.4 Criteria on Maintenance Optimization

The usual criteria on optimization of maintenance policies are based on maintenance cost measures only: expected maintenance costs per unit of time, total discounted costs, gain, *etc*. Hence, the optimal maintenance policies are the ones that minimize (maximize) a given cost (gain) criterion (Jensen 1996). A small portion of maintenance models has used reliability measures: availability, average up time, or average down time in optimization criteria. This book will consider both system maintenance cost measures and reliability measures to obtain global optimal maintenance policies. Section 1.5 will further explain this while Chapter 5 will demonstrate the necessity of considering system maintenance cost measures and reliability measures together through numerical examples.

1.5 Scope of this Book

1.5.1 General Methodologies

Repairable systems whose subsystems are subject to imperfect maintenance, economic dependence, correlated failure and repair, and failure dependence may be realistic in many applications, according to the previous sections. Therefore, for such systems a study of system reliability measures: availability, mean time between system failures (MTBSF), mean time between system repairs (MTBSR), of system maintenance cost measures, and of optimum PM policies would be necessary. The objectives of the book are to study the stochastic behavior of repairable multicomponent systems whose components are subject to imperfect maintenance, economic dependence, correlated failure and repair, and failure dependence, and to investigate the optimal system maintenance policies. In this book, to study the stochastic behavior of systems means mainly to:

a) Formulate and derive the system reliability measures: availability, MTBSF and MTBSR, *etc.*

b) Model and derive system maintenance cost per unit time, or cost rate.

c) Model and derive system warranty costs and their variance (for some system structures).

The optimal system maintenance policies mentioned above may be those which:

a) Minimize system maintenance cost rate.

b) Optimize the system reliability measures.

c) Minimize system maintenance cost rate while the system reliability requirements are satisfied.

d) Optimize the system reliability measures when the requirements for the system maintenance cost are satisfied.

Similarly, for software systems, we will first formulate and model their reliability and testing cost, and then discuss the optimal software testing policy which may:

a) Minimize software testing cost.

b) Optimize the software reliability measures.

c) Minimize software testing cost while the software reliability requirements are satisfied.

d) Optimize the software reliability measures when the requirements for the software testing cost are satisfied.

1.5.2 Directions

This book aims to discuss the stochastic behavior and optimal maintenance policies for typical reliability system architectures: single-unit systems, series systems, parallel systems, and k-out-of-n systems.

There exist many maintenance policies for one-unit hardware systems. This book uses the following practical maintenance policies for each subsystem or the entire system when applicable:

a) Age-dependent PM policy. We consider the situation in which either CM or PM or both are imperfect.

b) Periodic PM policy. We consider the cases of imperfect or perfect PMs, and minimal or imperfect CM at failures between PMs.

c) *T-N* policy. A subsystem is subject to (imperfect) PM T where T is a nonzero constant, or at the Nth failure ($N = 1,2,3,...$), whichever occurs first, and undergoes (imperfect) repair at failures between PMs.

d)   Repair limit policy.

Other maintenance policies will be formally described in related chapters, where their characteristics are also discussed.

For warranty policies, this book discusses three types of existing warranty policies: free repair warranty (FRPW), free replacement warranty (FRW), and pro-rata warranty (PRW), and two new warranty policies: renewable full service warranty (RFSW) and repair-limit risk-free warranty (RLRFW).

It is worthwhile to note that for a series system there exist various shut-off rules. For example, while a failed component in a series system is in repair, all other components remain in "suspended animation" (they do not age and do not fail). After the repair is completed, the system is returned to operation. At that instant, the components in "suspended animation" are as good as they were when the system stopped operating. This shut-off rule is used in Barlow and Proschan (1975). Obviously, it is practical and can be applicable in other system architectures. We will refer to it as shut-off rule 1 later. In shut-off rule 2, component A upon failure shuts off component B but not *vice versa*. The third shut-off rule is that components operate independently, and non-failed components continue to operate regardless of the failed components. Hudes (1979) and Khalil (1985) discuss various shut-off rules and system availability for the series system.

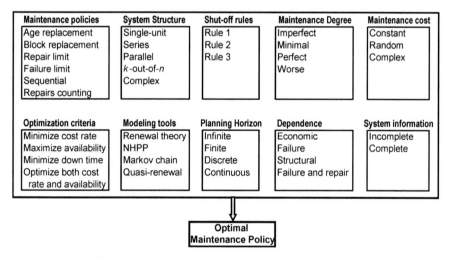

Figure 1.1. Maintenance policy and constituting factors

Figure 1.1 shows various factors which may affect an optimal maintenance policy of a system. An optimal maintenance policy should properly consider /incorporate various maintenance policies, system architectures, shut-off rules, maintenance restoration degrees, correlated failures and repairs, failure dependence, economic dependence, non-negligible maintenance time, *etc.* This book discusses optimal maintenance policies under various system architectures, maintenance policies, shut-off rules, imperfect maintenance, correlated failures and repairs, and economic dependence given that the planning horizon is infinite. Non-negligible maintenance time is also considered for some models in this book. It is worthwhile to mention the following points:

1. Because a unit is a building brick for a multicomponent system it is necessary to establish effective and efficient methods for modeling reliability measures and cost rates, and determining optimal maintenance policies for a single-unit system considering proper impact factors. All these methods for a single-unit system will be the basis for the analysis of a multicomponent system. For example, there are various modeling methods for imperfect repair for a single-unit system, and they may also be used to model imperfect repair for a multi-component systems.

2. Most work in the literature tends to determine optimal maintenance policies through minimizing the system maintenance cost rate. It is important to note that for multicomponent systems, minimizing the system maintenance cost rate may not result in optimal system reliability measures. Sometimes when the maintenance cost rate is minimized the system reliability measures are so low that they are not acceptable in practice. This is because various components in the system may have different maintenance costs and different reliability importance in the system. The details are demonstrated in Chapter 5 through numerical examples. Therefore, to achieve the best operating performance for multicomponent systems we need to consider both maintenance cost and reliability measures simultaneously.

3. The reliability architecture of the system must be considered to obtain optimal system reliability performance. For example, once a subsystem of a series system fails it is necessary to repair it at once. Otherwise, the system will have a longer downtime and worse reliability measures. However, when a subsystem of a parallel system fails, the system will still function even if this subsystem is repaired immediately. In fact, its repair can be delayed until it is time to do PM on the system, considering economic dependence; or repair can begin at a time such that only one subsystem operates and the other subsystems have failed and are awaiting repairs; or at the time that all subsystems fail and thus the system fails, if the system failure during actual operation is not very important.

4. In this book, maintenance cost measures, system reliability measures and optimal maintenance policies for imperfect maintenance will be compared with those for perfect maintenance because perfect maintenance may be a special case of imperfect maintenance, and the stochastic behavior and optimal maintenance policies for the perfect maintenance cases have been studied extensively. The comparisons will be helpful for verifying the results obtained under the imperfect maintenance cases.

5. Throughout this book, maintenance will be a general term and may represent PM or CM. Replacement is a perfect maintenance, preventive or corrective. Repair is an action made at component or system failure and has the same meaning as CM. CM and repair will be used alternatively when there can be no confusion.

6. Throughout this book, the planning horizon is assumed to be infinite.

7. In most existing literature, the maintenance time is assumed to be negligible

for reliability and maintenance models. This assumption makes availability, MTBF and MTBR modeling impossible or unrealistic. This book will consider maintenance time to obtain realistic system reliability measures whenever possible.

1.5.3 Framework

In this book, Chapter 2 will survey imperfect repair and dependence models in the literature and summarize approaches to model imperfect maintenance and dependence, while Chapter 3 will overview various maintenance policies in the literature and practice, such as age-dependent PM policy, repair limit policy.

Chapter 4 will introduce a new modeling tool for imperfect maintenance: a quasi-renewal process which includes ordinary renewal processes as a special case and discusses imperfect maintenance of one-unit systems by means of the quasi renewal process. Eleven imperfect maintenance models based on the quasi-renewal process are presented in this chapter.

In practice, many systems are series systems or can be simplified into series systems. Chapter 5 will investigate reliability and maintenance cost of a series system with n components and correlated failure and repair, and discuss some related optimal maintenance polices. Some important properties of reliability and maintenance cost of series systems are presented. Imperfect repairs are modeled through increasing and decreasing quasi-renewal processes.

Chapter 6 will discuss the opportunistic maintenance of a system with $(n+1)$ subsystems and economic dependence among them, in which whenever a subsystem fails, its repair is combined with PM of the functioning one having increasing failure rate (IFR) if the former reaches some age. Imperfect repair is assumed and two different imperfect modeling methods are used in this chapter.

In Chapter 7 we will look into a preparedness maintenance policy for a system with $n+1$ subsystems, economic dependency and imperfect maintenance. In this preparedness maintenance policy, the system is placed in storage and is called on to perform a given task only if a specific but unpredictable emergency occurs. Some maintenance actions resulting in optimal system "preparedness for field use" may be taken while the system is in storage.

A k-out-of-n system is one of the most important systems in reliability engineering and can include series and parallel systems as special cases. Chapter 8 will presents three new (τ, T) opportunistic maintenance models for a k-out-of-n:G system with economic dependency and imperfect maintenance. In these models, minimal repairs are performed on failed components before fixed time τ – a decision variable, and CM of all failed components is combined with PM of all functioning ones after τ. At time T – another decision variable, PM is performed if the system has not been subject to perfect maintenance before T. Applications to aircraft engine maintenance are also discussed in Chapter 8.

Chapter 9 will investigate multi-state degraded systems subject to multiple competing failure processes including two independent degradation processes and random shocks. We first discuss reliability model and then optimal condition-based inspection-maintenance for these systems. The system reliability model can be used not only to determine the reliability of the degraded systems but also to obtain the states of the systems by calculating the system state probabilities. A quasi-renewal process introduced in Chapter 4 is employed to establish the inter-inspection sequence. The PM thresholds for degradation processes and inspection sequence are the decision variables. An optimization algorithm to minimize the average long-run maintenance cost rate is discussed.

In current highly competitive markets, warranty policies become more and more complex. Chapter 10 will discuss warranty cost models of repairable complex systems from manufacturers' point of view by considering a comprehensive set of warranty cost impact factors, such as warranty policies, system structure, product failure mechanism, warranty service cost, impact of warranty service, value of time, warranty service time, and warranty claim related factors, under three types of existing warranty policies: free repair warranty (FRPW), free replacement warranty (FRW), and pro-rata warranty (PRW), and two new warranty policies: renewable full service warranty (RFSW) and repair-limit risk-free warranty (RLRFW). Imperfect or minimal repair is assumed. Monte Carlo simulation techniques and a new modeling tool: a truncated quasi-renewal process introduced in Chapter 4 will be used. The focus is on multi-component systems.

Chapter 11 will model software reliability and testing costs using the quasi-renewal process introduced in Chapter 4, and discusses optimal software testing and release policies. Several software reliability and cost models are presented in which successive error-free times form an increasing quasi-renewal process. It is assumed that the cost of fixing a fault during the software testing phase consists of deterministic and incremental random parts, and increases as the number of faults removed rises. The maximum likelihood estimates of parameters associated with these models are provided. Based on the valuable properties of quasi-renewal processes, the expected software testing and debugging cost, number of remaining faults in the software, and mean error-free time after testing are obtained. Optimum testing policies incorporating both reliability and cost measures are investigated.

To obtain the optimal maintenance policy for a complex system, we may first need to determine system availability or MTBF. Generally there are four major difficulties in evaluating complex large-scale system reliability, availability, and MTBF and their confidence limits using classical statistics: the system reliability structure may be very complicated; subsystems may follow different failure distributions; subsystems may have arbitrary failure and repair distributions for maintained systems; failure data of subsystems are sometimes not sufficient, sample size of life test or field population tends to be small. Therefore, it is difficult and often impossible to obtain s-confidence limits of the reliability measurements by classical statistics. It has been proven that Monte Carlo technique combined with Bayes method is a powerful tool to deal with this kind of complex systems. In Chapter 12, some existing Monte Carlo reliability, availability and MTBF simulation algorithms will be analyzed. Variance reduction methods, random variate generation techniques, commercial Monte Carlo reliability softwares, *etc.*, are addressed. The pros, cons, accuracy and computer execution time of Monte Carlo simulation in evaluating reliability, availability and MTBF of a complex network are discussed, and a general Monte Carlo reliability assessment method is presented.

2

Imperfect Maintenance and Dependence

Imperfect maintenance and dependence are major concerns of this book, as stated in Chapter 1. This chapter will present a detailed introduction to imperfect maintenance and dependence, and survey typical modeling methods, their characteristics and uses. Imperfect maintenance section of this chapter updates Pham and Wang (1996). Some modeling methods on imperfect maintenance and dependence will be used in the subsequent chapters in this book.

2.1 Imperfect Maintenance

Perfect maintenance assumes that the system is "as good as new" following maintenance. However, this assumption may not be true in practice. A more realistic assumption is that, upon maintenance, the system lies in a state somewhere between 'as good as new' and its pre-maintenance condition, *i.e.*, maintenance is imperfect, as mentioned in Chapter 1. Kay (1976), Ingle and Siewiorek (1977), Chaudhuri and Sahu (1977), and Chan and Downs (1978) are pioneers in imperfect maintenance study. Kay (1976) and Chan and Downs (1978) study the worst PM. Ingle and Siewiorek (1977) investigate imperfect maintenance. Chaudhuri and Sahu (1977) mention the concept of imperfect PM. An early work on imperfect repair can also be found in NAPS document No. 03476-A. In fact, NAPS document No. 03476-A plays a significant role in later imperfect maintenance research. Based on this work, other researchers have proposed some imperfect maintenance models.

In the existing imperfect maintenance literature, various methods for modeling imperfect maintenance have been used and most of them are on single-unit systems. It is necessary to summarize and compare these modeling methods because it will be helpful for later study in this area, especially for a multicomponent system which is the main concern in this book. It should be pointed out that although the existing literature is mainly on a single-unit system and the modeling methods will be summarized from it, they will also be useful for modeling a multi-component system. This is because individual subsystems can be regarded as single-unit systems, and thus the methods of treating imperfect maintenance for single-unit systems may also be effective for modeling imperfect maintenance of individual subsystems, based on which imperfect maintenance of a system will be investigated, possibly together with dependence (Pham and Wang 1996).

Methods for modeling imperfect, worse and worst maintenance can be classified into seven categories. Pham and Wang (1996) summarize these methods from related papers and technical reports throughout the literature, and these seven methods and some important results for them are presented next.

2.1.1 Modeling Methods for Imperfect Maintenance

2.1.1.1 Modeling Method 1 - (p,q) Rule

Nakagawa (1979) models imperfect PM in this way: after PM a unit is returned to the 'as good as new' state (perfect PM) with probability p and returned to the 'as bad as old' state (minimal PM) with probability $q = 1 - p$. Clearly, if $p = 1$ the PM coincides with perfect one and if $p = 0$ it corresponds to minimal PM. So in this sense, minimal and perfect maintenances are special cases of imperfect maintenance and imperfect maintenance is a general maintenance. Using such a study method for imperfect maintenance and assuming that PM is imperfect, Nakagawa (1979, 1980) succeeds in obtaining optimum PM policies minimizing the s-expected maintenance cost rate for one-unit system under age-dependent and periodic PM policies, respectively.

Similar to Nakagawa (1979a,b), Helvic (1980) states that, while the fault-tolerant system is usually renewed after PM with probability θ_2, its operating condition sometimes remains unchanged (as bad as old) with probability θ_1 where $\theta_1 + \theta_2 = 1$.

Brown and Proschan (1983) study the following model of the imperfect repair process. A unit is repaired each time it fails. The executed repair is either a perfect one with probability p or a minimal one with probability $1 - p$. Assuming that all repair actions take negligible time, they establish ageing preservation properties of this imperfect repair model and monotonicity of various parameters and random variables associated with the failure process. They obtain an important, useful result: if the life distribution of a unit is F and its failure rate is r, then the distribution function of the time between successive perfect repairs is $F_p = 1 - (1 - F)^p$ and the corresponding failure rate $r_p = pr$. Using this result, Fontenot and Proschan (1984), and Wang and Pham (1996b) obtain optimal imperfect maintenance policies for one-component system.

Later on, we will refer to this method for modeling imperfect maintenance as the (p,q) rule, that is, after maintenance (corrective or preventive) a system becomes "as good as new" with probability p and "as bad as old" with probability $1 - p$. In fact, this modeling method is getting popular: more and more imperfect maintenance models have used this rule in recent years.

Bhattacharjee (1987) obtains the same results as Brown and Proschan (1983), and some new results for Brown-Proschan model of imperfect repair via a shock model representation of the sojourn time.

Lim *et al.* (1998) extend the Brown and Proschan (1983) imperfect repair model, and propose a new Bayesian imperfect repair model where the probability of perfect repair, P, is considered to be a random variable. Assuming that P has a prior distribution $\text{II}(p)$, they obtain the distribution of waiting times between two successive perfect repairs and its corresponding failure rate. Lim *et al.* (1998) discuss the posterior distribution of P and its estimators, and study some preservation properties for certain nonparametric classes of life distributions and the monotonicity properties for several parameters. Cha and Kim (2001) model Bayesian availability where P is not fixed but a random variable with a prior distribution.

Li and Shaked (2003) equip the Brown and Proschan (1983) imperfect repair model with PM, and obtain stochastic maintenance comparisons for the numbers of failures under different policies via a point-process approach. They also obtain some results involving stochastic monotonicity properties of these models with respect to the unplanned complete repair probability.

2.1.1.2 Modeling Method 2 - $(p(t), q(t))$ Rule

Block *et al.* (1985) extend the above Brown-Proschan imperfect repair model with the (p, q) rule to the age-dependent imperfect repair for one-unit system: an item is repaired at failure (corrective maintenance). With probability $p(t)$, the repair is a perfect repair; with probability $q(t) = 1 - p(t)$, the repair is a minimal one, where t is the age of the item in use (the time since the last perfect repair). Block *et al.* (1985) prove that if the item's life distribution F is a continuous function and its failure rate is r, the successive perfect repair times form a renewal process with interarrival time distribution

$$F_p = 1 - \exp\left\{ \int_0^t p(x)[1 - F(x)]^{-1} F(dx) \right\}$$

and the corresponding failure rate $r_p(t) = p(t)r(t)$. In fact, similar results can be found in Beichelt and Fischer (1980), and NAPS Document No. 03476-A. Block *et al.* (1985) prove that the ageing preservation results of Brown and Proschan (1983) hold under suitable hypotheses on $p(t)$. Later on, we will call this imperfect maintenance modeling method as the $(p(t), q(t))$ rule.

Using this $(p(t), q(t))$ rule, Block *et al.* (1988) investigate a general age-dependent PM policy, where an operating unit is replaced when it reaches age T; if it fails at age $y < T$, it is either replaced by a new unit with probability $p(t)$, or it undergoes minimal repair with probability $q(t) = 1 - p(t)$. The cost of the ith minimal repair is a function, $c_i(y)$, of age y and number of repairs. After a perfect maintenance, planned or unplanned (preventive), the procedure is repeated.

Both Brown and Proschan (1983) model and Block *et al.* (1985) model assume that the repair time is negligible. It is worthwhile to mention that Iyer (1992) obtains availability results for imperfect repair using the $(p(t), q(t))$ rule given that the repair time is not negligible. His realistic treatment method will be helpful for later research.

Sumita and Shanthikumar (1988) propose and study an age-dependent counting process generated from a renewal process and apply that counting process to the age-dependent imperfect repair for the one-unit system.

Whitaker and Samaniego (1989) propose an estimator for the life distribution when the above model by Block *et al.* (1985) is observed until the time of the mth perfect repair. This estimator was motivated by a nonparametric maximum likelihood approach, and was shown to be a 'neighborhood MLE'. They derive large-sample results for this estimator. Hollander *et al.* (1992) take the more modern approach of using counting process and martingale theory to analyze these models. Their methods yield extensions of Whitaker and Samaniego's results to the whole line and provide a useful framework for further work on the minimal repair model.

The (p,q) rule and $(p(t),q(t))$ rule for imperfect maintenance seem practical and realistic. It makes imperfect maintenance be somewhere between perfect and minimal ones. The *degree* to which the operating conditions of an item is restored by maintenance can be measured by p or $p(t)$. Especially, in the $(p(t),q(t))$ rule, the *degree* to which the operating condition of an item is restored by maintenance is related to its age t. Thus, the $(p(t),q(t))$ rule seems more realistic but mathematical modeling of imperfect maintenance by using it will be more complicated. The two rules can be expected to be powerful in future imperfect maintenance modeling. In fact, both rules have received much attention and have been used in some imperfect repair models, as shown in the subsequent chapters.

Makis and Jardine (1992) consider a general treatment method for imperfect maintenance and model imperfect repair at failure in a way that repair returns a system to the "as good as new" state with probability $p(n,t)$ or to the "as bad as old" state with probability $q(n,t)$, or with probability $s(n,t) = 1 - p(n,t) - q(n,t)$ the repair is unsuccessful, the system is scrapped and replaced by a new one, where t is the age of the system and n is the number of failures since replacement. We will refer to this treatment method as $(p(n,t),q(n,t),s(n,t))$ rule later.

2.1.1.3 Modeling Method 3 - Improvement Factor Method

Malik (1979) introduces the concept of improvement factor in the maintenance scheduling problem. He believes that maintenance changes the system time of the failure rate curve to some newer time but not all the way to zero (not new), as shown in Figure 2.1 This treatment method for imperfect maintenance also makes the failure rate after PM lie between 'as good as new' and 'as bad as old'. The degree of improvement in failure rate is called improvement factor. Malik (1979) assumes that since systems need more frequent maintenance with increased age the successive PM intervals are decreasing in order to keep the system failure rate at or below a stated level (sequential PM policy), and proposes an algorithm to determine these successive PM intervals. Lie and Chun (1986) present a general expression to determine these PM intervals. Malik (1979) relies on an expert judgment to estimate the improvement factor, while Lie and Chun (1986) give a set

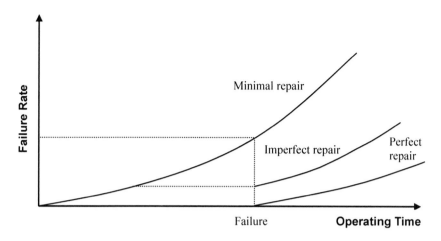

Figure 2.1. Minimal, perfect, imperfect repair *vs.* failure rate changes

of curves as a function of maintenance cost and the age of the system for the improvement factor.

Using the improvement factor and assuming finite planning horizon, Jayabalan and Chaudhuri (1992b) introduce a branching algorithm to minimize the average total cost for a maintenance scheduling model with assured reliability and they (1992c) discuss optimal maintenance policy for a system with increased mean down time and assured failure rate. It is worthwhile to note that using fuzzy set theory and improvement factor, Suresh and Chaudhuri (1994) establish a PM scheduling procedure to assure an acceptable reliability level or tolerable failure rate assuming finite planning horizon. They regard the starting condition, ending condition, operating condition, and type of maintenance of a system as fuzzy sets. Improvement factor is used to find out the starting condition of the system after maintenance.

Chan and Shaw (1993) think that failure rate is reduced after each PM and this reduction of failure rate depends on the item age and the number of PMs. Chan and Shaw propose two types of failure-rate reduction: (1) failure-rate with fixed reduction. After each PM, the failure rate is reduced such that all jump-downs of the failure rate are the same; (2) failure rate with proportional reduction. After PM, the failure rate is reduced such that each jump-down is proportional to the current failure rate. They obtain cycle-availability for single unit system and discuss the design scheme to maximize the probability of achieving a specified stochastic-cycle availability with respect to the duration of the operating interval between PMs .

In Doyen and Gaudoin (2004), the (conditional) failure intensity before the first repair is a continuous function of time. The repair effect is characterized by the change induced on the failure intensity before and after failure. Repair effect is expressed by a reduction of failure intensity. Several cases are studied, which take into account the possibility of a Markovian memory property.

This kind of study method for imperfect maintenance is in terms of failure rate and seems useful and practical in engineering where it can be used as a general treatment method for imperfect maintenance or even *worse* maintenance. Later on we call this treatment method *Improvement Factor Method.*

Besides, Canfield (1986) assumes that PM at time t restores the failure rate function to its shape at $t - \tau$, while the level remains unchanged where τ is less than or equal to the PM intervention interval.

2.1.1.4 Modeling Method 4 - Virtual Age Method

Kijima *et al.* (1988) develop an imperfect repair model by using the idea of the virtual age process of a repairable system. If the system has the virtual age $V_{n-1} = y$ immediately after the $(n-1)^{\text{th}}$ repair, the nth failure-time X_n is assumed to have the distribution function

$$\Pr\{X_n \leq x \mid V_{n-1} = y\} = \frac{F(x+y) - F(y)}{1 - F(y)}$$

where $F(x)$ is the distribution function of the time to failure of a new system. Let a be the degree of the nth repair where $0 \leq a \leq 1$. They construct such a repair model: the nth repair cannot remove the damage incurred before the $(n-1)^{\text{th}}$ repair. It reduces the additional age X_n to aX_n. Accordingly, the virtual age after the nth repair becomes:

$$V_n = V_{n-1} + aX_n$$

Obviously, $a = 0$ corresponds to a perfect repair while $a = 1$ to a minimal repair. Later Kijima (1989) extends the above model to the case that a is a random variable taking a value between 0 and 1 and proposes another imperfect repair model:

$$V_n = A_n(V_{n-1} + X_n)$$

where A_n is a random variable taking a value between 0 and 1 for $n = 1, 2, 3, \ldots$ For the extreme values 0 and 1, $A_n = 1$ means a minimal repair and $A_n = 0$ a perfect repair. Comparing this treatment method with Brown and Proschan's, we can see that if A_n is independently and identically distributed (*i.i.d.*) taking the two extreme values 0 and 1 they are the same. Therefore, the second treatment method by Kijima (1989) is general. He derives various monotonicity properties associated with the above two models.

In Doyen and Gaudoin (2004), repair effect is expressed by a reduction of the system virtual age. Several cases are studied, which take into account the possibility of a Markovian memory property.

This treatment method will be referred to as the *Virtual Age Method* later on.

It is worth mentioning that Uematsu and Nishida (1987) consider a more general model including the above two models by Kijima (1989) as special cases and obtain some elementary properties of the associated failure process. Let T_n

denote the time interval between the $(n-1)^{th}$ failure and the nth one, and X_n denote the degree of repair. After performing the nth repair, the age of the system becomes $q(t_1,...,t_n;x_1,...x_n)$ given that $T_i = t_i$ and $X_i = x_i$ $(i = 1,2,...n)$ where T_i and X_i are random variables. On the other hand, $q(t_1,...,t_n;x_1,...x_{n-1})$ represents the age of the system as just before the nth failure. The starting epoch of an interval is subject to the influence of all previous failure history, i.e., the nth interval is statistically dependent on $T_1 = t_1,....,T_{n-1} = t_{n-1}$, $X_1 = x_1,...,$ $T_{n-1} = t_{n-1}$. For example, if

$$q(t_1,...,t_n;x_1,...x_n) = \sum_{j=1}^{m} \sum_{i=j}^{m} x_i t_j,$$ then $X_i = 0$ ($X_i = 1$) represents that perfect repair (minimal repair) performs at the ith failure.

2.1.1.5 Modeling Method 5 - Shock Model

It is well-known that the time to failure of a unit can be represented as a first passage time to a threshold for an appropriate stochastic process that describes the levels of damage. Consider a unit which is subject to shocks occurring randomly in time. At time $t = 0$, the damage level of the unit is assumed to be 0. Upon occurrence of a shock, the unit suffers a non-negative random damage. Each damage, at the time of its occurrence, adds to the current damage level of the unit, and between shocks, the damage level stays constant. The unit fails when its accumulated damage first exceeds a specified level. To keep the unit in an acceptable operating condition, some PM is performed (Kijima and Nakagawa 1991).

Kijima and Nakagawa (1991) propose a cumulative damage shock model with imperfect periodic PM. The PM is imperfect in the sense that each PM reduces the damage level by $100(1-b)\%$, $0 \leq b \leq 1$, of total damage. Note that if $b = 1$ the PM is minimal and if $b = 0$ the PM coincides with a perfect PM. This research approach is similar to the one in treatment method 1. They derive a sufficient condition for the time to failure to have an IFR distribution and discuss the problem of finding the number of PMs that minimizes the expected maintenance cost rate.

Kijima and Nakagawa (1992) establish a cumulative damage shock model with a sequential PM policy assuming that PM is imperfect. They model imperfect PM in a way that the amount of damage after the kth PM becomes $b_k Y_k$ when it was Y_k before PM, i.e., the kth PM reduces the amount Y_k of damage to $b_k Y_k$ where b_k is called the improvement factor. They assume that a system is subject to shocks occurring according to a Poisson Process and, upon occurrence of shocks, it suffers a non-negative random damage which is additive. Each shock causes a system failure with probability $p(z)$ when the total damage is z at the shock. In this model, PM is done at fixed intervals x_k for $k = 1,2,3,...,N$ because more frequent maintenance is needed with age, and the Nth PM is perfect. If the system fails between PMs it undergoes only minimal repair. They derive the expected maintenance cost rate until replacement assuming that $p(z)$ is an exponential function and damage is independently and identically distributed and discuss the optimal replacement policies.

Finkelstein (1997) investigates the performance of a repairable system subject to shocks: each shock with a probability that depends on a virtual age of the effected system causes a breakdown that ends the process of system functioning. In Finkelstein (1997), various models of repair, ranging from minimal till perfect repair are studied. It is assumed that shocks occur according to the non-homogeneous Poisson process or a renewal process with identically distributed cycles. The probability of a system functioning without breakdowns for the mentioned models is derived.

This study approach for imperfect maintenance will be called *Shock Model* later in this book.

2.1.1.6 Modeling Method 6 - Quasi-renewal Process

Wang and Pham (1996b) treat imperfect repair in a way that, upon each repair, the lifetime of a system will be reduced to a fraction α of its immediately previous one where $0 < \alpha < 1$ and all lifetimes are independent, *i.e.*, the lifetime decreases with the number of repairs. In Wang and Pham (1996b), the successive lifetimes are defined to constitute a decreasing quasi-renewal process, whose details are given in Chapter 4. Assuming that the *pdf* of the lifetime of a system which has been subject to $(n-1)$ repairs, X_n, is $f_n(x)$ for $n = 1,2,3,...$, Wang and Pham (1996b) study this quasi-renewal process and prove that:

a.   If $f_1(x)$ belongs to IFR, DFR, IFRA, DFRA, and NBU (for definitions see Chapter 4 Appendix), then $f_n(x)$ is in the same category, $\forall n, n = 2,3,...$.

b.   The shape parameter of X_n are the same for $n = 1,2,3,...$ for a quasi-renewal process if X_1 follows the Gamma, Weibull or Lognormal distribution.

The second result means that after "renewal" the shape parameters of the interarrival time will not change. In reliability theory, the shape parameters of a lifetime of an item tend to relate to its failure mechanism. Usually, if a product possesses the same failure mechanism then its lifetimes will have the same shape parameters at different application conditions. Because most maintenance does not change the failure mechanism we can expect that the lifetime of a system will have the same shape parameters. Thus, in this sense, the quasi-renewal process will be plausible to model the imperfect maintenance process.

Wang and Pham (1996c) further assume that repair time is non-negligible, not as in most imperfect maintenance models, and upon repair the next repair time becomes a multiple β of its current one where $\beta > 1$ and all repair times are independent, *i.e.*, the time to repair increases with the number of repairs. In Wang and Pham (1996b), the successive repair times are defined to form an increasing quasi-renewal process. This method modeling imperfect maintenance will be referred to as the (α, β) *rule* or *quasi-renewal process* method. Yang and Lin (2005) apply the quasi-renewal process in age and block PM.

In investigating the optimal replacement problem, Lam (1988) uses the fixed life reduction idea after repair, referred to as the geometric process. Lam (1988, 1996) studies the geometric process by means of the ordinary renewal process. In

Chapter 4, the quasi-renewal process is introduced from defining the quasi-renewal function.

2.1.1.7 Modeling Method 7 – Multiple (p,q) Rule

Shaked and Shanthikumar (1986) introduce the multivariate imperfect repair concept. They consider a system whose components have dependent lifetimes and are subject to imperfect repairs respectively until they are replaced. For each component the repair is imperfect according to the (p, q) rule, i.e., at failure the repair is perfect with probability p and minimal with probability q. Assume that n components of the system start to function at the same time 0, and no more than one component can fail at a time. They establish the joint distribution of the times to next failure of the functioning devices after a minimal repair or perfect repair, and derive the joint density of the resulting lifetimes of the components and other probabilistic quantities of interest, from which the distribution of the lifetime of the system can be obtained. Sheu and Griffith (1992) further extend this work. Later we will call this treatment method the *multiple (p,q) rule*.

2.1.1.8 Others

Nakagawa (1979b) models imperfect PM in a way that, in the steady-state, PM reduces the failure rate of an item to a fraction of its value just before PM and during operation of the item the failure rate climbs back up. He believes that the portion by which the failure rate is reduced is a function of some resource consumed in PM and a parameter. That is, after PM the failure rate of the unit becomes $\lambda(t) = g(c_1,\theta) \cdot \lambda(t + T)$ where the fraction reduction of failure rate $g(c_1,\theta)$ lies between 0 and 1, T is the time interval length between PMs, c_1 is the amount of resource consumed in PM, and θ is a parameter. This treatment method is different from the improvement factor method in that, for improvement factor method, maintenance makes the system younger in terms of its age, i.e., its age becomes younger after maintenance.

Nakagawa (1986, 1988) uses two other methods to deal with imperfect PM for two sequential PM policies: (1) The failure rate after PM k becomes $a_k h(t)$ given that it was $h(t)$ in the previous period where $a_k \geq 1$. That is, the failure rate increases with the number of PMs; (2) The age, after PM k, reduces to $b_k t$ when it was t before PM where $0 \leq b_k < 1$. That is, PM reduces the age. Obviously, the second method is similar to the improvement factor method. Besides, in investigating periodic PM models, Nakagawa (1980) treats imperfect PM in that the age of the unit becomes x units of time younger by each PM and further suggests that x is in proportion to the PM cost where x is less than or equal to the PM interval. We will call it the *x Rule* later.

Nguyen and Murthy (1981) model imperfect PM in a way that, after PM, the unit has a different (worse) failure time distribution than after CM. Yak (1984) assumes that maintenance may result in its failure (the worst maintenance) in modeling the MTTF and the availability of a system.

Some typical work on imperfect maintenance are summarized in Table 2.1 by modeling methods. From this table we can see that the (p,q) rule and $(p(t),q(t))$

rule are popular in treating imperfect maintenance. This is partly because these two rules make imperfect maintenance modeling mathematically tractable, as demonstrated in Chapters 6 – 8.

Table 2.1. Summary of treatment methods for imperfect maintenance

Modeling method	References
(p, q) Rule	Chan and Downs (78), Helvic (80), Nakagawa (79, 80, 87), Brown and Proschan (82, 83), Fontenot and Proschan (84), Lie and Chun (86), Yun and Bai (87), Bhattacharjee (87), Rangan and Grace (89), Sheu and Liou (92), Srivastava and Wu (93), Wang and Pham (96a,b,c, 97b), Lim *et al.* (98), Pham and Wang (00), Cha and Kim (01), Kvam *et al.* (02), Li and Shaked (03)
$(p(t), q(t))$ Rule	Beichelt (80, 81), Block *et al.* (85, 88), Abdel-Hameed (87a), Whitaker and Samaniego (89), Sheu (91a, 92, 93), Makis and Jardine (91), Iyer (92), Hollander *et al.* (92), Sheu and Kuo (94), Sheu *et al.* (95), Wang *et al.* (01), Wang and Pham (99, 03)
Improvement factor	Malik (79), Canfield (86), Lie and Chun (86), Jayabalan and Chaudhuri (92a, b,c, 95), Chan and Shaw (93), Suresh and Chaudhuri (94), Doyen and Gaudoin (04)
Virtual age	Uematsu and Nishida (87), Kijima (88, 89), Makis and Jardine (93), Liu *et al.* (95), Gasmi *et al.* (03), Doyen and Gaudoin (04)
Shock model	Bhattacharjee (87), Kijima and Nakagawa (91,92), Sheu and Liou (92c), Finkelstein (97)
(α, β) Rule or quasi-renewal process	Lam (88, 96), Wang and Pham (96a,b,c, 97b, 99, 06), Pham and Wang (00, 01), Yang and Lin (05), Wu and Clements-Croome (05), Bai and Pham (06a)
Multiple (p, q) rule	Shaked and Shanthikumar (86), Sheu and Griffith (92)
Others	Nakagawa (79b, 80, 86, 88), Subramanian and Natarajan (80), Nguyen and Murthy (81), Yak (84), Yun and Bai (88), Dias (90), Subramanian and Natarajan (90), Zheng and Fard (91), Jack (91), Chun (92), Dagpunar and Jack (94)

The following further work on imperfect maintenance is necessary:

- Study optimal maintenance policy for multicomponent systems because previous work on imperfect maintenance was focused on one-unit systems.
- Construct statistical estimation methods for parameters of various imperfect maintenance models.
- Develop more and better methods for treating imperfect maintenance.
- Study more realistic imperfect maintenance models, for example, including non-negligible repair time, finite horizon.
- Use the reliability measures as the optimality criteria for maintenance policies instead of cost rates, or combine both, as stated in Section 1.5.1.

2.1.2 Typical Imperfect Maintenance Models by Maintenance Policies

2.1.2.1 Age-dependent PM Policy
In the age-dependent PM model, a unit is preventively maintained at predetermined age T, or repaired at failure, whichever comes first. For this policy there are various imperfect maintenance models according to the conditions that either or both of PM and CM is imperfect. The research under the age-dependent PM policy and its extensions is summarized in Table 2.2.

Table 2.2. Imperfect maintenance study under age-dependent PM policy

Study	PM	CM	Treatment method	Optimality criteria	Modeling tool	Planning horizon
Chan and Downs (1978)	Imperfect	Perfect	(p,q) rule	Availability cost rate	Semi-Markov	Infinite
Nakagawa (1979a)	Imperfect	Perfect	(p,q) rule	Cost rate	Renewal theory	Infinite
Beichelt (1980)	Perfect	Imperfect	$(p(t),q(t))$ rule	Cost rate	Renewal theory	Infinite
Fontenot and Proschan (1984)	Perfect	Imperfect	(p,q) rule	Cost rate	Renewal theory	Infinite
Block et al. (1988)	Perfect	Imperfect	$(p(t),q(t))$ rule	Cost rate total cost	Renewal theory	Infinite finite
Rangan and Grace (1989)	Perfect	Imperfect	(p,q) rule	Total cost	Renewal theory	Finite
Sheu (1991a)	Perfect	Imperfect	$(p(t),q(t))$ rule	Cost rate (random cost)	Renewal theory	Infinite finite
Sheu and Kuo (1993)	Perfect	Imperfect	$(p(t),q(t))$ rule	Cost rate (random cost)	Renewal theory	Infinite
Sheu et al. (1995)	Perfect	Imperfect	$(p(t),q(t))$ rule	Cost rate (random cost)	Renewal theory	Infinite
Wang and Pham (1996a)	Imperfect	Imperfect	(p,q) rule (α,β) rule	Cost rate Availability	Renewal theory	Infinite

One of the pioneer imperfect maintenance models for the age-dependent PM policy is due to Nakagawa (1979a) and NAPS Document No. 03476-A. Nakagawa (1979a) investigates three age-dependent PM models with imperfect PM and perfect or minimal repair at failure using the (p, q) rule. He derives the expected maintenance cost rate and discusses the optimal maintenance policies in terms of PM time interval T.

Using the $(p(t),q(t))$ rule, Block et al. (1988) discuss an age-dependent PM policy where CM is imperfect and the cost of the ith minimal repair is a function, $c_i(y)$, of age and number of repairs. Sheu et al. (1993) generalized the age-dependent PM policy where if a system fails at age $y < t$, it is subject to perfect repair with $p(y)$, or undergoes minimal repair with probability $q(y) = 1 - p(y)$.

Otherwise, a system is replaced when the first failure after t occurs or the total operating time reaches age T ($0 \le t \le T$), whichever occurs first. They discussed the optimal policy (t^*, T^*) to minimize the expected cost rate. This is a realistic PM model. Sheu et al. (1995) further extend this model. They assume that a system has two types of failures when it fails at age z and is replaced at the nth type 1 failure or first type 2 failure or at age T, whichever occurs first. Type 1 failure occurs with probability $p(z)$ and is corrected by minimal repair. Type 2 failure occurs with probability $q(z) = 1 - p(z)$ and is corrected by perfect repair (replacement). Using the ($p(t), q(t)$) rule and random minimal repair costs, they derive the expected cost rate and a numerical example is presented.

Table 2.3. Imperfect maintenance study under periodic PM policy

Study	PM	CM	Treatment method	Optimality criterion	Modeling tool	Planning horizon
Nakagawa (1979)	Imperfect	Minimal	(p,q) rule	Cost rate	Renewal theory	Infinite
Nakagawa (1980)	Imperfect	Perfect minimal	(p,q) rule x rule	Cost rate	Renewal theory	Infinite
Beichelt (1981a, b)	Perfect	Imperfect	$p(t), q(t)$	Cost rate	Renewal theory	Infinite
Fontenot and Proschan (1984)	Perfect	Imperfect	(p,q) rule	Cost rate	Renewal theory	Infinite
Nakagawa (1986)	Imperfect	Minimal	Different failure rates	Cost rate	Renewal theory	Infinite
Abdel-Hameed (1987a)	Perfect	Imperfect	$p(t), q(t)$	Cost rate	Stochastic process	Infinite
Nakagawa and Yasui (1987)	Imperfect	Perfect	(p,q) rule	Availability	Renewal theory	Infinite
Kijima et al. (1988)	Perfect	Imperfect	Virtual age	Cost rate	Renewal theory	Infinite
Kijima and Nakagawa (1991)	Imperfect	Perfect	Shock model	Cost rate	Renewal theory	Infinite
Jack (1991)	Perfect	Imperfect	Others	Total cost	Renewal theory	Finite
Chun (1992)	Imperfect	Minimal	x rule	Total cost	Probability	Finite
Sheu (1992)	Perfect	Imperfect	$p(t), q(t)$	Cost rate	Renewal theory	Infinite
Liu et al. (1995)	Imperfect	Minimal	Virtual age	Cost rate	Renewal theory	Infinite
Wang and Pham (1996c)	Imperfect	Imperfect	(p,q) rule (α, β) rule	Cost rate Availability	Quasi-renewal theory	Infinite
Wang and Pham (1999)	Imperfect	Imperfect	$p(t), q(t)$	Cost rate Availability	Renewal theory	Infinite

2.1.2.2 Periodic PM Policy

In the periodic PM policy, a unit is preventively maintained at fixed time intervals and repaired at intervening failures. Liu *et al.* (1995) investigate an extended periodical PM model using the notation of the virtual age. They assume that a unit receives (imperfect) PM every T time unit, intervening failures are subject to minimal repairs and the unit is replaced every fixed number of PMs. Nakagawa (1986) studies a similar model but he assumes that PM is imperfect in the sense that after PM the failure rate will be changed. The research under the periodic PM policy and its extensions are summarized in Table 2.3.

Table 2.4. Imperfect maintenance study under failure limit policy

Study	PM	CM	Measure improved	Optimality criterion	Modeling tool	Planning horizon
Malik (1979)	Imperfect	None	Reliability	Reliability	Probability	Infinite
Canfield (1986)	Imperfect	None	Failure rate	Cost rate	Renewal theory	Infinite
Lie and Chun (1986)	Imperfect	Imperfect	Failure rate	Cost rate	Renewal theory	Infinite
Jayabalan (1992a)	Imperfect	Minimal	Failure rate	Total cost	Probability	Finite
Jayabalan and Chaudhuri (1992c)	Imperfect	Minimal	Age Others	Cost rate	Probability	Infinite
Jayabalan and Chaudhuri (1992d)	Imperfect	None	Age	Total cost	Probability	Finite
Chan and Shaw (1993)	Imperfect	Perfect	Failure rate	Availability	Probability	Infinite
Suresh and Chaudhuri (94)	Imperfect		Reliability and failure rate	Total cost	Probability	Finite
Jayabalan and Chaudhuri (1995)	Imperfect	Minimal	Age	Total cost	Renewal theory	Finite
Monga *et al.* (1996)	Imperfect	Minimal	Reduction (age and failure rate)	Cost rate	Renewal theory	Infinite

2.1.2.3 Failure Limit Policy

This policy assumes that PM is performed only when the failure rate or reliability of a unit reaches a predetermined level. Malik (1979) derives the PM schedule points so that a unit works at or above the minimum acceptable level of reliability. Lie and Chun (1986) formulate a maintenance cost model where PM is performed whenever the unit reaches the predetermined maximum failure rate. Jayabalan and Chaudhuri (1992a) obtain the optimal maintenance policy for a specific period of time given that downtime for installation and for PM are negligible. In other work, Jayabalan and Chaudhuri (1992b) consider downtime for replacement as a nonzero constant. As a unit ages, the successive downtime for PM interventions is expected to consume more time. To incorporate this point, Jayabalan and Chaudhuri (1992b) assume that PM time follows exponential distribution and is increasing with age. Jayabalan and Chaudhuri (1995) present an algorithm to obtain optimal maintenance policies which require less computational time. The research under the failure limit policy and its extensions is summarized in Table 2.4.

2.1.2.4 Sequential PM Policy

When a system is maintained at unequal intervals, the PM policy is known as *Sequential PM policy*. Nakagawa (1986,1988) discusses a sequential PM policy where PM is done at fixed intervals x_k where $x_k \leq x_{k-1}$ for $k = 2,3...$ This policy is practical because most units need more frequent maintenance with increased age. This PM policy is different from the failure limit policy in that it controls x_k lengths directly but the failure limit policy controls failure rate, age, reliability, *etc.*, directly. In Wu and Clements-Croome (2005), the PM is sequentially executed with τ_n time units after the $(n-1)^{th}$ PM, where $n = 1,2,...$ Between two adjacent PMs, a CM is carried out immediately on failure. Both PM and CM are imperfect. τ_n is dependent on n and determined through minimizing the maintenance cost rate. The research under the sequential PM policy and its extensions is summarized in Table 2.5.

Table 2.5. Sequential PM policy

Study	PM	CM	Treatment	Optimality criterion	Modeling tool	Planning horizon
Nakagawa (1986)	Imperfect	Minimal	Different failure rates	Cost rate	Renewal theory	Infinite
Nakagawa (1987)	Imperfect	Minimal	Reduction (age and failure rate)	Cost rate	Renewal theory	Infinite
Kijima and Nakagawa (1992)	Imperfect	Minimal	Shock model	Cost rate	Renewal theory	Infinite
Monga *et al.* (1996)	Imperfect	Minimal	Reduction (age and failure rate)	Cost rate	Renewal theory	Infinite
Wu and Clements-Croome (2005)	Imperfect	Imperfect	(α, β) rule	Cost rate	Renewal theory	Infinite

2.1.2.5 Repair Limit Policy

When a unit fails, the repair cost is estimated and repair is undertaken if the estimated cost is less than a predetermined limit; otherwise, the unit is replaced. This is called the *Repair Cost Limit Policy* in the literature. Yun and Bai (1987) study the optimal repair cost limit policies under an imperfect maintenance assumption.

The *Repair Time Limit Policy* is proposed by Nakagawa and Osaki (see Nguyen and Murthy 1981) in which a unit is repaired at failure: if the repair is not completed within a specified time T, it is replaced by a new one; otherwise the repaired unit is put into operation again where T is called the *repair limit time*. Nguyen and Murthy (1981) study the repair time limit replacement policies with imperfect repair in which there are two types of repair – local and central repair. The local repair is imperfect while the central repair is perfect. The optimal policies are derived to minimize the expected cost rate for an infinite time span. The research under the repair limit policy and its extensions is summarized in Table 2.6.

Table 2.6. Imperfect maintenance study under repair limit policy

Study	CM before cost limit	CM after cost limit	Treatment method	Optimality criterion	Modeling tool	Planning horizon
Beichelt (1978, 1981b)	Minimal	Perfect	$(p(t), q(t))$	Cost rate	Renewal theory	Infinite
Nguyen and Murthy (1981)	Imperfect	Perfect	Others	Cost rate	Renewal theory	Infinite
Yun and Bai (1987)	Imperfect	Perfect	(p, q) rule	Cost rate	Renewal theory	Infinite
Yun and Bai (1988)	Minimal	Perfect	Others	Cost rate	Renewal theory	Infinite
Wang and Pham (1996c)	Imperfect	Imperfect	(p, q) rule/ (α, β) rule	Cost rate Availability	Quasi-renewal theory	Infinite

2.1.2.6 Multicomponent Systems

Imperfect maintenance models for multi-unit systems are summarized in Table 2.7. For series system Zhao (1994) establishes a series system availability model in which either minimal repair or perfect repair of all components can be modeled based on Barlow and Proschan's work (1975). He assumes that the repaired component might not be as good as new and its lifetime may follow any distribution which can be different from that of old one after repair and obtain mean limiting availability and mean system down and up time. In this model of series system, repair time is not negligible and thus it is practical. This treatment method for imperfect repair is similar to the one by Subramanian and Natarajan (1980). Besides, Sheu *et al.* have done some work on this problem. The related research is summarized in Table 2.7.

2.1.2.7 Others

Jack (1991) investigates a maintenance policy involving imperfect repairs on failure with replacement upon the Nth failure. Dagpunar and Jack (1994) determine the optimal number of imperfect PM during a finite horizon given that the minimal repairs are made at any failures between PMs and the ith PM makes the age of a unit x_i units of time younger (*x rule*). Chun (1992) studies determination of the optimal number of periodic PMs under a finite planning horizon using the *x rule*.

Makis and Jardine (1992) contemplate a replacement policy without PMs in which a unit can be replaced at any time at a cost c_0, and at the nth failure the unit is either replaced at the cost c_0 or undergoes an imperfect repair at a cost $c(n,t)$ where t is the age of the unit. They use the $(p(n,t),q(n,t),s(n,t))$ rule to model imperfect repair. Makis and Jardine (1991, 1993) discuss the optimal replacement policy with imperfect repair at failure: a unit is replaced each time at the first failure after some fixed time using the ($p(t),q(t),s(t)$) rule and the virtual age method, respectively.

Table 2.7. Multicomponent systems subject to imperfect maintenance

Study	PM	CM	Treatment	Optimality criterion	Modeling tool	Horizon / architecture / policy
Shaked and Shanthikumar (1986)	None	Imperfect	Multiple (p,q) rule	None	Renewal	Infinite / Arbitrary/
Subramanian and Natarajan (1990)	None	Imperfect	Other	Reliability Availability	Stochastic process	Infinite / Two-unit standby/
Zheng and Fard (1991)	Perfect	Imperfect	Other	Cost rate	Probability	Infinite / Arbitrary / Failure limit
Sheu and Griffith (1991b)	None	Imperfect	Multiple $(p(t),q(t))$	None	Renewal theory	Infinite / Arbitrary / Age-dependent
Sheu and Liou (1992c)	Perfect	Imperfect	$(p_1(t),...,p_n(t))$	Cost rate	NHPP	Infinite / k-out-of-n / Age-dependent
Zhao (1994)	None	Imperfect	Other	Availability	Probability	Infinite/ Series
Sheu and Kuo (1994)	Perfect	Imperfect	$(p(t),q(t))$	Cost rate (random cost)	Renewal theory	Infinite / k-out-of-n / Age-dependent
Wang and Pham (2006)	None	Imperfect	(α,β)	Availability	Quasi-renewal theory	Infinite / Series /
Wang (1997)	Perfect	Imperfect	$(p(t),q(t))$	Availability Cost rate	Renewal theory	Infinite / Arbitrary / Age-dependent
Pham and Wang (2000)	Im-perfect	Imperfect	(p,q) rule	Availability cost rate	Renewal theory	Infinite / k-out-of-n / Periodic
Wang et al. 2001)	Im-perfect	Perfect	$(p(t),q(t))$	Availability Cost rate	Renewal theory	Infinite / Arbitrary / Age-dependent

Block et al. (1992) introduce a generalized age replacement policy – repair replacement policy where units are preventively maintained when a certain time has elapsed since their last repair. If the last repair was a perfect repair, this policy is essentially the same as an age replacement policy.

Srivastava and Wu (1993) present an imperfect-inspection model in which failures can only be detected with probability p, and they discuss the estimation of parameter p. Sheppard (1983) and Nicolescu (1985) study imperfect testing which may result in reduction of availability of a unit. Ebrahimi (1985) derives the mean time to keep a failure-free operation with imperfect repair which either restore a unit to "as good as new" or to "as bad as old". Guo and Love (1992), and Love and Guo (1993) contemplate the statistical analysis for imperfect repair models.

Fontenot and Proschan (1984) explore four imperfect maintenance models using the (p,q) rule under various maintenance policies. Helvic (1980) investigates maintenance of the fault-tolerant system using the (p,q) rule. Besides, Abdel-Hameed (1987b, 1995), Makis and Jardine (1992), Murthy(1991), Nguyen and Murthy (1981), Natvig (1990), and Zheng and Fard (1991) also discuss the imperfect repair problem.

2.2 Dependence

There are three kinds of dependencies: Economic Dependence, Correlated Failures and Repairs, and Failure Dependence, as stated in Chapter 1. McCall (1963), and Radner and Jorgenson (1963) address the economic dependence in a system with n components. Ozekici (1988) studies the effects of failure and economic dependencies on periodic replacement policies and provide useful characterizations of the optimal replacement policy.

Goel *et al.* (1992, 1993, 1996) investigate the *correlated failure and repair* for some repairable systems: a two-unit standby system, two unit priority redundant system, warm standby system, two-unit cold standby system, two server two unit cold standby system, *etc.* He uses a bivariate exponential distribution to model the joint distribution of failure and repair times of components. Gupta (1999) discusses profit analysis of a two non-identical unit cold standby system with correlated failure and repair and switchover.

Harris (1968) utilizes a bivariate exponential to describe *correlated failures* of two components and derives the mean time to system failure by using the supplementary variable technique for an arbitrary repair time distribution. Osaki (1970) extends this analysis to obtain the availability of the system by using a variant of a semi-Markov process with some non-regeneration points. Pijinenburg (1993) obtains reliability, mean time to system failure, pointwise and steady-state availability and joint availability and interval reliability using the imbedded renewal process. Shaked and Shanthikumar (1986), and Sheu and Griffith (1992) model failure dependence in a system with n components using the joint distribution of the lifetimes of n components.

Albin *et al.* (1992) investigates the PM policies for the series system with failure dependence and economic dependence using Markov chain and presents a brief summary of previous work on failure dependence. Nakagawa and Murthy (1993) consider a two-unit system with two kinds of failure dependence: when unit 1 fails, (1) unit 2 fails with probability α_j where j represents the jth failure of unit 1 and (2) unit 1 causes damage with distribution $G(z)$ to unit 2, and the damage is cumulative and unit 2 fails once the total damage exceeds some specified level. Nakagawa and Murthy (1993) derive expected maintenance cost rates of the two models assuming that the system is replaced at failure of unit 2 or at the nth failure of unit 1 and discuss the optimum maintenance policies. Murthy and Wilson (1994) consider the parameter estimation problem for failure dependence models.

Pham (1992) studies a high voltage system reliability with dependent failures and treats the failure dependence in the way that the failure of one component causes the failure rate of the other working component to increase.

3

Maintenance Policies and Analysis

In the past several decades, maintenance and replacement problems of deteriorating systems have been extensively studied in the literature. Hundreds of maintenance and replacement models have been created. However, all these models can fall into certain categories of maintenance policies: age replacement policy, random age replacement policy, block replacement policy, periodic PM policy, failure limit policy, sequential PM policy, repair cost limit policy, repair time limit policy, repair number counting policy, reference time policy, mixed age policy, preparedness maintenance policy, group maintenance policy, opportunistic maintenance policy, *etc.* Each kind of policy has different characteristics, advantages, disadvantages, and relationship with others. This chapter summarizes, classifies, and compares various existing maintenance policies in the maintenance literature and practice for both single unit and multi-unit systems, following Wang (2002). Relationships among different maintenance policies are also addressed.

3.1 Introduction

Systems used in the production of goods and delivery of services constitute the vast majority of most industry's capital. These systems are subject to deterioration with usage and age (Valdez-Flores and Feldman 1989). Most of them are maintained or repairable systems. Therefore, maintenance on them may be necessary since it can improve reliability. The growing importance of maintenance has generated an increasing interest in the development and implementation of optimal maintenance strategies for improving system reliability, preventing the occurrence of system failures, and reducing maintenance costs of deteriorating systems.

As mentioned earlier in this chapter, maintenance, inspection, and replacement problems have been extensively investigated in the past several decades. In this chapter, a classification scheme of maintenance models that is amenable to current theoretical development is presented. This classification is intended to serve as guidance to both practitioners and researchers. The idea is to classify maintenance models such that a decision maker can recognize those that best fit his maintenance problem. Although thousands of maintenance models have been published, there are a limited number of maintenance policies which all maintenance models can be based on. For example, hundreds of maintenance models fall into the age replacement policy, and many fall into the failure limit policy. Therefore, this chapter examines existing maintenance models in terms of maintenance policies that they belong to. It is organized into two sections reflecting the classification scheme: maintenance policies of single unit systems and multi-unit systems. Since maintenance polices for single-unit systems are more established, and are the basis for maintenance policies of multi-unit systems, this chapter discusses single-unit systems in larger space. Note that maintenance policies can also be classified into time-based and system condition-based. For example, periodic PM policy is time-based, and failure limit maintenance policy is condition-based.

3.2 Maintenance Policies for One-unit Systems

As mentioned earlier, although thousands of maintenance models have been developed they can be classified into certain kinds of maintenance policies. This section summarizes, classifies, and compares maintenance policies of one-unit systems. The characteristics, advantages, and drawbacks for each kind of policy will be addressed. Maintenance models with different maintenance cost structures and/or different maintenance restoration degrees (minimal, imperfect, perfect) under the same maintenance policy will be classified into the same policy. The first five subsections of this section discuss maintenance policies with PMs and another subsection contemplates those without PMs. The last subsection provides a summary of them.

The basic assumption for single-unit systems under all PM polices is that the system lifetime has increasing failure rate (IFR).

3.2.1 Age-dependent PM Policy

The most common and popular maintenance policy might be the age-dependent PM policy. Studies on this type of policy went back to as early as Morse (1958). In some early work, the *age replacement policy* was extensively studied. Under this policy, a unit is always replaced at its age T or failure, whichever occurs first, where T is a constant (Barlow and Hunter 1960). Later, as concepts of minimal repair and especially imperfect maintenance (Pham and Wang 1996) became more and more established, various extensions and modifications of the age replacement policy have been proposed. This class of policies, *i.e.*, the age replacement policy and its extensions, is referred to as the *age-dependent PM policy* in this chapter since their PM times are based on the age of the unit. Under this type of policy, a unit is preventively maintained at some predetermined age T, or repaired at failure, until a perfect maintenance, preventive or corrective, is received. Note that PM at T and CM at failure might be either minimal, imperfect, or perfect. Thus, for this class of policies various maintenance models can be constructed according to different types of PMs (minimal, imperfect, perfect), CMs (minimal, imperfect, perfect), cost structures, *etc.* For example, PM at T might be a replacement or imperfect, CM at failure might be minimal or imperfect, maintenance cost may be a constant or a function of unit age or number of repairs, *etc*. Details can be found in Pham and Wang (1996) and Valdez-Flores and Feldman (1989), and Chapter 2 of this book. If T is a random variable, the policy is referred to as the random age-dependent maintenance policy that is in force when it is impractical to maintain a unit in a strictly periodic fashion. For example, a given unit may have a variable work cycle so that maintenance in midcycle is impossible or impractical. In this eventuality, the maintenance policy would have to be a random one, taking advantage of any free time available to perform maintenance. In the age replacement policy, items are replaced if they reach a certain age. This age is measured from the time of the last replacement. If only minimal repair is undertaken upon failure, the age replacement policy amounts to the "Periodic replacement with minimal repair at failure" policy (see Section 3.2.2).

Some researchers have produced many interesting and significant results for variations of the age replacement model. Tahara and Nishida (1975) introduce the maintenance policy "Replace the unit when the first failure after t_0 hours of operation or when the total operating time reaches T ($0 \le t_0 \le T$), whichever occurs first; failures in $[0, t_0]$ are removed by minimal repair." Note that if $t_0 \equiv 0$, it becomes the age replacement policy, and if $t_0 \equiv T$ it reduces to the "Periodic replacement with minimal repair at failure" policy. Observe that t_0 is a *reference time*, and maintenance actions are not performed exactly at that moment t_0 (unlike PM time).

Nakagawa (1984) extends the age replacement policy to replacing a unit at time T or at number N of failures, whichever occurs first, and undergoes minimal repair at failure between replacements. The decision variables for this policy are T and N. Note that this policy combines the fixed age and the repair number counting ideas. Clearly, if $N \equiv 1$, this policy reduces to the age replacement policy. Herein this policy is called T-N policy. A more general policy is that a subsystem is subject to imperfect PM at T, or CM at the Nth failure ($N = 1,2,3,...$), whichever occurs first, and undergoes imperfect repair at failure between replacements.

Two other expansions of the age replacement policy are provided by Sheu *et al*. (1993, 1995). Sheu *et al*. (1993) examine a generalized age replacement policy by using the idea similar to Tahara and Nishida (1975). In this policy if a unit fails at age $y < t$, it is subject to a perfect repair with $p(y)$, or undergoes a minimal repair with probability $q(y) = 1 - p(y)$. Otherwise, the unit is replaced when the first failure after t occurs or the total operating time reaches age T ($0 \le t \le T$), whichever occurs first. The policy decision variables are t and T. Obviously, if $t \equiv 0$ then this policy becomes the age replacement policy. If $t \equiv T$ and $q(y) \equiv 1$, it becomes the "Periodic replacement with minimal repair at failure" policy (see Section 3.2.2). Therefore, this policy is also general since it includes both age replacement policy and "Periodic replacement with minimal repair at failure", which are in two different categories of this chapter. Sheu *et al*. (1995) make another extension to the age replacement policy. They assume that a unit has two types of failures at age z, and is replaced at either the nth Type 1 failure or first

Type 2 failure, or at age T, whichever occurs first. Type 1 failure occurs with probability $p(z)$ and is corrected by minimal repair. Type 2 failure occurs with probability $q(z) = 1 - p(z)$ and is corrected by perfect repair. Clearly, if $p(z) = 0$ this policy becomes the age replacement policy. If $p(z) \equiv 1$ and $n \equiv \infty$, it becomes the "Periodic replacement with minimal repair at failure" policy (see Section 3.2.2). The policy decision variables are n and T. Again, this policy is quite general since it includes both the age replacement policy and the "Periodic replacement with minimal repair at failure" policy.

Block *et al.* (1993) introduce another generalized age replacement policy, *repair replacement policy*, where units are preventively maintained when a certain time has elapsed since their last repair. That is, items are repaired if they fail and are replaced only if they survive beyond a certain fixed time from the last repair or replacement. Units are either minimally or perfectly repaired at failure or they are replaced if they survive a certain fixed time from the last repair without suffering a CM. If at failure only perfect repair is allowed, then the repair replacement policy reduces to the age replacement policy. Consequently, the concept of a repair replacement policy is a more general type of replacement policy than the age replacement policy. This policy seems convenient, since, at repair, a schedule to maintain the item is in place and so the bookkeeping to start the maintenance policy can also be undertaken at this time. Furthermore, it seems reasonable, especially for an item which is aging and has undergone minimal repairs, to have some replacement policy rather than to do nothing.

Wang and Pham (1999) make another extension of age replacement policy, called "*Mixed age PM policy*". In this policy, after the nth imperfect repair, there are two types of failures. A Type 1 failure might be total breakdowns, while a Type 2 failure can be interpreted as a slight and easily fixed problem. When a failure occurs, it is a Type 1 failure with probability $p(t)$ and a Type 2 failure with probability $q(t) = 1 - p(t)$. Type 1 failures are subject to perfect repairs and Type 2 failures are subject to minimal repairs. Therefore, each repair is a perfect repair with probability $p(t)$ and a minimal one with probability $q(t) = 1 - p(t)$. After the first n imperfect repairs, the unit will be subject to a perfect maintenance at age T or at the first Type 1 failure, whichever occurs first. This process continues along an infinite time horizon. The policy decision variables are T and n. Obviously, if $p(t) \equiv 0$ and $n \equiv 0$, it becomes the "Periodic replacement with minimal repair at failures" policy. If $p(t) \equiv 1$ and $n \equiv 0$, it becomes the age replacement policy. Chapter 4 will further discuss this policy by investigating the maintenance cost rate, availability and optimal maintenance policy.

The age-dependent PM policy has probably received most of the attention in the literature. In the age-dependent PM policy, the failure rate is increasing with age. Various age-dependent PM policies, summarized from numerous existing maintenance models, are listed in Table 3.1. Table 3.1 shows that age replacement policy is the basic one and most extended policies are general and can include the age replacement policy and/or the "Periodic replacement with minimal repair at failure" policy as special cases. Note also that most of them are proposed based on

Table 3.1. Summary of age-dependent PM policies

Maintenance policy	Typical reference	PM time points	Decision variables	Special cases
Age replacement	Barlow and Hunter (1960)	Fixed age T	T	
Repair Replacement	Block *et al.* (1993)	Time since last maintenance	Fixed time	Age replacement
T-N	Nakagawa (1984)	Fixed age T or time	T, N	Age replacement Periodic PM
(T, t)	Sheu *et al.* (1993)	Fixed Age T or time	T, t	Age replacement Periodic PM
(t_0, T)	Tahara and Nishida (1975)	Fixed age T	t_0, T	Age replacement Periodic PM
Mixed age	Wang and Pham (1999)	Fixed age T or time	k, T	Age replacement Periodic PM
(T, n)	Sheu *et al.* (1995)	Fixed Age T	T, n	Age replacement Periodic PM

imperfect maintenance concepts. Most extended policies have more than one decision variables.

3.2.2 Periodic PM Policy

In the *periodic PM policy*, a unit is preventively maintained at fixed time intervals kT ($k = 1,2,...$) independent of the failure history of the unit, and repaired at intervening failures where T is a constant. In some early research, the *block replacement policy* was examined in which a unit is replaced at pre-arranged times kT ($k = 1,2,...$) and at its failures. The block replacement policy derives its name from the commonly employed practice of replacing a block or group of units in a system at prescribed times kT ($k = 1,2,...$) independent of the failure history of the system and is often used for multi-unit systems. Early research on the block replacement policy can be found in Welker (1959) and Drenick (1960) (see Barlow and Proschan 1965). Another basic periodic PM policy in this class is *"Periodic replacement with minimal repair at failures"* policy under which a unit is replaced at pre-determined times kT ($k = 1,2,...$) and failures are removed by minimal repair (Barlow and Hunter 1960, Policy II). This policy is good for large systems where minimal repair are plausible at failures. The third basic periodic PM policy: *no failure replacement*, is that a unit is always replaced at times kT ($k = 1,2,...$) but it is not replaced at failure.

As the concepts of minimal repair and especially imperfect maintenance (Pham and Wang 1996) became more and more established, various extensions and variations of these two policies were proposed. One expansion of the "Periodic replacement with minimal repair at failure" policy is the one where a unit receives imperfect PM every T time unit, intervening failures are subject to minimal repairs, and it is replaced after its age has reached $(O+1)T$ time units where O is the number of imperfect PMs which have been done (Liu *et al.* 1995). $O = 0$ is allowed in this policy, which means the unit will be replaced whenever it has operated for T time units and there will be no imperfect PM for it. The policy decision variables are O and T. Obviously, if $O = 0$, this policy becomes the "Periodic replacement with minimal repair at failure" policy.

Cox (1962) extends the block replacement policy to one where if a failure occurs just before a preventive replacement at T, it will be left down until the following preventive replacement. In particular, if a failure occurs in an interval $(kT - \delta, kT)$, $\forall k, k = 0,1,2,...$, the replacement will be made instantaneously but at kT. Obviously, if $\delta = 0$, this policy reduces to the block replacement policy. if $\delta = T$, this policy reduces to the third basic periodic PM policy.

Berg and Epstein (1976) have modified the block replacement policy by setting an age limit. Under this modified policy, a failed unit is replaced by a new one; however, units whose ages are less than or equal to t_0 $(0 \le t_0 \le T)$ at the scheduled replacement times kT $(k = 1,2,...)$ are not replaced, but remain working until failure or the next scheduled replacement time point. Obviously, if $t_0 = T$, it reduces to the block replacement policy. In Berg and Epstein (1976), this modified block replacement policy is shown to be superior to the block replacement policy in terms of the long-run maintenance cost rate.

Tango (1978) suggests that some failed units be replaced by used ones, which have been collected before the scheduled replacement times. Under this extended block replacement policy, units are replaced by new ones at periodic times kT $(k = 1,2,...)$. The failed units are, however, replaced by either new ones or used ones based on their individual ages at the times of failures. A time limit r is set in this policy, similar to t_0 in Berg and Epstein (1976). Under this policy, if a failed unit' age is less than or equal to a predetermined time limit r, it is replaced by a new one; otherwise, it is replaced by a used one. This policy is different from Berg and Epstein's (1976) because it modifies the ordinary block replacement policy by considering rules on the failed units rather than on the working ones. Obviously, if $r = T$, this policy becomes the block replacement policy.

Nakagawa (1981a, b) presents three modifications to the "Periodic replacement with minimal repair at failure" policy. The modifications give alternatives that emphasize practical considerations. The three policies all establish a reference time T_0 and periodic time T^*. If failure occurs before T_0, then minimal repair occurs. If the unit is operating at time T^*, then replacement occurs at time T^*. If failure occurs between T_0 and T^*, then: **(Policy I)** the unit is not repaired and remains failed until T^*; **(Policy II)** the failed unit is replaced by a spare unit as many times as needed until T^*; **(Policy III)** the failed unit is replaced by a new one. In all these three policies, the policy decision variables are T_0 and T^*. Clearly, if $T_0 \equiv T^*$, Policies I, II, and II all become the "Periodic replacement with minimal

Table 3.2. Summary of periodic PM policies

Maintenance policy	Typical reference	PM time points	Decision variables	Special cases
Block replacement	Barlow and Hunter (1960)	Periodic time	Periodic time	
Periodic replacement with minimal repair	Barlow and Hunter (1960)	Periodic time	Periodic time	
Overhaul and Minimal repair	Liu *et al.* (1995)	Periodic time and its multiples	Fixed number of PMs / Periodic time	Periodic replacement with minimal repair
(T_0, T^*) Policy I	Nakagawa (1981)	Periodic time	Periodic time/ reference time	Periodic replacement with minimal repair
(T_0, T^*) Policy II	Nakagawa (1981)	Periodic time	Periodic time/ reference time	Periodic replacement with minimal repair
(T_0, T^*) Policy III	Nakagawa (1981)	Periodic time	Periodic time/ reference time	Periodic replacement with minimal repair/ Block replacement
(n, T)	Nakagawa (1986)	Periodic time	Periodic time /number of failures	Periodic replacement with minimal repair
(r, T)	Tango (1978)	Periodic time	Periodic time/ reference age	Block replacement
(N, T)	Wang and Pham (1999)	Periodic time and its multiples	Periodic time /number of repairs	Block replacement/ Periodic replacement with minimal repair
(δ, T)	Cox (1962)	Periodic time	Periodic time/ reference age	Block replacement/ No failure repair
(t_0, T)	Berg and Epstein (1976)	Periodic time	Periodic time/ reference age	Block replacement

repair at failure" policy. If $T_0 \equiv 0$, Policy III becomes the block replacement policy.

Nakagawa (1980) also makes an expansion to the block replacement policy. In his policy, a unit is replaced at times kT ($k = 1, 2, ...$) independent of the age of the unit. A failed unit remains failed until the next planned replacement. Another variant of the "Periodic replacement policy with minimal repair" policy is also due to Nakagawa (1986), in which the replacement is scheduled at periodic times kT ($k = 1, 2, ...$) and failure is removed by minimal repair. If the total number of failures is equal to or greater than a specified number n, the replacement should be done at the next scheduled time; otherwise, no maintenance should be done. The decision variable is n and T. In this policy, if $n = \infty$, this policy becomes the "Periodic replacement with minimal repair at failure" policy.

Chun (1992) studies determination of the optimal number of periodic PMs under a finite planning horizon. Dagpunar and Jack (1994) determine the optimal number of imperfect PMs for a finite horizon given that the minimal repair is made at any failure between PMs.

Wang and Pham (1999) extend the block replacement policy to a general case. In their policy, a unit is imperfectly repaired at failure if the number of repairs is less than N (a positive integer). The repair is imperfect in the sense that the unit has shorter and shorter lifetime upon each repair. Upon the Nth imperfect repair at failure, the unit is preventively maintained at kT ($k = 1,2,...$) where the constant $T > 0$. The PM is imperfect in the sense that after PM the unit is "as good as new" with probability p and "as bad as old" with $(1 - p)$. Upon a perfect PM, the maintenance process repeats. The decision variables are N and T. The justification of this policy is that when a new unit is put into operation, the first N repairs at failure will be performed at a low cost. This is because the unit is young at those times and these repairs turn out to be imperfect. Usually, these repairs are just minor repairs because it is in good operating condition. After the Nth imperfect repair at failure, the unit may be in worse operating condition due to usage, aging and imperfectness of repairs, and then a major maintenance is necessary at a higher cost. If the repair at failure and PM are perfect and $N \equiv \infty$, this policy reduces to the block replacement policy. If the repair at failure is minimal and PM is perfect and $N \equiv \infty$, this policy amounts to the "Periodic replacement with minimal repair at failure" policy. Chapter 4 will further discuss this policy.

Maintenance schedules under the periodic PM policy, summarized from various existing maintenance models, are listed in Table 3.2, which shows that block replacement policy and periodic replacement with minimal repair are the basic ones with one decision variable. Other policies have more than one decision variables.

3.2.3 Failure Limit Policy

Under the *failure limit policy*, PM is performed only when the failure rate or other reliability indices of a unit reach a predetermined level and intervening failures are corrected by repairs. This PM policy makes a unit work at or above the minimum acceptable level of reliability. For example, Lie and Chun (1986) formulate a maintenance cost policy where PM is performed whenever a unit reaches the predetermined maximum failure rate, and failures are corrected by minimal repair. Bergman (1978) investigates a failure limit policy in which replacements are based on measurements of some increasing state variable, *e*.g., wear, accumulated damage or accumulated stress, and the proneness to failure of an active unit is described by an increasing state-dependent failure rate function. The optimal replacement rule in terms of average long-run maintenance cost rate is shown to be a failure limit rule, *i.e.*, it is optimal to replace either at failure or when the state variable has reached some threshold value, whichever occurs first. Bergman's model includes the age replacement policy as a special case.

Other research on the failure limit policy can be found in Malik (1979), Canfield (1986), Jayabalan and Chaudhuri (1992a), Jayabalan and Chaudhuri (1992c), Jayabalan and Chaudhuri (1995), Chan and Shaw (1993), Suresh and Chaudhuri (1994), Monga et al. (1997), and Pham and Wang (1996). In addition, Love and Guo (1996) study failure limit policy for PM decisions under Weibull failure rates.

Generally, the problem for this class of policies is that it requires much computing effort to determine maintenance schedules and wasteful to implement.

The failure limit policy and its extensions are summarized in Table 3.3, that shows that most failure limits are measured by failure rates, and maintenance cost rate is most used to determine the optimal maintenance policies. Several policies consider finite planning horizon.

Table 3.3. Summary of failure limit policies

Typical reference	Reliability index monitored	Optimality criterion	Planning horizon
Bergman (1978)	Failure rate through wear, accumulated damage or stress	Cost rate	Infinite
Malik (1979)	Reliability	Reliability	Infinite
Canfield (1986)	Failure rate	Cost rate	Infinite
Zheng and Fard (1991)	Failure rates	Cost rate	Infinite
Lie and Chun (1986)	Failure rate	Cost rate	Infinite
Jayabalan and Chaudhuri (1992a)	Failure rate	Total cost	Finite
Jayabalan and Chaudhuri (1992c)	Age others	Cost rate	Infinite
Jayabalan and Chaudhuri (1992d)	Age	Total cost	Finite
Chan and Shaw (1993)	Failure rate	Availability	Infinite
Suresh and Chaudhuri (1994)	Reliability and failure rate	Total cost	Finite
Jayabalan and Chaudhuri (1995)	Age	Total cost	Finite
Monga et al. (1997)	Reduction (age and failure rate)	Cost rate	Infinite
Love and Guo (1996)	Weibull failure rate	Cost rate	Infinite

3.2.4 Sequential PM Policy

Unlike the periodic PM policy, a unit is preventively maintained at unequal time intervals under the *sequential PM policy*. Usually, the time intervals become shorter and shorter as time passes, considering that most units need more frequent maintenance with increased ages. An early sequential PM policy is designed for a finite span (Barlow and Proschan 1962). Under this sequential policy, the age for which PM is scheduled is no longer the same following successive PMs, but depends on the time still remaining. Clearly the added flexibility permits the achievement of an optimum sequential PM policy having lower cost than that of the corresponding optimum age replacement policy. Under sequential PM policy, the next PM interval is selected to minimize the expected expenditure during the remaining time. Thus, this policy does not specify at the beginning of the original time span each future PM interval; rather, after each PM, it specifies only the next PM interval. This gain in flexibility leads to reduction in expected cost.

Nguyen and Murthy (1981b) introduce a sequential policy which calls for a PM if a failure has not occurred by some reference time t_i, where t_i is the maximum time that a unit should be left without maintenance after the $(i\text{-}1)^{th}$ repair (time from the last repair or replacement). In this policy, a unit is replaced after $(k-1)$ repairs. It is repaired (or replaced at the kth repair) at the time of failure or at age t_i, whichever occurs first. The decision variables are k and t_i for $i = 1,...,k$, given that each PM increases the failure rate of the unit. If $k = 1$, this sequential policy reduces to the age replacement policy.

Nakagawa (1986, 1988) discusses a sequential PM policy where PM is done at fixed intervals x_k for $k = 1,2,,...,N$. The unit is replaced at the Nth PM and failures between PMs are corrected by minimal repairs, given the unit has different failure distributions between PMs (the failure rate of the unit increases with the number of PMs, or its age is reduced (1988), *i.e.*, the first $(N-1)$ PMs are imperfect). The policy decision variables are N and x_k ($k = 1,2,,...,N$). Nakagawa (1986, 1988) also presents two numerical examples indicating that the optimal policy satisfies $x_k \leq x_{k-1}$ for $k = 2,...,N$. Nguyen and Murthy (1981b) study this policy (Policy II in their paper). If $N = 1$, this sequential policy reduces to the "Periodic replacement with minimal repair at failure" policy.

These sequential policies are practical because most units need more frequent maintenance with increased age. They are different from the failure limit policy in that it controls x_k lengths directly but the failure limit policy controls failure rate, reliability, *etc.*, directly. Moreover, Kijima and Nakagawa (1992) develop a sequential PM policy using an accumulated damage concept.

In Wu and Clements-Croome (2005), both PM and CM are carried out. The PM is sequentially executed with τ_n time units after the $(n\text{-}1)^{th}$ PM, where $n = 1,2,...$ Between two adjacent PMs, a CM is carried out immediately on failure. τ_n is dependent on n and determined through minimizing the maintenance cost rate. Obviously, if τ_n is independent on n, the sequential PM policy becomes a periodic PM policy.

3.2.5 Repair Limit Policy

The repair limit policies and their extensions are summarized in Table 3.4. Note that in the existing literature, there are two types of repair limit policies: *repair cost limit policy* and *repair time limit policy*.

When a unit fails, the repair cost is estimated and repair is undertaken if the estimated cost is less than a predetermined limit; otherwise, the unit is replaced. This is called *repair cost limit policy* in the literature, investigated by Gardent and Nonant (1963) and Drinkwater and Hastings (1967).

A drawback of the repair cost limit policy is that the replacement or repair decision depends only on the cost of a single repair. Long-lasting situations characterized by frequent repairs whose costs are below the corresponding limit do not directly influence the time of replacement, although the repair cost rate might justify a replacement. Thus, further financial savings seem possible if the replacement decision depends on the whole history of the repair process. Considering this drawback, Beichelt (1982) examines repair cost limit policy and uses the repair cost rate (repair cost per unit time) as a criterion of replacement or repair: a unit is replaced as soon as the repair cost rate reaches or exceeds a fixed level, otherwise, it is repaired. In this policy (Beichelt, 1982), the replacement intervals are independently and identically distributed random variables. Yun and Bai (1987) propose a repair cost limit policy in which when a unit fails, the repair cost is estimated and repair is undertaken if the estimated cost is less than a predetermined limit L, where the repair is imperfect. Otherwise, the unit is replaced. This policy by Yun and Bai (1987) is generalized from the one by Hastings (1967). In addition, Kapur *et al.* (1989) extend the repair cost limit policy to incorporate the number of repairs as a policy decision variable.

Table 3.4. Summary of repair limit policies

Reference	CM before limit	CM after limit	Limit type	Optimality criterion	Planning horizon
Hastings (1969)	Minimal	Perfect	Cost	Cost rate	Infinite
Kapur et al. (1989)	Minimal	Perfect	Cost	Cost rate	Infinite
Beichelt (1982)	Perfect	Perfect	Cost rate	Cost rate	Infinite
Beichelt (1978,1981b)	Minimal	Perfect	Cost rate	Cost rate	Infinite
Nguyen and Murthy (1981)	Imperfect	Perfect	Time	Cost rate	Infinite
Yun and Bai (1988)	Minimal	Perfect	Cost	Cost rate	Infinite
Koshimae et al. (1996)	Perfect	Perfect	Time	Cost rate	Infinite
Nguyen and Murthy (1980)	Minimal	Perfect	Time	Cost rate	Infinite
Dohi et al. (1997)	Minimal	Imperfect	Time	Cost rate	Infinite
Park (1983)	Minimal	Perfect	Cost	Cost rate	Infinite
Nakagawa and Osaki (1974)	Minimal	Perfect	Time	Cost rate	Infinite
Yun and Bai (1987)	Imperfect	Perfect	Cost	Cost rate	Infinite
Wang and Pham (1996c)	Imperfect	Imperfect	Cost	Availability/ Cost rate	Infinite

The *repair time limit policy* is proposed by Nakagawa and Osaki (1974) in which a unit is repaired at failure: if the repair is not completed within a specified time T, it is replaced by a new one; otherwise the repaired unit is put into operation again, where T is called repair time limit. Nguyen and Murthy (1980) study a repair time limit replacement policy with imperfect repair in which there are two types of repair – local and central repair. The local repair is imperfect while the central repair is perfect, which may take a longer time. Dohi *et al.* (1997) consider a generalized repair time limit replacement problem with lead time and imperfect repair, which is subject to a time constraint, and propose a nonparametric solution procedure to estimate the optimal repair time limit. Koshimae *et al.* (1996) consider another repair time limit policy. Under this policy, when the original unit fails, the repair is started immediately. If the repair is completed in a time limit t_0, then the repaired unit is installed as soon as the repair is finished. On the other hand, if the repair time is greater than the time limit t_0, the failed unit is scrapped and a spare is ordered immediately. It is delivered and installed after a lead time. The policy decision variable is the repair time limit t_0.

3.2.6 Repair Number Counting and Reference Time Policy

Morimura and Makabe (1963a) introduce a policy where a unit is replaced at the kth failure. The first $(k-1)$ failures are removed by minimal repair. Upon replacement, the process repeats. This policy is called *repair number counting policy* in this chapter. The policy decision variable is k. Later, Morimura (1970) extends this policy by introducing another policy variable T – critical *reference time* which is a positive number. Under this extended policy, all failures before the kth failure are corrected only with minimal repair. If the kth failure occurs before an accumulated operating time T, it is corrected by minimal repair and the next failure induces replacement. But if the kth failure occurs after T, it induces replacement of the unit. Obviously, this policy combines the ideas of counting the number of repairs and recording the elapsed time. The policy decision variables are k and T. If the policy decision variable T is zero, this policy reduces to the repair number counting policy. An imperfect repair version of the repair number counting policy is examined by Jack (1991): performing imperfect repair on failure, and replacement upon the kth failure. A policy similar to the repair number counting policy is also investigated by Park (1979) in which a unit is replaced at the kth failure and minimal repairs are performed for the first $(k-1)$ failures. Later, Lam (1988) and Stadje and Zuckerman (1990) investigate the repair number counting policy, given that the lengths of the operating intervals decrease whereas the durations of the repair increase in different ways.

Muth (1977) examines a replacement policy, similar to the reference time idea of the extended policy by Morimura and Makabe (1970), in which a unit is minimally repaired up to time T and replaced at the first failure after T. This policy is referred to as the *reference time policy* later in this review. Note that in this policy the maintenance action is not undertaken exactly at the reference time point T (unlike PM time). Makis and Jardine (1991, 1993) discuss an optimal replacement policy with imperfect repair at failure: a unit is replaced at the first failure after some fixed time. Makis and Jardine (1992) also introduce a general policy in which a unit can be replaced at any time and at the nth failure the unit can be either replaced or can undergo an imperfect repair. Under different conditions, this policy can reduce to the repair number counting policy, reference time policy, and "Periodic replacement with minimal repair at failure" policy, respectively. Therefore, it is a quite general policy.

In general, the repair number counting policy is effective when the total operating time of a unit is not recorded or it is time consuming and costly to replace a unit in operation. It has been proven (Muth 1977) that the reference time policy yields a lower long-run expected cost per unit time than the periodic PM policy given that the mean residual life function of the unit is strictly decreasing after some age t_0. With this condition, called positive aging, the unit deteriorates and eventually reaches a condition where it is no longer economically justifiable to perform minimal repair after repair. It is shown that the repair number counting policy yields lower asymptotic expected cost rate than the age replacement policy. Also the number of failures before replacement in the repair number counting policy is less than that in the age replacement policy. However, all these results are proven numerically for the Weibull distribution (*i.e.*, for some specific Weibull distribution parameter values).

Phelps (1981) compares the "Periodic replacement with minimal repair at failure" policy (Barlow and Hunter 1960), the repair number counting policy (Morimura and Makabe 1963a,b; Park 1979), and the reference time policy (Muth 1977), given an increasing failure rate. Phelps (1981) shows that the reference time policy, replacing after the first failure that occurs after reference time T, is the optimal of the three policies in terms of the long-run cost rate; the repair number counting policy is more economical than the "Periodic replacement with minimal repair at failure" policy.

Note that generally there are no PMs scheduled for this type of policy. These policies are mainly based on counting the number of repairs and/or reference time, but the age-dependent PM policy and periodic PM policy rely on PM times, at which maintenance actions are performed. In the repair number counting and reference time policy, maintenance actions are not undertaken precisely at the reference time point T. In the repair number counting and reference time policy, number of repairs and/or reference time are policy decision variable(s). In the age-dependent PM policy and periodic PM policy, PM time is one of the policy decision variables.

3.2.7 On the Maintenance Policies for Single-unit Systems

The age-dependent PM policy and periodic PM policy have received much more attention in the literature. Hundreds of papers and models have been published under these two kinds of maintenance policies, as summarized in McCall (1963), Barlow and Proschan (1965, 1975), Pieskalla and Voelker (1976), Osaki and Nakagawa (1976), Sherif and Smith (1981), Jardine and Buzacott (1985), Valdez-Flores and Feldman (1989), Pham and Wang (1996), and Wang (2002). Detailed mathematical comparisons on the age and block replacement policies can be found in Barlow and Proschan (1965, 1975) in which the general conclusion is that the age replacement policy is an economical way to the block replacement policy. They prove that more unfailed components under the block replacement policy are removed than under age replacement policy, and the total number of removals for both failed and unfailed components under the block replacement policy is larger. Berg and Epstein (1978) compare three types of replacement policies: age, block, failure replacement policies and provide a heuristic rule for choosing the best one. Berg (1976a) and Bergman (1980) prove that an age replacement policy is optimal among all reasonable maintenance policies. In Block *et al.* (1990), comparisons are made between the block replacement policy and "Periodic replacement with minimal repair at failure" policy. In Block *et al.* (1993), comparisons are made among the age replacement policy, block replacement policy, and repair replacement policy. The periodic PM policy is perhaps more practical than the age-dependent PM policy since it does not require keeping records on unit usage. The block replacement policy is more wasteful than the age replacement policy since a unit of "young" age might be replaced at periodic times. Generally, the same argument may hold for the age-dependent PM and periodic PM policy.

The failure limit policy, repair limit policy, and sequential policy are more practical, but there has been much less research done on it. The failure limit policy is also directly consistent with the maintenance objectives: improving reliability and reducing failure frequency. One of the disadvantages of the failure limit policy and sequential policy is that their PM intervals are not equal and thus it is wasteful to implement them.

Note that maintenance policies have become more and more general because they include some previous policies as special cases. This is reflected in Tables 3.1 and 3.2, and described in Sections 3.2.1 – 3.2.6. In general, optimal maintenance plans obtained from these general policies may result in some cost savings since the optimal maintenance schedules under them might be "globally" optimal (optimal in a larger range). However, as they become more and more complicated, these general policies may also cause inconvenience in implementation in practice. Similarly, the maintenance cost is no longer a constant and becomes more and more general. For example, it may be a function of unit age and number of repairs already performed on the unit (Note that Frenk *et al.* (1997) establish a general method for modeling complicated maintenance costs, which is also convenient for this case).

Generally, each maintenance policy for one-unit systems either depends on counting/recording of the number of repairs, PM time, or reference time. In practice, counting number of repairs and recording PM time, and reference time are all possible ways. The current research seems to intend to use two or more of them as policy decision variables in a single policy.

Note that in some policies no PMs are involved. For example, in Gasmi *et al.* (2003), a system is observed to operate in alternative states. The most common mode is loaded (or regular) operation. Occasionally the system is placed in an unloaded state, while the system is mechanically still operating; it is assumed that the failure intensity is lower due to this reduction in operating intensity. They use a proportional hazard model to capture this reduction in failure intensity due to switching of operating modes. In either operating state, the system is occasionally shut down for repair and upon failure, one of the three actions is taken: a) minimal repair, b) minor repair, or c) major repair.

3.3 Maintenance Policies of Multi-unit Systems

The six types of maintenance policies in Section 3.2 are designed for a system composed of a single stochastically deteriorating subsystem. A natural generalization of these maintenance policies is to consider a system with a number of subsystems. Optimal maintenance policies for such systems reduce to those for systems with a single subsystem only if there exists neither economic dependence, failure dependence nor structural dependence. In this case, maintenance decisions are independent, and the "optimal" maintenance policy is to employ one of the six classes of maintenance policies described in Section 3.2 for each separate subsystem. However, if there exists dependence, for example, economic dependence, then the optimal maintenance policy is not one of considering each subsystem separately and maintenance decisions will not be independent. Obviously, the optimal maintenance action for a given subsystem at any time point depends on the states of all subsystems in the system: the failure of one subsystem results in the possible opportunity to undertake maintenance on other subsystems (opportunistic maintenance). In this chapter, economic dependence means that performing maintenance on several subsystems jointly costs less money and/or time than on each subsystem separately. Failure dependence means that failure distributions of several subsystems are stochastically dependent.

Economic dependency is common in most continuous operating systems. Examples of such systems include aircraft, ships, power plants, telecommunication systems, chemical processing facilities, and mass production lines. For this type of system, the cost of system unavailability (one-time shut-down) may be much higher than maintenance costs. Therefore, there is often a great potential for cost savings by implementing an opportunistic maintenance policy.

Currently, there is an increasing interest in multicomponent maintenance policies and models. As pointed out in van der Duyn Schouten (1995), one of the reasons that is often put forward to explain the lack of success in applications of maintenance and replacement models is the simplicity of the models compared to the complex environment where the applications occur. In particular, the fact that up to ten years ago the vast majority of the maintenance models were concerned with one single piece of equipment operating in a fixed environment was considered an intrinsic barrier for applications. Next we summarize maintenance policies for multi-unit systems. Cho and Parlar (1991) survey the multi-unit system maintenance models created before 1991, and Dekker *et al.*'s review (1997) is focused on economic dependence models published after 1991. This chapter emphasizes classifications and characteristics of maintenance policies though sometimes it cites the same existing maintenance models as the previous survey papers. The basic assumptions for multi-unit systems under all maintenance polices are that there are virtually infinitely many disposable identical units with *i.i.d.* lifetimes for all items; salvage values of all units are negligible.

3.3.1 Group Maintenance Policy

The problem of establishing group maintenance policies, which are best from the point of view of the system's reliability or operational cost, has received significant attention in the maintenance literature. One class of problem for group maintenance policies has been to establish categories of units that should be replaced when a failure occurs. This is particularly important when there are varying access costs associated with disassembly and reassembly, and simultaneous PM of categories of parts may be appropriate. A second class of group replacement studies has been concerned with reducing costs by including redundant parts into systems design. A third class of papers has been concerned with establishing group maintenance policies for systems of independently operating machines, all of which are subject to stochastic failures from the same distribution (Ritchken and Wilson 1990). For this class of problems, there are three existing group maintenance policies. The first policy, referred to as a T-age group replacement policy, calls for a group replacement when the system is of age T. A second policy, referred to as an m-failure group replacement policy, calls for a system inspection after m failures have occurred. The third policy combines the advantages of the m-failure and T-age policies. This policy, referred to as an (m, T) group replacement policy, calls for a group replacement when the system is of age T, or when m failures have occurred, whichever comes first. The (m, T) group replacement policy requires inspection at either the fixed age T or the time when m machines have failed, whichever comes first. At an inspection, all failed units are replaced with new ones and all functioning units are serviced so that they become as good as new. The policy decision variables are m and T.

Gertsbakh (1984) introduces a policy in which a system has n identical units with exponential lifetimes, and it is repaired when the number of failed units reaches some prescribed number k, the policy decision variable. Vergin and Scriabin (1977) propose an (n, N) policy. Under this group maintenance policy, a unit undergoes preventive replacement if it has operated for N periods, and undergoes a group replacement if it has operated n periods and if either another unit fails or another unit reaches its preventive replacement age (where $n < N$). Love et al. (1982) establish another group replacement policy for a fleet of vehicles. Under this group maintenance policy a vehicle is replaced when repair cost for the vehicle exceeds a pre-set repair limit; otherwise, it is repaired. Sheu and Jhang (1997) propose a 2-phase group maintenance policy for a group of identical repairable items. The time interval $(0, T)$ is defined as the first phase, and the timer interval $(T, T+W)$ is defined as the second phase. As individual units fail, individual units have two types of failures. Type I failures are removed by minimal repairs, whereas Type II failures are removed by replacements or are left idle. A group of maintenance is conducted at time $T+W$ or upon the kth idle, whichever comes first. The policy decision variables are T, W, and k.

Wildeman et al. (1997) discuss a group maintenance policy considering that a maintenance activity carried out on a technical system involves a system-dependent set-up cost that is the same for all maintenance activities carried out on that system, and grouping activities thus saves costs since execution of a group of activities requires only one set-up. Under this policy, a rolling-horizon approach is proposed that takes a long-term tentative plan as a basis for a subsequent adaptation according to information that becomes available in the short term. This policy makes it easy to incorporate short-term circumstances such as opportunities or a varying use of components because these are either not known beforehand or make the problem intractable.

Assaf and Shanthikumar (1987) propose a group preparedness policy for a set of N machines having exponential lifetimes with constant rate. A failed machine can be repaired at any time, and the repair is perfect. The number of failed machines in the system is unknown unless an inspection is carried out. Upon an inspection, a decision will be made on whether to repair the failed machines or not, based on the number of failed machines in the system, a policy decision variable.

3.3.2 Opportunistic Maintenance Policies

As pointed out earlier, maintenance of a multicomponent system differs from that of a single-unit system because there exists dependence in multicomponent systems. One of the dependencies is economic dependence. For example, it is possible to do PM to non-failed subsystems at a reduced additional cost while failed subsystems are being repaired. Another dependence is failure dependence, or correlated failures. For example, the failure of one subsystem may affect one or more of the other functioning subsystems, and times to failures of different units are then statistically dependent (Nakagawa and Murthy 1993). Berg (1976, 1977) suggests a preventive replacement policy for a machine with two identical components which are subject to exponential failure. Under this policy, upon a component failure the other component as well as the failed one is also replaced by a new one if its age exceeds a pre-determined control limit L. Later, Berg (1978) extends it to such an policy: both units are replaced either when one of them fails and the age of the other unit exceeds the critical control limit L, or when any of them reaches a predetermined critical age S. A unit is replaced at age T or at failure, whichever occurs first. Note that this policy will become two independent age replacement policies if $L \equiv \infty$.

Zheng and Fard (1991) examine an opportunistic maintenance policy based on failure rate tolerance for a system with k different types of units. A unit is replaced (active replacement) either when the hazard rate reaches L or at failure with the failure rate in a predetermined interval $(L - u, L)$. When a unit is replaced due to the hazard rate reaching L, all of the operating units with their hazard rate falling in the interval $(L - u, L)$ are replaced (passive replacement) at that time. A unit is subject to minimal repair at failure when the hazard rate is in interval $(0, L - u)$. The policy decision variables are L and u.

Kulshrestha (1968) investigates an opportunistic maintenance policy in which there are two classes of units, 1 and 2. Class 1 contains M standby redundant units so that upon the failure of the currently operating Class-1 units, a standby takes over. When all the Class-1 standbys have failed, the system suffers catastrophic failure. The Class-2 units, on the other hand, form a series system; if one of them should fail, the system suffers a minor breakdown. When a minor breakdown occurs, there is a possible chance for opportunistic repair of those Class-1 units which have failed.

Pham and Wang (2000) propose two new (τ, T) opportunistic maintenance policies for a k-out-of-n system. In these two policies, minimal repairs are performed on failed components before time τ - a policy decision variable, and CM of all failed components is combined with PM of all functioning ones after τ. At time T, another policy decision variable, PM is performed if the system has not been subject to a perfect maintenance before $T > \tau$. The policy decision variables are τ and T. Pham and Wang (2000) also extend these two policies to the one including the third decision variable – the number of failed components to start CM, considering the k-out-of-n system may still operate even if some of its components have failed. Chapter 8 will further discuss this policy.

Dagpunar (1996) introduces a general maintenance policy where replacement of a component within a system is available at an opportunity. An opportunity arises if the failure of some other part of the system allows the component in question to be replaced. It is assumed that the opportunity process is Poisson, which is reasonable if the system consists of a large number of components which are regularly maintained. In this policy the component will be replaced if its age at an opportunity exceeds a specified control limit.

Radner and Jorgenson (1963) and Wang $et\ al.$ (2001) investigate an opportunistic preparedness maintenance of multi-unit systems with ($n + 1$) subsystems. Wang $et\ al.$ (2001) examine such a preparedness policy:

i)   If subsystem i fails when the age of subsystem 0 is in the time interval $[0, t_i)$, replace subsystem i alone at a cost of C_i and at a time of w_i, $\forall i, i = 1, 2, ..., n$.

ii)   If subsystem i fails when the age of subsystem 0 is in the time interval $[t_i, T)$, replace subsystem i and do perfect PM on subsystem 0, $\forall i, i = 1, 2, ..., n$. The total maintenance cost is C_{0i} and total maintenance time is w_{0i}.

iii) If subsystem 0 survives until its age $x = T$, perform PM on subsystem 0 alone at a cost of C_0 and at a maintenance time of w_0 at $x = T$. PM is imperfect.

iv)   If subsystem 0 has not received a perfect PM at T, perform PM on it alone at time jT ($j = 2, 3, ...$) until it gets a perfect PM; if subsystem 0 has not experienced a perfect maintenance and subsystem i fails after some PM, replace subsystem i and do perfect PM on subsystem 0, $\forall i, i = 1, 2, ..., n$. The total maintenance cost is still C_{0i} and total maintenance time is w_{0i}. This process continues until subsystem 0 gets a perfect maintenance.

Chapter 7 will further discuss the above preparedness maintenance policy by investigating its reliability and maintenance cost measures. Chapter 6 will discuss the following opportunistic maintenance policy for multi-unit systems with $(n+1)$ subsystems and arbitrary reliability architecture:

(i)   If subsystem 0 fails at any time before T, perform imperfect repair on it at a cost C_{00}.

(ii)   If subsystem i fails when the age of subsystem 0 is in the time interval $0 \le x < t_i$, replace subsystem i alone at a cost C_i and at a time w_i, $\forall i, i = 1,2,...,n$.

(iii)   If subsystem i fails when the age of subsystem 0 is in the time interval $t_i \le x < T$, replace subsystem i and do perfect PM on subsystem 0, $\forall i, i = 1,2,...,n$. The total maintenance cost is C_{0i} and total maintenance time is w_{0i}.

(iv)   If subsystem 0 survives until $x = T$, perform PM on subsystem 0 at a cost C_0 and at a maintenance time w_0.

4

A Quasi-renewal Process and Its Applications

Renewal theory had its origin in the study of strategies for replacement of technical components (Cox 1962). In a renewal process, the times between successive events are supposed to be independent and identically distributed (*i.i.d*). Most maintenance models using renewal theory were actually based on this assumption, *i.e.*, "as good as new" after each maintenance. As pointed out in Chapters 1 and 2, "as good as new" represents one extreme type of maintenance results and usually a system may not be good as new but younger after maintenance, *i.e.*, maintenance is imperfect. Therefore, it is useful to establish some directly effective theories which can be used to model imperfect maintenance.

In this chapter, a general renewal process, a quasi-renewal process, is introduced and its usefulness in treating imperfect maintenance of one-unit systems is demonstrated. Section 4.1.1 presents the definition of the quasi-renewal process and discusses its properties. The quasi-renewal function of the quasi-renewal process is derived in Section 4.1.2. The justification of the quasi-renewal process in modeling imperfect maintenance and the problems of statistical hypothesis testing are discussed in Section 4.1.3. Using the quasi-renewal theory, eleven imperfect maintenance models with periodic PM, age-dependent PM, and cost limit replacement policies for one-component systems are developed. For each of the 11 maintenance models, the expected maintenance cost rate and/or availability are derived, and optimum maintenance policies as well as related optimization problems are discussed. The developed models show that the quasi-renewal process and its basic idea are an effective tool to deal with imperfect maintenance problems. This chapter is based on Wang and Pham (1996a, b, c, 1999). In Chapter 5, the quasi-renewal process will be used to study imperfect repair of the series system. Chapter 9 uses it to establish inspection time sequences and Chapter 10 uses the truncated quasi-renewal process to model warranty cost. Software reliability growth and testing cost by using quasi-renewal process is discussed in Chapter 11.

NOTATION

T Fixed age at which a unit is subject to PM, $T > 0$
A Asymptotic average availability

c_p PM cost at fixed age T

c_f Fixed part of imperfect repair cost at failure for each of the first $(k-1)$ failures

c_v Incremental part of imperfect repair cost at failure for each of the first $(k-1)$ failures

c_{fr} Repair (perfect or imperfect) cost at failure after the first $(k-1)$ imperfect repairs

$(k-1)$ Number of imperfect repairs in the sense of the fixed lifetime reduction rule

L Expected total maintenance cost per unit time or cost rate

D Expected renewal cycle length

$f_1(t), F_1(t)$ Probability density function (*pdf*) and cumulative distribution function (*cdf*) of the failure time of a new unit

$r_1(t), R_1(t)$ Failure rate and cumulative failure rate function of the failure time of a new unit

$s_1(t)$ Survival function (*sf*) or reliability function of a new unit, $s_1(t) = 1 - F_1(t)$

p Probability that maintenance is perfect

q Probability that maintenance is minimal where $q = 1 - p$

μ, η Expected lifetime and expected repair time of a new unit

α Lifetime reduction factor

β Incremental factor for repair time

ACRONYMS

NBU   New better than used
NWU   New worse than used
NBUE   New better than used in expectation
NWUE   New worse than used in expectation
IFRA   Increasing failure rate in average
DFRA   Decreasing failure rate in average
IFR   Increasing failure rate
DFR   Decreasing failure rate
i.i.d.   Independently and identically distributed

4.1 A Quasi-renewal Process

The quasi-renewal process is defined and studied in Wang and Pham (1996b). In this section, we will introduce its definition, properties, quasi-renewal function, parameter estimation problems, and truncated quasi-renewal processes.

4.1.1 Definition

Let $\{N(t), t > 0\}$ be a counting process, and X_n denote the time between the $(n-1)^{th}$ and the nth event of this process, $n \geq 1$.

DEFINITION 4.1 Observe the sequence of nonnegative random variables $\{X_1, X_2, X_3, ...\}$. The counting process $\{N(t), t \geq 0\}$ is said to be a quasi-renewal process with parameter α and the first interarrival time X_1, if $X_1 = Z_1$, $X_2 = \alpha Z_2$, $X_3 = \alpha^2 Z_3$, ..., where Z_i s are $i.i.d.$ and $\alpha > 0$ is a constant.

When $\alpha = 1$ this quasi-renewal process becomes the ordinary renewal process. We will see that this quasi-renewal process can be used to model hardware maintenance process when $0 < \alpha \leq 1$. It can also be utilized to model software reliability growth process in testing or operation phase, hardware repair times, and hardware reliability growth in burn-in stage for $\alpha > 1$. Later on, a quasi-renewal process with parameter $\alpha > 1$ will be known as an increasing quasi-renewal process, and a quasi-renewal process with $0 < \alpha < 1$ as a decreasing quasi-renewal process.

Assuming that the probability density function (*pdf*), cumulative density function (*cdf*), survival function (*sf*), and failure rate of random variable X_1 are $f_1(x), F_1(x), s_1(x)$ and $r_1(x)$ respectively. Wang and Pham (1996b) show that the *pdf*, *cdf*, *sf*, failure rate, mean and variance of random variable X_n for $n = 2, 3, 4, ...$ are given by Equation (4.1):

$$\begin{cases} f_n(x) = \alpha^{1-n} f_1(\alpha^{1-n} x) & F_n(x) = F_1(\alpha^{1-n} x) \\ s_n(x) = s_1(\alpha^{1-n} x) & r_n(x) = \alpha^{1-n} r_1(\alpha^{1-n} x) \\ E(X_n) = \alpha^{n-1} E(X_1) & Var(X_n) = \alpha^{2n-2} Var(X_1) \end{cases} \qquad (4.1)$$

Because the nonnegativity of X_1 and the fact that X_1 is not identically 0, we conclude that $E(X_1) = \mu_1 \neq 0$. Now we investigate some properties of the quasi-renewal process.

THEOREM 4.1 *If $f_1(x)$ belongs to IFR, DFR, IFRA, DFRA, NBU, or NWU, then $f_n(x)$ is in the same category for $n = 2, 3, ...$*

Proof. For mathematical definitions of the above terms: IFR, DFR, IFRA, DFRA, NBU, and NWU, see Appendix at the end of this chapter. Suppose that the failure rate of X_n is differentiable with respect to time x. From Equation (4.1) the derivative of the failure rate of X_n is given by

$$r_n'(x) = \frac{1}{\alpha^{2n-2}} r_1'(\frac{1}{\alpha^{n-1}} x)$$

From the above equation we can see that if $r_1(x)$ is increasing (decreasing) then $r_n(x)$ is also increasing (decreasing). Therefore, for the first two categories: IFR or DFR, the conclusion follows.

Next assume that

$$s_1(x+y) \leq (\geq) s_1(x) s_1(y).$$

Then it follows that

$$s_n(x+y) = s_1\left(\frac{x+y}{\alpha^{n-1}}\right)$$

$$\leq (\geq) s_1\left(\frac{x}{\alpha^{n-1}}\right) s_1\left(\frac{y}{\alpha^{n-1}}\right)$$

$$= s_n(x) s_n(y).$$

Therefore, if $s_1(x)$ is NBU (NWU) then $s_n(x)$ is also in the same category. Finally, note that the derivatives with respect to x:

$$\left[s_n^{1/x}(x)\right]_x' = \left[\exp\left(\frac{1}{x}\ln s_1\left(\frac{x}{\alpha^{n-1}}\right)\right)\right]_x'$$

$$= -\frac{1}{\alpha^{2n-2}} s_1^{1/x}\left(\frac{x}{\alpha^{n-1}}\right)\left[\left(\frac{x}{\alpha^{n-1}}\right)^{-2} \ln s_1\left(\frac{x}{\alpha^{n-1}}\right) + \left(\frac{x}{\alpha^{n-1}}\right)^{-1} \frac{f_1(\alpha^{1-n}x)}{s_1(\alpha^{1-n}x)}\right]$$

and

$$\left[s_1^{1/x}(x)\right]_x' = -s_1^{1/x}(x)\left[x^{-2}\ln s_1(x) + x^{-1}\frac{f_1(x)}{s_1(x)}\right].$$

Note also that

$$s_1^{1/x}(x) \geq 0 \qquad s_1^{1/x}(\alpha^{1-n}x) \geq 0 \qquad \text{for } x \geq 0.$$

From above it follows that, if $\left[s_1^{1/x}(x)\right]_x'$ is increasing or decreasing respectively, $\left[s_n^{1/x}(x)\right]_x'$ is also increasing or decreasing respectively. Thus, for the last two categories: NBU or NWU, the conclusion holds. This completes the proof of Theorem 4.1. ◆

It is worthwhile to note that if $f_1(x)$ is NBUE or NWUE, then $f_n(x)$ may not be in the same category for $n = 2,3,...$

The following result is due to Wang and Pham (1996b):

THEOREM 4.2 *The shape parameter of random variable X_n is the same, $\forall n, n = 1,2,3,...$ for a quasi-renewal process if X_1 follows the Gamma, Weibull or Lognormal distribution.*

Remark. This means after "renewal" the shape parameters of the interarrival time will not be changed. In reliability theory, the shape parameters of a lifetime of a hardware product tend to relate to its failure mechanism and modes. Usually, if it possesses the same failure mechanism then a product will have the same shape parameters of its lifetimes at different environments. Therefore, the use of a quasi-renewal process is generally justified in the maintenance process of a hardware system and hardware burn-in stage.

It is worth noting that

$$\lim_{n \to \infty} \frac{E(X_1 + X_2 + \cdots + X_n)}{n} = \lim_{n \to \infty} \frac{\mu_1(1-\alpha^n)}{(1-\alpha)n} = \begin{cases} 0 & \text{when} \quad \alpha < 1 \\ +\infty & \text{when} \quad \alpha > 1 \end{cases}$$

Therefore, if the interarrival time represents the failure-free time of a hardware system with imperfect maintenance the average failure-free time goes to zero when the planning horizon is infinite. This is because the operating condition of the hardware system becomes generally worse and worse as time goes on if it is subject to imperfect maintenance. If the interarrival time represents the error-free time of a software system the average error-free time goes to infinity when its debugging process goes on for a very long time. This conclusion seems reasonable because the faults in the software become generally less and less when it is subject to testing and debugging. When the debugging time is infinite we can expect that there exist no faults with this software and thus average error-free time and the error-free time at the infinite time point is infinite.

4.1.2 Quasi-renewal Function

Consider a quasi-renewal process with parameter α and the first interarrival time X_1. Clearly, the total number $N(t)$ of "renewals" that has occurred up to time t and the arrival time of the nth renewal, SS_n, has the following relationship:

$$N(t) \geq n \quad \Leftrightarrow \quad SS_n \leq t.$$

That is, $N(t)$ is at least n if and only if the nth renewal occurs prior to or at time t. It is easily seen that

$$SS_n = \sum_{i=1}^{n} X_i = \sum_{i=1}^{n} \alpha^{i-1} Z_i \qquad n \geq 1$$

Take

$$SS_0 = 0$$

Thus, we have

$$P\{N(t) = n\} = P\{N(t) \geq n\} - P\{N(t) \geq n+1\}$$
$$= P\{SS_n \leq t\} - P\{SS_{n+1} \leq t\}$$
$$= G^{(n)}(t) - G^{(n+1)}(t)$$

where $G^{(n)}(t)$ is the convolution of the interarrival times $F_1, F_2, ..., F_n$.

In Wang and Pham (1996b), the mean value of $N(t)$ is defined as the quasi-renewal function $M(t)$. Therefore,

$$M(t) = E[N(t)]$$

$$= \sum_{n=1}^{\infty} P\{N(t) \geq n\}$$

$$= \sum_{n=1}^{\infty} P\{SS_n \leq t\}$$

$$= \sum_{n=1}^{\infty} G^{(n)}(t)$$

The derivative of $M(t)$ is known as quasi-renewal density:

$$m(t) = M'(t)$$

In renewal theory, random variables representing the interarrival distributions assume nonnegative values only, and the Laplace transform of distribution $F_1(t)$ is defined by

$$\widetilde{F}_1(s) = \int_0^{\infty} e^{-sx} dF_1(x)$$

Thus,

$$\widetilde{F}_n(s) = \int_0^{\infty} e^{-\alpha^{n-1}st} dF_1(t) = \widetilde{F}_1(\alpha^{n-1}s)$$

$$\widetilde{M}(s) = \sum_{n=1}^{\infty} \widetilde{G}^{(n)}(s)$$

$$= \sum_{n=1}^{\infty} \widetilde{F}_1(s) \cdot \widetilde{F}_1(\alpha s) \cdots \widetilde{F}_1(\alpha^{n-1}s)$$

Since there is a one-to-one correspondence between distribution functions and its Laplace transforms, it follows that:

THEOREM 4.3 *The first interarrival distribution of a quasi-renewal process uniquely determines its quasi-renewal function.*

In investigating optimal hardware replacement problem, Lam (1988) uses the fixed life reduction idea after repair, referred to as a geometric process. Lam (1988, 1996) studies the geometric process by means of the ordinary renewal process. As shown in this section, the quasi-renewal process is investigated from defining the quasi-renewal function, not from the ordinary renewal process.

4.1.3 Associated Statistical Testing Problems

Ascher and Feingold (1984) observe that the interarrival times between successive failures of a deteriorating system tend to become smaller and smaller based on some actual examples. One of the actual examples is the average failure-free time lengths of some bus engines between successive failures: 9400, 7000, 5400, 4100, 3300 miles. In this example, there were 191 engines run to first failure and a sample size of approximately 100 was for each of the other four interarrival miles. Since the sample sizes are large there is overwhelming evidence that the miles between successive failures were decreasing (p.70, Ascher and Feingold 1984). Generally, whether decreases in failure-free times in maintenance processes have geometric reduction patterns needs statistical hypothesis testing. However, since there may not be many failure data available in practice, geometric decay of failure-free times implied by a quasi-renewal process is a good choice as it can, at least, approximate the failure process. Besides, the quasi-renewal process makes imperfect repair modeling mathematically tractable as it can be seen later in this chapter.

In the above example, the estimates of parameter α are respectively

$$\hat{\alpha}_1 = 7000/9400 = .745 \qquad \hat{\alpha}_2 = 5400/7000 = .771$$
$$\hat{\alpha}_3 = 4100/5400 = .759 \qquad \hat{\alpha}_4 = 3300/4100 = .805$$

which are very close to each other. Therefore, we have no strong evidence to reject the hypothesis

$$H_0 : \alpha_1 = \alpha_2 = \alpha_3 = \alpha_4 = \alpha$$
$$H_a : \alpha_1, \alpha_2, \alpha_3 \text{ and } \alpha_4 \text{ are not all equal}$$

To discuss this problem further, assume that n units operate at time zero independently with lifetimes $X_1^1, X_1^2, ..., X_1^n$, respectively. The times between failures for each of them are recorded as follows:

To 1st failure	To 2nd failure	...	To $(m_i + 1)^{th}$ failure
x_1^1	x_2^1		$x_{m_1+1}^1$
x_1^2	x_2^2	...	$x_{m_2+1}^2$
$\vdots$	$\vdots$		$\vdots$
x_1^n	x_2^n	...	$x_{m_n+1}^n$

where x_j^i represents the lifetime of unit i after the $(j-1)^{th}$ repair, and i, j, n and m_i are all integers.

From these data we can obtain the following estimates of parameter α:

$$\begin{array}{cccc} \hat{\alpha}_1^1 & \hat{\alpha}_2^1 & ... & \hat{\alpha}_{m_1}^1 \\ \hat{\alpha}_1^2 & \hat{\alpha}_2^2 & ... & \hat{\alpha}_{m_2}^2 \\ & \vdots & & \\ \hat{\alpha}_1^n & \hat{\alpha}_2^n & ... & \hat{\alpha}_{m_n}^n \end{array}$$

where $\hat{\alpha}_j^i = x_{j+1}^i / x_j^i$.

If α_i is the parameter related to the $(i-1)^{th}$ failure and the ith failure, the following hypothesis:

$$H_0 : \alpha_1 = \alpha_2 = \cdots = \alpha_k = \alpha$$

$$H_a : \alpha_1, \alpha_2, \ldots, \text{and } \alpha_k \text{ are not all equal}$$

can be tested by Analysis of Variance (ANOVA) techniques with the normality assumption where $k = \max(m_l \mid l = 1, 2, \ldots, n)$. Note that this is generally an unbalanced experimental design. When $m_1 = m_2 = \cdots = m_n$ it becomes a balanced design. For pairwise comparisons, the pairwise t-test can be used. These techniques are available in standard textbooks on statistical design of experiments and so there will be no further discussion in this book. If we have no evidence to reject the null hypothesis then parameter α can be estimated by

$$\hat{\alpha} = \frac{\sum_{j=1}^{m^*} (\sum_{i=1}^{n} \alpha_j^i)/n}{m^*}$$

where $m^* = \min(m_l \mid l = 1, 2, \ldots, n)$

For other estimation methods of parameter α, we refer to Wang and Pham (1996e) about acceleration factor estimation.

A quasi-renewal process may be characterized by several parameters, including α. Their estimation can be carried out by using the maximum likelihood estimate (MLE). Assume that we observe one unit and its maintenance process follows a quasi-renewal process. Denote by t_i the ith time to failure since the unit operate at time 0. Assume that $0 = t_0 < t_1 \cdots < t_n$. The likelihood function of this maintenance model is, noting Equation (4.1),

$$L(t_1, t_2, \ldots, t_n) = \prod_{i=1}^{n} f_1(t_1 \mid \Theta) f_2(t_2 \mid \Theta) \cdots f_n(t_n \mid \Theta)$$

$$= \prod_{i=1}^{n} f_1(t_1 \mid \Theta) \alpha^{-1} f_1(\alpha^{-1} t_2 \mid \Theta) \cdots \alpha^{1-n} f_1(\alpha^{1-n} t_n \mid \Theta)$$

$$= \alpha^{-n(n-1)/2} \prod_{i=1}^{n} f_1(t_1 \mid \Theta) f_1(\alpha^{-1} t_2 \mid \Theta) \cdots f_1(\alpha^{1-n} t_n \mid \Theta)$$

where Θ represents the parameter family including parameter α, specified by the parameter space Ω.

From the above likelihood function, the parameters associated with the quasi-renewal process can be estimated by the maximum likelihood method. For example, assume that the first failure time, X_1, of a new unit follows the normal distribution with mean μ and variance σ^2, that is

$$f_1(x) = \sigma^{-1} (2\pi)^{-\frac{1}{2}} e^{-(x-\mu)^2/2\sigma^2}$$

and that its maintenance process can be modeled by the quasi-renewal process. Given the observed failure times $\{t_1, t_2, ..., t_n\}$ we can estimate quasi-renewal process parameter α and normal distribution parameters μ and σ. The likely-hood function becomes

$$L(t_1, t_2, \cdots, t_n) = \alpha^{-n(n-1)/2} \frac{1}{(\sigma\sqrt{2\pi})^n} \exp\left\{-\frac{1}{2\sigma^2}\sum_{i=1}^{n}(\alpha^{1-i}t_i - \mu)^2\right\} \quad \text{and}$$

$$\ln L(t_1, t_2, \cdots, t_n) = -\frac{n(n-1)}{2}\ln\alpha - n\ln(\sigma\sqrt{2\pi}) - \frac{1}{2\sigma^2}\sum_{i=1}^{n}(\alpha^{1-i}t_i - \mu)^2$$

Taking derivatives of $\ln L$ with respect to α, μ and σ result in the MLE of parameters α, μ, and σ, which can be obtained by solving following simultaneous equations:

$$\begin{cases} \hat{\mu} = \frac{1}{n}\sum_{i=1}^{n}\hat{\alpha}^{1-i}t_i \\ n = \frac{1}{2\hat{\sigma}^3}\sum_{i=1}^{n}(\hat{\alpha}^{1-i}t_i - \hat{\mu})^2 \\ \frac{n(n-1)}{2} = \frac{1}{\hat{\sigma}^2}\sum_{i=1}^{n}(\hat{\alpha}^{1-i}t_i - \hat{\mu})\frac{t_i}{\hat{\alpha}^{2i-1}} \end{cases}$$

If failure times of many identical units are recorded, similar MLE procedure can be used to estimate the associated parameters.

It is noted that Whitaker and Samaniego (1989) discuss parameter estimation problems associated with the (p, q) rule for the imperfect repair. It should be pointed out that statistical hypothesis testing and parameter estimation of maintenance models need more attention and study while most work on reliability and maintenance in the literature is focused on probabilistic modeling.

4.1.4 Truncated Quasi-renewal Processes

Bai and Pham (2005) introduce the truncated quasi-renewal processes through omitting a range of possible values for the total number $N(t)$ of "renewals". As one can see in Section 10.3 of Chapter 10, truncated quasi-renewal processes arise naturally in warranty cost study for repairable products. There are also potential applications in reliability and maintenance modeling. Depending on the values omitted from a quasi-renewal process, there are three types of truncations: truncation above m, truncation below m, and double truncation, where m is a fixed non-negative number. Here we focus on the discussion of truncation above m, which will be used in warranty cost models in Chapter 10.

Model I

A quasi-renewal process truncated above m means that for a given t, the total number of "renewals" $N(t)$ can only take values of $0, 1, \cdots, m$. For such $N(t)$, let $P_i(t) \equiv P[N(t) = i]$. Assume that the probability law governing $P_i(t)$ does not change by the truncation, or in other words, all these probabilities $P_i(t)$ are standardized, then

$$P_i(t) = \frac{G^{(i)}(t) - G^{(i+1)}(t)}{1 - G^{(m+1)}(t)}, \qquad \forall i, i = 0, 1, \cdots, m$$

As a result, the first and second moments of $N(t)$ are given by

$$E[N(t)] = \sum_{i=0}^{m} i[G^{(i)}(t) - G^{(i+1)}(t)]/[1 - G^{(m+1)}(t)]$$

$$= \frac{1}{1 - G^{(m+1)}(t)}\left[\sum_{i=1}^{m} iG^{(i)}(t) - \sum_{j=2}^{m+1}(j-1)G^{(j)}(t)\right] \qquad (\text{let } j = i+1)$$

$$= \frac{1}{1 - G^{(m+1)}(t)}\left[\sum_{i=1}^{m} iG^{(i)}(t) - \sum_{j=2}^{m+1} jG^{(j)}(t) + \sum_{j=2}^{m+1} G^{(j)}(t)\right]$$

$$= \frac{\sum_{i=1}^{m+1} G^{(i)}(t) - (m+1)G^{(m+1)}(t)}{1 - G^{(m+1)}(t)}$$

$$= \frac{\sum_{i=1}^{m} G^{(i)}(t) - mG^{(m+1)}(t)}{1 - G^{(m+1)}(t)}$$

and

$$E[N^2(t)] = \frac{1}{1 - G^{(m+1)}(t)}\sum_{i=0}^{m} i^2[G^{(i)}(t) - G^{(i+1)}(t)]$$

$$= \frac{1}{1 - G^{(m+1)}(t)}\left[\sum_{i=1}^{m} i^2 G^{(i)}(t) - \sum_{j=2}^{m+1}(j-1)^2 G^{(j)}(t)\right] \qquad (\text{let } j = i+1)$$

$$= \frac{\sum_{i=1}^{m} i^2 G^{(i)}(t) - (\sum_{j=2}^{m+1} j^2 G^{(j)}(t) - \sum_{j=2}^{m+1} 2jG^{(j)}(t) + \sum_{j=2}^{m+1} G^{(j)}(t))}{1 - G^{(m+1)}(t)}$$

$$= \frac{\sum_{i=1}^{m+1}(2i-1)G^{(i)}(t) - (m+1)^2 G^{(m+1)}}{1 - G^{(m+1)}(t)}$$

$$= \frac{\sum_{i=1}^{m}(2i-1)G^{(i)}(t) - m^2 G^{(m+1)}(t)}{1 - G^{(m+1)}(t)}$$

Model II

Sometimes truncation may change the relative magnitude of $P_i(t)$. In particular, let $N(t)$ be a quasi-renewal process truncated above m. Suppose that for $i \in \{0,1,\cdots,m-1\}$, $P_i(t)$ is the same for that without truncation, so for $i = m$, $P_m(t) = 1 - \sum_{j=0}^{m-1} P_j(t)$. That is:

$$P_i(t) = G^{(i)}(t) - G^{(i+1)}(t), \qquad \text{for } i = 0,1,\cdots,m-1$$
$$P_m(t) = G^{(m)}(t)$$

Consequently, the first and second moments of $N(t)$ are

$$E[N(t)] = \sum_{i=0}^{m-1} i[G^{(i)}(t) - G^{(i+1)}(t)] + mG^{(m)}(t)$$

$$= \sum_{i=1}^{m-1} iG^{(i)}(t) - \sum_{j=2}^{m}(j-1)G^{(j)}(t) + mG^{(m)}(t) \qquad \text{(let } j = i+1\text{)}$$

$$= \sum_{i=1}^{m} G^{(i)}(t)$$

and

$$E[N^2(t)] = \sum_{i=0}^{m-1} i^2[G^{(i)}(t) - G^{(i+1)}(t)] + m^2 G^{(m)}(t)$$

$$= \sum_{i=1}^{m-1} i^2 G^{(i)}(t) - \sum_{j=2}^{m}(j-1)^2 G^{(j)}(t) + m^2 G^{(m)}(t) \qquad \text{(let } j = i+1\text{)}$$

$$= \sum_{i=1}^{m-1} i^2 G^{(i)}(t) - (\sum_{j=2}^{m} j^2 G^{(j)}(t) - \sum_{j=2}^{m} 2jG^{(j)}(t) + \sum_{j=2}^{m} G^{(j)}(t)) + m^2 G^{(m)}(t)$$

$$= \sum_{i=1}^{m}(2i-1)G^{(i)}(t)$$

Chapter 10 will discuss modeling of warranty cost by using truncated quasi-renewal processes.

The next three sections will model imperfect maintenance of a single-unit system by using the quasi-renewal process. The following three assumptions are made in this chapter:

i)   A new unit begins to operate at time 0.

ii) The failure rate $r_1(t)$ of the new unit is continuous and monotonously increasing and differentiable.

iii) *cdf* $F_1(t)$ of the unit is absolutely continuous and $F_1(0) = 0$.

4.2 Periodic PM with Imperfect Maintenance

4.2.1 Model 1: Imperfect Repair and Perfect PM

Suppose that a unit is preventively maintained at times $T, 2T, 3T, ...$ at a cost c_p, independently of the unit's failure history where the constant $T > 0$ and PM is perfect. The unit undergoes imperfect repair at failures between PMs at cost c_f in the sense that upon each repair lifetime (random variable) will be reduced to a fraction α of its immediately previous one and all successive lifetimes are independent, i.e., the lifetimes follow a decreasing quasi-renewal process with parameter α. Thus we can apply the quasi-renewal theory to model this maintenance process. We consider T as a decision variable, α as a parameter in this section. The following result is from Wang and Pham (1996b):

PROPOSITION 4.1 *The long-run expected maintenance cost per unit time, or maintenance cost rate, is*

$$L(T;\alpha) = \frac{c_p + c_f M(T)}{T} \qquad (4.2)$$

where $M(T)$ is the quasi-renewal function of a quasi-renewal process with parameter α.

We can see the form of cost rate $L(T;\alpha)$ is the same as the well-known result obtained from the ordinary renewal theory based on the perfect repair assumption (Barlow and Proschan, 1965). However, the renewal functions are different.

PROPOSITION 4.2 *There exists an optimum T^* which minimizes $L(T;\alpha)$ where $0 < T^* \leq \infty$ and the resulting minimum value of $L(T;\alpha)$ is $c_f m(T^*)$.*

Proof. Note that $L(T;\alpha)$ is continuous for $0 < T < \infty$ because we assume that $F_1(t)$ is continuous. It is easy to see that $L(T,\alpha) \to \infty$ when $T \to 0$ from Equation (4.2). If we explain PM at interval $T = \infty$ as maintenance only at failure, i.e., no PM, it follows that $L(T;\alpha)$ has a minimum for $0 < T \leq \infty$. A necessary condition that a finite value T^* minimizes $L(T;\alpha)$ is that it must satisfy the following equation, obtained by differentiating $L(T;\alpha)$ with respect to T and setting the derivative equal to 0:

$$T^* m(T^*) - M(T^*) = c_p / c_f$$

where $m(\cdot)$ is the renewal density. Substituting this equation into Equation (4.2) it follows that the minimum value of $L(T;\alpha)$ is $c_f m(T^*)$. $\blacklozenge$

4.2.2 Model 2: Imperfect Repair and Imperfect PM

This model is identical to Model 1 in Section 4.2.1 except that the unit is imperfectly preventively maintained at times $T, 2T, 3T, \dots$ at a cost c_p where the constant $T > 0$. Imperfect PM is treated by the (p, q) rule, that is, after PM the unit is 'as good as new' with probability p and is restored to 'as bad as old' with probability $q = 1 - p$.

PROPOSITION 4.3 *The long-run expected maintenance cost per unit time, or cost rate, is*

$$L(T; \alpha, p) = \frac{c_p + c_f p^2 \left[M(T) + \sum_{i=2}^{\infty} q^{i-1} M(iT) \right]}{T} \qquad (4.3)$$

where $M(iT)$ is the quasi-renewal function of a quasi-renewal process with parameter α.

Proof. For detailed proof, see Wang and Pham (1996b). The times between consecutive perfect PM constitute a renewal cycle. Note that the expected duration of a renewal cycle $D(T; \alpha, p)$ is

$$D(T; \alpha, p) = \sum_{i=1}^{\infty} q^{i-1} p(iT).$$

Equation (4.3) follows.   ◆

Note that if $p = 1$ (corresponding to perfect PM), the above equation coincides with the result of Model 1. In this section T is considered as a decision variable, α and p are considered as parameters.

PROPOSITION 4.4 *There exists an optimum T^* which minimizes $L(T; \alpha, p)$ where $0 < T^* \leq \infty$ and the resulting minimum value of expected cost rate $L(T; \alpha, p)$ is*

$$c_f p^2 \sum_{i=1}^{n} q^{i-1} \left[iT^* m(iT^*) \right].$$

Proof. Note that $L(T; \alpha, p)$ is continuous for $0 < T < \infty$ because we assume that $F_1(t)$ is continuous and that $L(T; \alpha, p) \to \infty$ as $T \to 0$ from Equation (4.3). If we explain PM at an interval $T = \infty$ as maintenance only at failure, that is, no PM, it follows that $L(T; \alpha, p)$ has minimum for $0 < T \leq \infty$.

A necessary condition that a finite value T^* minimizes $L(T; \alpha, p)$ is that it satisfies the following equation, from Wang and Pham (1996b):

$$\sum_{i=1}^{n} q^{i-1} \left[iT^* m(iT^*) - M(iT^*) \right] = c_p / (c_f p^2)$$

From this it follows that the optimum maintenance cost rate $L(T;\alpha,p)$ is $c_f p^2 \sum_{i=1}^{n} q^{i-1} \left[iT^* m(iT^*) \right]$, where $m(\cdot)$ is the quasi-renewal density of the quasi-renewal process. ◆

4.2.3 Model 3: Imperfect Repair and Imperfect PM

Assume that a unit start working at time 0; upon failure i it is imperfectly repaired at a cost $c_f + (i-1)c_v$ if and only if $i \le k-1$ where $i = 1,2,3,...$, c_f and c_v are fixed repair cost and incremental cost respectively. The repair is imperfect in the sense that upon each repair the time to failure will be reduced to a fraction α of its immediate predecessor and be independent of all previous ones; the repair time will increase to a multiple β of its immediately previous one and be independent of all previous ones. In other words, the successive times to failure constitute a decreasing quasi-renewal process with parameter α and the successive repair times follow an increasing quasi-renewal process with parameter β. Notice that the repair cost increases by c_v for each next imperfect repair. Boland and Proschan (1982) introduce this kind of increasing maintenance cost notation for minimal repair.

Given that the lifetime X_1 of the new unit and the first imperfect repair time Y_1 are independent random variables with means μ and η, respectively, and suppose that Z_i s and ζ_i s are respectively *i.i.d.* random variable sequences. The lifetime of the unit upon the first imperfect repair and the second imperfect repair time will become respectively αZ_1 with mean $\alpha\mu$ and $\beta\zeta_1$ with means $\beta\eta$, where the constant $\beta \ge 1$ means that the repair time is increasing as the number of imperfect repairs increases and $0 < \alpha \le 1$ means that the lifetime is decreased at each imperfect repair. Note that the lifetime of the unit upon the $(k-2)^{th}$ repair and the $(k-1)^{th}$ repair time are $\alpha^{k-2} Z_{k-1}$ and $\beta^{k-2} \zeta_{k-1}$ with means $\alpha^{k-2}\mu$ and $\beta^{k-2}\eta$ respectively.

After the $(k-1)^{th}$ imperfect repair at failure, the unit is imperfectly preventively maintained at times $T,2T,3T,...$ at a cost c_p (independently of the unit's failure history), and imperfect PM is treated by the (p,q) rule. If there is a failure between PMs an imperfect repair is performed at a cost c_{fr} with negligible repair time; and the repair is imperfect in the sense that upon each repair the lifetime of this unit will be reduced to a fraction λ of its immediately previous one and the successive lifetimes are independent where $0 < \lambda < 1$, *i.e.*, the lifetimes constitute a decreasing quasi-renewal process with parameter λ. This section considers the case that the PM time is a random variable W with mean w. This maintenance process will repeat itself once a perfect PM is incurred.

One possible interpretation of this model is: when a new unit is put into operation, the first $(k-1)$ repairs at failures, because the unit is young at that time, will be performed at a low cost $c_f + (i-1)c_v$ for $i = 1,2,...k-1$, and the repairs turn out to be imperfect. Usually, these repairs are just minor repairs because the unit is in a good operating condition. For example, if a new car is put into use, it will be in a good operating state and should not need any major repairs for some short period, to say, in the first half year. After the $(k-1)^{th}$ imperfect (minor) repair at failure, this car will be in a bad condition and then a better or perfect (major) maintenance (preventive or unplanned, especially preventive) is necessary at a higher cost of c_p or c_{fr}.

4.2.3.1 Maintenance Cost Rate and Availability

The following result is due to Wang and Pham (1996b), and we provide it here without the proof.

PROPOSITION 4.5 *The long-run expected maintenance cost per unit time, or maintenance cost rate, is*

$$L(T,k;\alpha,\beta,\lambda,p) = \frac{(k-1)c_f + \dfrac{(k-1)(k-2)}{2}c_v + c_p p^{-1} + pc_{fr}\sum_{i=1}^{\infty}q^{i-1}M(iT)}{\dfrac{\mu(1-\alpha^{k-1})}{1-\alpha} + \dfrac{\eta(1-\beta^{k-1})}{1-\beta} + \dfrac{T}{p} + w} \quad (4.4)$$

and the asymptotic average availability is given by

$$A(T,k;\alpha,\beta,\lambda,p) = \frac{\dfrac{\mu(1-\alpha^{k-1})}{1-\alpha} + \dfrac{T}{p}}{\dfrac{\mu(1-\alpha^{k-1})}{1-\alpha} + \dfrac{\eta(1-\beta^{k-1})}{1-\beta} + \dfrac{T}{p} + w} \quad (4.5)$$

where $M(t)$ is the quasi-renewal function of a quasi-renewal process with parameter λ and the first interarrival time distribution $F_1(\alpha^{1-k}t)$.

In this section, T and k are decision variables; α,β,λ and p are parameters. $C(T,k;\alpha,\beta,\lambda,p)$ implies that the expected maintenance cost per renewal cycle C is a function of variables T and k with parameters of α,β,λ and p. We will use similar notation throughout this book.

Wang and Pham (1996b) prove that a necessary condition that finite values (T^*,k^*) minimize $L(T,k;\alpha,\beta,p)$ is that they satisfy the following simultaneous equations:

$$\left[\frac{\mu(1-\alpha^{k-1})}{1-\alpha}+\frac{\eta(1-\beta^{k-1})}{1-\beta}+\frac{T}{p}+w\right]pc_{fr}\sum_{i=1}^{\infty}iq^{i-1}m(iT)-c_{fr}\sum_{i=1}^{\infty}q^{i-1}M(iT)$$

$$=(k-1)c_f p^{-1}+\frac{(k-1)(k-2)}{2}c_v p^{-1}+c_p p^{-2}$$

$$\left[\frac{\mu(1-\alpha^{k-1})}{1-\alpha}+\frac{\eta(1-\beta^{k-1})}{1-\beta}+\frac{T}{p}+w\right]\left[c_f+\frac{(2k-3)}{2}c_v\right]=\left[\frac{\mu\alpha^{k-1}\ln\alpha}{\alpha-1}-\frac{\eta\beta^{k-1}\ln\beta}{1-\beta}\right]$$

$$\times\left[(k-1)c_f+\frac{(k-1)(k-2)}{2}c_v+c_p p^{-1}+pc_{fr}\sum_{i=1}^{\infty}q^{i-1}M(iT)\right]$$

4.2.3.2 Optimization and Numerical Example

Sometimes it may be required that while some reliability requirements are satisfied the optimum maintenance policy is obtained. For maintenance Model 3, noting the asymptotic average availability in Equation (4.5), the following optimization problem can be formulated in terms of decision variables T and k as well as parameters α, β and p:

Minimize

$$L(T,k;\alpha,\beta,p)=\frac{(k-1)c_f+\frac{(k-1)(k-2)}{2}c_v+c_p p^{-1}+pc_{fr}\sum_{i=1}^{\infty}q^{i-1}M(iT)}{\frac{\mu(1-\alpha^{k-1})}{1-\alpha}+\frac{\eta(1-\beta^{k-1})}{1-\beta}+\frac{T}{p}+w}\qquad(4.6)$$

Subject to

$$\begin{cases}A(T,k;\alpha,\beta,p)=\dfrac{\dfrac{\mu(1-\alpha^{k-1})}{1-\alpha}+\dfrac{T}{p}}{\dfrac{\mu(1-\alpha^{k-1})}{1-\alpha}+\dfrac{\eta(1-\beta^{k-1})}{1-\beta}+\dfrac{T}{p}+w}\geq A_0\\[4mm]k=2,3,\dots\\T>0\end{cases}$$

where constant A_0 is the predetermined availability requirement.

Similarly, some other optimization models can be formulated and these models can be solved by using any nonlinear programming software. To illustrate the optimal maintenance model (4.6) we now present a numerical example. Note that the normal distribution is IFR. Assume that the lifetime, X_1, of a new unit follows the normal distribution with mean μ and variance σ^2. Then upon the $(k-1)^{th}$ imperfect repair at failure, the *pdf* of the lifetime of this unit will become

$f_k(x) = \alpha^{1-k} f_1(\alpha^{1-k} x)$. From Section 4.1.2, the quasi-renewal function of the quasi-renewal process with parameter λ and the first interarrival time distribution $\alpha^{1-k} f_1(\alpha^{1-k} x)$ is

$$M(t) = \sum_{n=1}^{\infty} G^n(t)$$

It is easy to obtain that

$$G^{(n)}(t) = P\{SS_n \leq t\}$$

where random variable SS_n follows the normal cumulative distribution function with mean $\mu\alpha^{k-1}(1-\lambda^n)/(1-\lambda)$ and variance $\sigma^2\alpha^{2k-2}(1-\lambda^{2n})/(1-\lambda^2)$. Therefore, the quasi-renewal function is given by

$$M(t) = \sum_{n=1}^{\infty} P\{SS_n \leq t\} = \sum_{n=1}^{\infty} \Phi\left(\left[t - \frac{\mu\alpha^{k-1}(1-\lambda^n)}{1-\lambda}\right] \middle/ \sqrt{\frac{\sigma^2\alpha^{2k-2}(1-\lambda^{2n})}{1-\lambda^2}}\right)$$

where $\Phi(\cdot)$ is the standard normal cdf. Now assume that

$$\mu = 10 \qquad \sigma = 1 \qquad \eta = 0.9$$
$$c_f = \$1 \qquad c_p = \$3 \qquad c_{fr} = \$4 \qquad c_v = \$0.06$$
$$\alpha = .95 \qquad \beta = 1.05 \qquad p = 0.95$$
$$w = 0.2 \qquad \lambda = 0.95 \qquad A_0 = 0.94$$

Substituting the above parameter values and the quasi-renewal function into the optimization model (4.6) yields:

Minimize
$L(T,k;0.95,1.05,0.95) =$

$$\frac{k + 0.03(k-1)(k-2) + \frac{41}{19} + 3.8\sum_{i=1}^{\infty} 0.05^{i-1} \sum_{n=1}^{\infty} \Phi\left(\dfrac{iT - 200 \times 0.95^{k-1}(1-0.95^n)}{\sqrt{\dfrac{0.95^{2k-2}(1-0.95^{2n})}{0.0975}}}\right)}{200(1-0.95^{k-1}) + 18(1.05^{k-1}-1) + \frac{T}{0.95} + 0.2}$$

Subject to

$$\begin{cases} A(T,k;0.95,1.05,0.95) = \dfrac{200(1-0.95^{k-1}) + \frac{T}{0.95}}{200(1-0.95^{k-1}) + 18(1.05^{k-1}-1) + \frac{T}{0.95} + 0.2} \geq 0.94 \\ \\ k = 2,3,... \\ T > 0 \end{cases}$$

Various kinds of approximations for the standard normal $\Phi(\cdot)$ have been developed and a simple approximation with high accuracy is by Zelen and Severo (see Johnson and Kotz, 1970):

$$\Phi(x) \approx 1 - \left(0.4361836t - 0.1201676t^2 + 0.9372980t^3\right)\left(\sqrt{2\pi}\right)^{-1}\exp(-\tfrac{1}{2}x^2) \quad (4.7)$$

where $t = (1 + 0.33267x)^{-1}$. The error in $\Phi(x)$, for $x \ge 0$, is less than 1×10^{-5}.

Note that $\Phi(x) = 1 - \Phi(-x)$. Thus, for $x < 0$ we can use this relationship and Equation (4.7) to approximate $\Phi(x)$.

Using the above approximation and nonlinear integer programming software we can find the optimal solution (T^*, k^*) that minimizes the maintenance cost rate given that the availability is at least 0.94:

$$T^* = 7.6530 \qquad\qquad k^* = 3$$

and the corresponding minimum cost rate and availability are respectively

$$L(T^*, k^*; 0.95, 1.05, 0.95) = \$0.2332 \qquad A(T^*, k^*; 0.95, 1.05, 0.95) = 0.9426$$

The results show that the optimal maintenance policy is to perform repair at the first two failures of the unit at a cost of \$1 and \$1.06 respectively, and then perform PM every 7.6530 time units at a cost of \$3 and repair the unit upon failure between PMs at a cost of \$4.

4.2.4 Model 4: Imperfect Repair and Imperfect PM

In the periodic PM Model 3 in Section 4.2.3, if we further assume that after the first $(k-1)$ imperfect repairs the unit will be subject to imperfect PMs at times $T, 2T, 3T,\ldots$ and repairs at failure. The repair at failure is perfect and the corresponding repair time is a random variable Q with mean η_2. The PM is imperfect in the sense that after PM the unit will be as good as new with probability p_1 and as bad as old with probability p_2 and will fail (worst repair) and need repair with probability p_3 where $p_1 + p_2 + p_3 = 1$. Let us further assume that the perfect and worst PM times have means η_4 and η_5 respectively and the repair time upon failure caused by PM is a random variable V with mean η_3. The reason that PM may result in a unit failure is stated in Nakagawa (1987) and Chapter 1 of this book.

According to the ordinary renewal reward theory, the limiting average availability A is

$$A(T, k; \alpha, \beta, p_1, p_2, p_3) = \frac{U(T, k; \alpha, \beta, p_1, p_2, p_3)}{U(T, k; \alpha, \beta, p_1, p_2, p_3) + D(T, k; \alpha, \beta, p_1, p_2, p_3)}$$

where $U(T, k; \alpha, \beta, p_1, p_2, p_3)$ and $D(T, k; \alpha, \beta, p_1, p_2, p_3)$ are the accumulating failure-free time and repair time in one renewal cycle. It is easy to obtain

$$U(T,k;\alpha,\beta,p_1,p_2,p_3) =$$

$$\frac{\mu_1(1-\alpha^{k-1})}{1-\alpha} + \left[\int_0^T tdF_1(\alpha^{1-k}t) + p_2\int_T^{2T} tdF_1(\alpha^{1-k}t) + p_2^2\int_{2T}^{3T} tdF_1(\alpha^{1-k}t) + \cdots\right]$$

$$+ (p_1+p_3)\left[T\cdot s(\alpha^{1-k}T) + 2T\cdot p_2\cdot s(2\alpha^{1-k}T) + 3T\cdot p_2\cdot s(3\alpha^{1-k}T) + \cdots\right]$$

$$= \frac{\mu_1(1-\alpha^{k-1})}{1-\alpha} + (1-p_2)\sum_{i=1}^{\infty} p_2^{i-1}\int_0^{iT} s_1(\alpha^{1-k}t)\,dt$$

$$D(T,k;\alpha,\beta,p_1,p_2,p_3) = \frac{\eta_1(1-\beta^{k-1})}{1-\beta} + (\eta_3+\eta_5)p_3\sum_{i=1}^{\infty} p_2^{i-1}s_1(\alpha^{1-k}iT)$$

$$+ \eta_2(1-p_2)\sum_{i=1}^{\infty} p_2^{i-1}F_1(\alpha^{1-k}iT) + \eta_4 p_1\sum_{i=1}^{\infty} p_2^{i-1}s_1(\alpha^{1-k}iT)$$

Now let

$$CL(T,k;\alpha,\beta,p_1,p_2,p_3) = U(T,k;\alpha,\beta,p_1,p_2,p_3) + D(T,k;\alpha,\beta,p_1,p_2,p_3)$$

$$= \frac{\mu_1(1-\alpha^{k-1})}{1-\alpha} + \frac{\eta_1(1-\beta^{k-1})}{1-\beta} + (1-p_2)\sum_{i=1}^{\infty} p_2^{i-1}\int_0^{iT} s_1(\alpha^{1-k}t)\,dt$$

$$+ (\eta_3+\eta_5)p_3\sum_{i=1}^{\infty} p_2^{i-1}s_1(\alpha^{1-k}iT) + \eta_2(1-p_2)\sum_{i=1}^{\infty} p_2^{i-1}F_1(\alpha^{1-k}iT)$$

$$+ \eta_4 p_1\sum_{i=1}^{\infty} p_2^{i-1}s_1(\alpha^{1-k}iT)$$

PROPOSITION 4.6 *The unit's asymptotic average availability is*

$$A(T,k;\alpha,\beta,p_1,p_2,p_3) = \frac{\dfrac{\mu_1(1-\alpha^{k-1})}{1-\alpha} + (1-p_2)\sum_{i=1}^{\infty} p_2^{i-1}\int_0^{iT} s_1(\alpha^{1-k}t)\,dt}{CL(T,k;\alpha,\beta,p_1,p_2,p_3)} \qquad (4.8)$$

In this section, T and k are decision variables; α,β,p_1,p_2 and p_3 are parameters. From Equation (4.8) we can see that $A(T,k;\alpha,\beta,p_1,p_2,p_3)$ is uniquely determined by variables T and k as well as parameters α,β,p_1,p_2 and p_3. The optimal T and k which maximizes $A(T,k;\alpha,\beta,p_1,p_2,p_3)$ satisfies the following simultaneous equations if they exist:

$$\left[(1-p_2)\sum_{i=1}^{\infty} ip_2^{i-1}s_1(\alpha^{1-k}iT)\right] \times \left[\frac{\mu_1(1-\alpha^{k-1})}{1-\alpha} + \frac{\eta_1(1-\beta^{k-1})}{1-\beta}\right.$$

$$+ (1-p_2)\sum_{i=1}^{\infty} p_2^{i-1}\int_0^{iT} s_1(\alpha^{1-k}t)\,dt + (\eta_3+\eta_5)p_3\sum_{i=1}^{\infty} p_2^{i-1}s_1(\alpha^{1-k}iT)$$

$$\left. + \eta_2(1-p_2)\sum_{i=1}^{\infty} p_2^{i-1}F_1(\alpha^{1-k}iT) + \eta_4 p_1\sum_{i=1}^{\infty} p_2^{i-1}s_1(\alpha^{1-k}iT)\right]$$

$$-\left[\frac{\mu_1(1-\alpha^{k-1})}{1-\alpha}+(1-p_2)\sum_{i=1}^{\infty}p_2^{i-1}\int_0^{iT}s_1(\alpha^{1-k}t)\,dt\right]\times$$

$$\left\{\sum_{i=1}^{\infty}ip_2^{i-1}\left[(1-p_2)s_1(\alpha^{1-k}iT)+\alpha^{1-k}f_1(\alpha^{1-k}iT)\left(\eta_2(1-p_2)-(\eta_3+\eta_5)p_3-\eta_4p_1\right)\right]\right\}=0$$

and

$$\left[-\frac{\mu_1\alpha^{k-1}\ln\alpha}{1-\alpha}-(1-p_2)\alpha^{1-k}\ln\alpha\sum_{i=1}^{\infty}p_2^{i-1}\int_0^{iT}tf_1(\alpha^{1-k}t)\,dt\right]$$

$$\times\left[\frac{\mu_1(1-\alpha^{k-1})}{1-\alpha}+\frac{\eta_1(1-\beta^{k-1})}{1-\beta}+(1-p_2)\sum_{i=1}^{\infty}p_2^{i-1}\int_0^{iT}s_1(\alpha^{1-k}t)\,dt\right.$$

$$\left.+\left((\eta_3+\eta_5)p_3+\eta_4p\right)\sum_{i=1}^{\infty}p_2^{i-1}s_1(\alpha^{1-k}iT)+\eta_2(1-p_2)\sum_{i=1}^{\infty}p_2^{i-1}F_1(\alpha^{1-k}iT)\right]-$$

$$\left[\frac{\mu_1(1-\alpha^{k-1})}{1-\alpha}+(1-p_2)\sum_{i=1}^{\infty}p_2^{i-1}\int_0^{iT}s_1(\alpha^{1-k}t)\,dt\right]\times\left\{-\frac{\mu_1\alpha^{k-1}\ln\alpha}{1-\alpha}-\frac{\eta_1\beta^{k-1}\ln\beta}{1-\beta}+\right.$$

$$\left.\sum_{i=1}^{\infty}\left[(p_2-1)\int_0^{iT}tf_1(\alpha^{1-k}t)\,dt+iT\alpha^{1-k}f_1(\alpha^{1-k}iT)\left(\eta_2(1-p_2)-(\eta_3+\eta_5)p_3-\eta_4p_1\right)\right]\right\}$$

$$\times p_2^{i-1}\alpha^{1-k}\ln\alpha=0$$

The above two equations are obtained by differentiating asymptotic average availability $A(T,k;\alpha,\beta,p_1,p_2,p_3)$ with respect to T and k respectively, and setting the derivatives equal to 0. Note that k is an integer and is regarded as a real number temporarily when differentiating $A(T,k;\alpha,\beta,p_1,p_2,p_3)$ with respect to k. Note also that α,β,p_1,p_2 and p_3 are all parameters.

4.2.5 Model 5: Imperfect Repair and Imperfect PM

This model is exactly like Model 3 in Section 4.2.3 except that the imperfect PM is treated by the x rule, *i.e.*, the age of the unit becomes x units of time younger upon PM (see Chapter 2); that the unit undergoes minimal repair at failures between PMs at cost c_{fm} instead of imperfect repairs in terms of parameter λ in Model 3. Assume that the Nth PM since the last perfect PM is perfect, where N is a positive integer. A cost c_{Np} and an independent replacement time V with mean v is suffered for the perfect PM at time NT. Given that imperfect PM at other times takes W time with mean w and imperfect PM cost is c_p. Suppose that $c_{Np}>c_p$, $v\geq w$, and W and V are independent of the previous failure history of the unit.

PROPOSITION 4.7 *The long-run expected maintenance cost per unit time is*

$$L(T,k,N;\alpha,\beta,p) =$$

$$\frac{(k-1)c_f + \dfrac{(k-1)(k-2)}{2}c_v + c_{Np} + c_p(N-1) + c_{fm}\displaystyle\sum_{i=0}^{N-1}\int_{i(T-x)}^{T+i(T-x)} r(\alpha^{1-k}t)dt}{\dfrac{\mu(1-\alpha^{k-1})}{1-\alpha} + \dfrac{\eta(1-\beta^{k-1})}{1-\beta} + NT + v + (N-1)w}$$

and the asymptotic average availability is

$$A(T,k,N;\alpha,\beta,p) = \frac{\dfrac{\mu(1-\alpha^{k-1})}{1-\alpha} + NT}{\dfrac{\mu(1-\alpha^{k-1})}{1-\alpha} + \dfrac{\eta(1-\beta^{k-1})}{1-\beta} + NT + v + (N-1)w} \qquad (4.9)$$

Proof. The times between consecutive perfect maintenance constitute a renewal cycle. From the classical renewal reward theory we have:

$$L(T,k,N;\alpha,\beta,p) = \frac{C(T,k,N;\alpha,\beta,p)}{D(T,k,N;\alpha,\beta,p)}$$

$$A(T,k,N;\alpha,\beta,p) = \frac{U(T,k,N;\alpha,\beta,p)}{D(T,k,N;\alpha,\beta,p)}$$

where $C(T,k,N;\alpha,\beta,p)$ is the expected maintenance cost per renewal cycle, $U(T,k,N;\alpha,\beta,p)$ is the accumulated operating time in a renewal cycle, and $D(T,k,N;\alpha,\beta,p)$ is the expected duration of a renewal cycle. Following Wang and Pham (1996c),

$$C(T,k,N;\alpha,\beta,p) = c_f + \cdots + [c_f + (k-2)c_v] + c_{Np} + c_p(N-1)$$
$$+ c_{fm}\sum_{i=0}^{N-1}\int_{i(T-x)}^{T+i(T-x)} r(\alpha^{1-k}t)dt$$
$$= (k-1)c_f + \frac{(k-1)(k-2)}{2}c_v + c_{Np} + c_p(N-1)$$
$$+ c_{fm}\sum_{i=0}^{N-1}\int_{i(T-x)}^{T+i(T-x)} r(\alpha^{1-k}t)dt$$

$$D(T,k,N;\alpha,\beta,p) = E[\sum_{i=1}^{k-1}(\alpha^{i-1}X_1 + \beta^{i-1}Y_1)] + [NT + v + (N-1)w]$$
$$= \frac{\mu(1-\alpha^{k-1})}{1-\alpha} + \frac{\eta(1-\beta^{k-1})}{1-\beta} + NT + v + (N-1)w$$

$$U(T,k,N;\alpha,\beta,p) = \frac{\mu(1-\alpha^{k-1})}{1-\alpha} + NT$$

Hence, Proposition 4.7 follows.   ◆

In this section, T, k and N are decision variables; α, β and p are parameters. From Equation (4.9) we can see that maintenance cost rate L and availability A are uniquely determined by variables T, k and N as well as parameters α, β and p.

From Proposition 4.7, the optimum solution (T^*, k^*, N^*) which minimizes $L(T, k, N; \alpha, \beta, p)$ or maximizes $A(T, k, N; \alpha, \beta, p)$ or optimizes both can be obtained by using a nonlinear integer programming software.

4.2.6 Model 6: Imperfect Repair and Imperfect PM

Model 6 is identical to Model 3 in Section 4.2.3 except that upon the $(k-1)^{\text{th}}$ imperfect repair we assume that there are two types of failures (see Beichelt 1980, 1981), and that PMs at times T, $2T$, $3T$,... are perfect. Type 1 failure might be total breakdowns while Type 2 failure can be interpreted as a slight and easily fixed problem. Type 1 failures are subject to perfect repairs and Type 2 failures are subject to minimal repairs. When a failure occurs it is a Type 1 failure with probability $p(t)$ and a Type 2 failure with probability $q(t) = 1 - p(t)$ where t is the age of the unit. Thus, the repair at failure can be modeled by the $(p(t), q(t))$ rule described in Chapter 2. Assume that the failure repair time is negligible, and PM time is a random variable V with mean v.

Consider T and k as decision variables in this section. For this maintenance model, the times between consecutive perfect PMs constitute a renewal cycle. The long-run expected maintenance cost per system time, or cost rate, is

$$L(T, k) = \frac{C(T, k)}{D(T, k)} \tag{4.10}$$

where $C(T, k)$ is the expected maintenance cost per renewal cycle and $D(T, k)$ is the expected duration of a renewal cycle.

After the $(k-1)^{\text{th}}$ imperfect repair, let Y_p denote the time until the first perfect repair without PM since last perfect repair, i.e., the time between successive perfect repairs. As mentioned in Chapter 2 the survival distribution of Y_p is given by

$$\overline{S}(t) = \exp\left\{-\int_0^t p(x) r_k(x) dx\right\}$$
$$= \exp\left\{-\alpha^{1-k} \int_0^t p(x) r(\alpha^{1-k} x) dx\right\}$$

which is proved in Block et al. (1985) and NAPS 03476-A, and utilized in Beichelt (1980, 1981a) and Sheu et al. (1995). Block et al. (1985) further prove that $\overline{S}(t)$ has IFR if $r(t)$ is IFR.

Assume that Z_t represents the number of minimal repairs during the time interval $\left(0, \min\{t, Y_p\}\right)$ and $S(t) = 1 - \overline{S}(t)$. Using the results shown in NAPS 03476-A and used in Beichelt (1981a), we have that:

$$E\{Z_t \mid Y_p < t\} = \frac{1}{S(t)} \int_0^t \int_0^y q(x) \cdot r_k(x) \cdot dx \cdot dS(y)$$

$$= \frac{\alpha^{1-k}}{S(t)} \int_0^t \int_0^y q(x) r_1(\alpha^{1-k} x) dx dS(y)$$

(4.11)

$$E\{Z_t \mid Y_p \geq t\} = \int_0^t q(x) \cdot r_k(x) \cdot dx$$

$$= \alpha^{1-k} \int_0^t q(x) r_1(\alpha^{1-k} x) dx$$

(4.12)

Let $N_1(t)$ and $N_2(t)$ denote s-expected number of perfect repairs and minimal repairs in $(0, t)$ respectively; c_1, c_2, and c_p denote costs of perfect repair, minimal repair, and PM, respectively. Wang and Pham (1999) obtain

$$D(T,k) = \frac{\mu(1-\alpha^{k-1})}{1-\alpha} + \frac{\eta(1-\beta^{k-1})}{1-\beta} + T + v$$

(4.13)

$$C(T,k) = (k-1)c_f + \frac{(k-1)(k-2)}{2} c_v + c_1 N_1(T) + c_2 N_2(T) + c_p$$

(4.14)

Obviously, $N_1(t)$ is the renewal function for the renewal process with the interarrival time distribution $S(t)$ and can be determined by the solution method to the renewal function in renewal theory. According to Beichelt (1981a), we have for $t \leq T$,

$$N_2(t) = E\left[Z_t \mid Y_p \geq t\right] \cdot \bar{S}(t) + \int_0^t \left[E\{Z_x \mid Y_p = x\} + N_2(t-x)\right] dS(x)$$

Note that $E\{Z_t \mid Y_p < t\} \cdot S(t) = \int_0^t E\{Z_x \mid Y_p = x\} dS(x)$ and Equations (4.11) and (4.12). It follows that

$$N_2(t) = E\{Z_t \mid Y_p \geq t\} \cdot \bar{S}(t) + E\{Z_t \mid Y_p < t\} S(t) + \int_0^t N_2(t-x) dS(t)$$

$$= E(Z_t) + \int_0^t N_2(t-x) dS(t)$$

$$= \alpha^{1-k} \int_0^t \bar{S}(t) r_1(\alpha^{1-k} x) dx - S(t) + \int_0^t N_2(t-x) dS(t)$$

(4.15)

Therefore, $N_2(t)$ can be obtained by the Laplace transform or by solving the integral Equation (4.15) using numerical computation. Substituting Equations (4.13) and (4.14) into Equation (4.10), it follows that:

PROPOSITION 4.8 *The long-run expected maintenance cost rate is given by*

$$L(T,k) = \frac{(k-1)c_f + \frac{(k-1)(k-2)}{2}c_v + c_1 N_1(T) + c_2 N_2(T) + c_p}{\frac{\mu(1-\alpha^{k-1})}{1-\alpha} + \frac{\eta(1-\beta^{k-1})}{1-\beta} + T + v}$$

(4.16)

Again, the optimal maintenance policy (T^*, k^*) to minimize the expected cost rate can be obtained from Equation (4.16) in the same manner as in Section 4.2.3.

4.3 Cost Limit Replacement Policy - Model 7

This model is the same as Model 3 in Section 4.2.3 except that at next failures after the $(k-1)^{th}$ imperfect repair since time zero, repair cost is estimated by perfect inspection to determine whether to replace or imperfectly repair it. Assume that the repair cost has a cumulative distribution function $C(x)$ which is independent of the age of the unit. If the estimated cost does not exceed a constant cost limit Q, then this unit is imperfectly repaired at an expected repair cost not exceeding Q. Otherwise, it is replaced by a new one at a higher fixed cost c_2 and the replacement time is W with mean w. Imperfect repair is modeled by the (p,q) rule. Given that the repair time is V with mean v, and that W and V are independent of the previous failure history of the unit. Upon a perfect repair or replacement the process repeats.

This section considers k and Q as decision variables, and α, β and p as parameters.

PROPOSITION 4.9 *The long-run expected maintenance cost per unit time is*

$L(k,Q;\alpha,\beta,p) =$

$$\frac{(k-1)c_f + \frac{(k-1)(k-2)}{2}c_v + \frac{c_2[1-C(Q)] + \bar{c}_1 C(Q)}{1-pC(Q)}}{\frac{\mu(1-\alpha^{k-1})}{1-\alpha} + \frac{\eta(1-\beta^{k-1})}{1-\beta} + \frac{[1-C(Q)]w + pC(Q)v}{1-qC(Q)} + \int\limits_0^\infty \exp\{-H(\alpha^{1-k}t)[1-qC(Q)]\}dt}$$

(4.17)

and the asymptotic average availability is

$A(k,Q;\alpha,\beta,p) =$

$$\frac{\frac{\mu(1-\alpha^{k-1})}{1-\alpha} + \int_0^\infty \exp\{-H(\alpha^{1-k}t)[1-qC(Q)]\}dt}{\frac{\mu(1-\alpha^{k-1})}{1-\alpha} + \frac{\eta(1-\beta^{k-1})}{1-\beta} + \frac{[1-C(Q)]w + pC(Q)v}{1-qC(Q)} + \int\limits_0^\infty \exp\{-H(\alpha^{1-k}t)[1-qC(Q)]\}dt}$$

where $H(\alpha^{1-k}t) = \int_0^t \alpha^{1-k} r_1(\alpha^{1-k}x)dx$ *is the cumulative hazard of the unit right after* the $(k-1)^{th}$ imperfect repair and $\bar{c}_1 = C^{-1}(Q)\int_0^L t\,dC(t)$ is the mean of repair costs less than Q.

Proof. The times between consecutive perfect maintenance, either replacement or perfect repair, constitute a renewal cycle. From the classical renewal reward theory we have

$$L(k,Q;\alpha,\beta,p) = \frac{C(k,Q;\alpha,\beta,p)}{D(k,Q;\alpha,\beta,p)}$$

$$A(k,L;\alpha,\beta) = \frac{U(k,Q;\alpha,\beta,p)}{D(k,Q;\alpha,\beta,p)}$$

where $C(k,Q;\alpha,\beta,p)$ is the expected maintenance cost per renewal cycle, $U(k,Q;\alpha,\beta,p)$ is the accumulated operating time in a renewal cycle, and $D(k,Q;\alpha,\beta,p)$ is the expected duration of a renewal cycle. Denote by $Z_0,Z_1,Z_2,...,$ the failure times of the unit before a replacement or a perfect repair where $Z_0 = 0$. Wang and Pham (1996c) show

$$C(k,Q;\alpha,\beta,p) = (k-1)c_f + \frac{(k-1)(k-2)}{2}c_v$$
$$+ \sum_{i=1}^{\infty}\left\{q^{i-1}p[C(Q)]^i i\bar{c}_1 + [C(Q)]^{i-1}[1-C(Q)]q^{i-1}[(i-1)\bar{c}_1 + c_2]\right\}$$
$$= (k-1)c_f + \frac{(k-1)(k-2)}{2}c_v + \frac{c_2[1-C(Q)]+\bar{c}_1 C(Q)}{1-pC(Q)}$$

$$D(k,Q;\alpha,\beta,p) = \frac{\mu(1-\alpha^{k-1})}{1-\alpha} + \frac{\eta(1-\beta^{k-1})}{1-\beta}$$
$$+ \sum_{i=1}^{\infty}[qC(Q)]^{i-1}\left\{[1-C(Q)][E(Z_i + W)] + pC(Q)[E(Z_i + V)]\right\}$$

$$U(Q,k;\alpha,\beta,p) = \frac{\mu(1-\alpha^{k-1})}{1-\alpha} + \sum_{i=1}^{\infty}[qC(Q)]^{i-1}[1-C(Q)+C(Q)p]E(Z_i)$$

Noting the results of Nakagawa and Kowada (1983):

$$E(Z_i) = \sum_{n=0}^{i-1}\int_0^{\infty}H(\alpha^{1-k}t)^n e^{-H(\alpha^{1-k}t)}/n! \qquad \forall i, i = 1,2,3,...$$

it follows that

$$D(k,Q;\alpha,\beta,p) = \frac{\mu(1-\alpha^{k-1})}{1-\alpha} + \frac{\eta(1-\beta^k)}{1-\beta} + \frac{[1-C(Q)]w + pC(Q)v}{1-qC(Q)}$$

$$+ \sum_{i=1}^{\infty} [qC(Q)]^{i-1} [1 - C(Q) + pC(Q)] E(Z_i)$$

$$= \frac{\mu(1 - \alpha^{k-1})}{1 - \alpha} + \frac{\eta(1 - \beta^{k-1})}{1 - \beta} + \frac{[1 - C(Q)]w + pC(Q)v}{1 - qC(Q)}$$

$$+ [1 - qC(Q)] \sum_{i=1}^{\infty} [qC(Q)]^{i-1} \sum_{n=0}^{i-1} \int_0^{\infty} H(\alpha^{1-k}t)^n e^{-H(\alpha^{1-k}t)} / n! \, dt$$

$$= \frac{\mu(1 - \alpha^{k-1})}{1 - \alpha} + \frac{\eta(1 - \beta^{k-1})}{1 - \beta} + \frac{[1 - C(Q)]w + pC(Q)v}{1 - qC(Q)}$$

$$+ \int_0^{\infty} \exp\{-H(\alpha^{1-k}t)[1 - qC(Q)]\} dt$$

$$U(k, Q; \alpha, \beta, p) = \frac{\mu(1 - \alpha^{k-1})}{1 - \alpha} + \int_0^{\infty} \exp\{-H(\alpha^{1-k}t)[1 - qC(Q)]\} dt$$

Proposition 4.9 follows from the above equations.   ◆

The optimum maintenance policy (k^*, Q^*) which minimizes $L(k, Q; \alpha, \beta, p)$ or maximizes $A(k, Q; \alpha, \beta, p)$ can be obtained using any nonlinear programming software.

4.4 Age-dependent PM Policies with Imperfect Maintenance

4.4.1 Model 8: Imperfect Repair

The model in this section is identical to Model 3 in Section 4.2.3 except that after the $(k-1)^{th}$ repair at failure the unit will be either replaced at next failure at a cost of c_{fr}, or preventively replaced at age T at a cost c_p, whichever occurs first. That is, after time zero a unit is imperfectly repaired at failure i at a cost $c_f + (i-1)c_v$ for $i \leq k-1$ where c_f and c_v are constants. The repair is imperfect in the sense that upon each repair the lifetime will be reduced to a fraction α of the immediate previous lifetime, and the repair time will be increased to a multiple β of the immediately previous one, and the successive lifetimes and repair times are independent. In other words, the successive times to failure constitute a decreasing quasi-renewal process with parameter α and the successive repair times form an increasing quasi-renewal process with parameter β. Note that the lifetime of the unit after the $(k-2)^{th}$ repair and the $(k-1)^{th}$ repair time are $\alpha^{k-2} Z_{k-1}$ and

$\beta^{k-2}\zeta_{k-1}$ with means $\alpha^{k-2}\mu$ and $\beta^{k-2}\eta$ respectively, where Z_is and ζ_is are *i.i.d.* random variable sequences respectively.

Similar to Model 3, one possible interpretation of this model is: when a new system is put into operation, the first $(k-1)$ repairs at failures, because the system is young at that time, will be performed at low cost $c_f + (i-1)c_v$, and the repairs turn out to be imperfect. Usually, these repairs are just minor repairs because the system is in a good operating state. After the first $(k-1)$ repairs at failures, the system will be in a worse operating condition and a perfect maintenance, especially a PM, is necessary at a higher cost.

We consider T and k as decision variables, α and β as parameters.

4.4.1.1 Cost Rate

PROPOSITION 4.10 *The long-run expected maintenance cost per unit time, or maintenance cost rate is*

$$L(T,k;\alpha,\beta) = \frac{(k-1)c_f + \frac{(k-1)(k-2)}{2}c_v + c_p \cdot s_1(\alpha^{1-k}T) + c_{fr} \cdot F_1(\alpha^{1-k}T)}{\frac{\mu(1-\alpha^{k-1})}{1-\alpha} + \frac{\eta(1-\beta^{k-1})}{1-\beta} + \int_0^T s_1(\alpha^{1-k}x)dx} \qquad (4.18)$$

Proof. The times between consecutive perfect maintenances, preventive or unscheduled at failure, constitute a renewal cycle. From the ordinary renewal reward theory we have

$$L(T,k;\alpha,\beta) = \frac{C(T,k;\alpha,\beta)}{D(T,k;\alpha,\beta)}$$

where $C(T,k;\alpha,\beta)$ is the expected total maintenance cost per renewal cycle and $D(T,k;\alpha,\beta)$ is the expected duration of a renewal cycle.

Wang and Pham (1996a) show

$$C(T,k;\alpha,\beta) = c_f + (c_f + c_v) + \cdots + [c_f + (k-2)c_v] + c_p \cdot s_1(\frac{1}{\alpha^{k-1}}T)$$

$$+ c_{fr} \cdot F_1(\frac{1}{\alpha^{k-1}}T)$$

$$= (k-1)c_f + \frac{(k-1)(k-2)}{2}c_v + c_p \cdot s_1(\frac{1}{\alpha^{k-1}}T) + c_{fr} \cdot F_1(\frac{1}{\alpha^{k-1}}T)$$

and

$$D(T,k;\alpha,\beta) = E\left[\sum_{i=1}^{k-1}(\alpha^{i-1}X_1 + \beta^{i-1}Y_1)\right] + T \cdot s_1(\frac{1}{\alpha^{k-1}}T) + \alpha^{1-k}\int_0^T f_1(\frac{1}{\alpha^{k-1}}x)dx$$

$$= \frac{\mu(1-\alpha^{k-1})}{1-\alpha} + \frac{\eta(1-\beta^{k-1})}{1-\beta} + \int_0^T s_1(\frac{1}{\alpha^{k-1}}x)dx$$

From them Proposition 4.10 follows.   ◆

Let

$$M(T) = r_1(\alpha^{1-k}T)\alpha^{1-k}\left[\frac{\mu(1-\alpha^{k-1})}{1-\alpha} + \frac{\eta(1-\beta^{k-1})}{1-\beta} + \int_0^T s_1(\alpha^{1-k}x)dx\right] - F_1(\alpha^{1-k}T)$$

and
$$E = \left[(k-1)c_f + \frac{(k-1)(k-2)}{2}c_v + c_p\right]\Big/(c_{fr} - c_p)$$

We now determine the optimum maintenance policies that minimize the expected maintenance cost rate. The result is summarized in Proposition 4.11.

PROPOSITION 4.11 *For a fixed value of k, an optimum value of PM time T, say* T^*, *to minimize the maintenance cost rate exists, and is perhaps infinite, if*

$$M(0) \le E ; \tag{4.19}$$

if $r_1(t)$ *is continuous and strictly increasing to infinity and Inequality (4.19) is satisfied, the optimal solution* T^* *is unique and finite, given by*

$$T^* = M^{-1}(E) \tag{4.20}$$

where $M^{-1}(E)$ *is the inverse function of* $M(T)$; *if* $M(0) > E$ *the optimal solution is* $T^* = 0^+$.

Proof. Following Wang and Pham (1996c), the derivative of $L(T,k;\alpha,\beta)$ with respect to T is given by

$$\frac{\partial L(T,k;\alpha,\beta)}{\partial T} = r_1(\alpha^{1-k}T)\alpha^{1-k}\left[\frac{\mu(1-\alpha^{k-1})}{1-\alpha} + \frac{\eta(1-\beta^{k-1})}{1-\beta} + \int_0^T s_1(\alpha^{1-k}x)dx\right]$$

$$- F_1(\alpha^{1-k}T) - \left[(k-1)c_f + \frac{(k-1)(k-2)}{2}c_v + c_p\right]\Big/(c_{fr} - c_p)$$

$$= M(T) - E$$

where $r_1(t)$ represents the failure rate of a new component.

A necessary condition for T^* to minimize $L(T,k;\alpha,\beta)$ can be obtained by setting the derivative of $L(T,k;\alpha,\beta)$ with respect to T equal to zero:

$$M(T) - E = 0 \tag{4.21}$$

The derivative of $M(T)$ is

$$\frac{dM(T)}{dT} = r_1'(\alpha^{1-k}T)\alpha^{2-2k}\left[\frac{\mu(1-\alpha^{k-1})}{1-\alpha} + \frac{\eta(1-\beta^{k-1})}{1-\beta} + \int_0^T s_1(\alpha^{1-k}x)dx\right]$$

Since $r_1(t)$ is continuous and increasing, $M(T)$ is also continuous and increasing in T. Note also that $M(0) \geq 0$. Thus, there exists a unique and finite solution for Equation (4.21) only if $r_1(t)$ is strictly increasing to infinity and Inequality (4.19) is satisfied. If Inequality (4.19) is satisfied and $r_1(t)$ is increasing, a solution to Equation (4.20) exists, and is perhaps infinite. If $M(0) > E$ there is a unique optimal solution:

$$T^* = \lim_{\substack{T \to 0 \\ T > 0}} T = 0^+$$

since we assume that $T > 0$. Therefore, Proposition 4.11 follows.   ◆

When repair number k and the PM interval T are both decision variables, we can find the optimal values for k and T by solving the following simultaneous equations if they exist and k is taken as a real number temporarily:

$$\frac{\partial L(T,k;\alpha,\beta)}{\partial T} = 0 \qquad\qquad \frac{\partial L(T,k;\alpha,\beta)}{\partial k} = 0$$

That is,

$$\left[\frac{\mu(1-\alpha^{k-1})}{1-\alpha} + \frac{\eta(1-\beta^{k-1})}{1-\beta} + \int_0^T s_1\left(\frac{1}{\alpha^{k-1}}x\right)dx \right]$$
$$\times \left[c_f + (k-\tfrac{3}{2})c_v + (c_p - c_{fr})T\alpha^{1-k}f_1(\alpha^{1-k}T)\ln\alpha \right]$$
$$= \left[(k-1)c_f + \frac{(k-1)(k-2)}{2}c_v + c_p \cdot s_1\left(\frac{1}{\alpha^{k-1}}T\right) + c_{fr} \cdot F_1\left(\frac{1}{\alpha^{k-1}}T\right) \right] \times$$
$$\left[-\frac{\mu\alpha^{k-1}\ln\alpha}{1-\alpha} - \frac{\eta\beta^{k-1}\ln\beta}{1-\beta} + \alpha^{1-k}\ln\alpha\int_0^T xf_1(\alpha^{1-k}x)dx \right] \qquad (4.22)$$

and Equation (4.21). One can determine the optimal values for T and k from Equations (4.21) and (4.22) by numerical computation methods. Note that k is taken as a real number in Equation (4.21) but the final optimal k value should be rounded to an integer.

4.4.1.2 Numerical Example

This section illustrates the results from Section 4.4.1.1 by a numerical example. Given the lifetime of a unit follows the Weibull distribution with a scale parameter $\lambda = 1$ and a shape parameter $\theta = 2$, that is, $F_1(t) = 1 - e^{-(\lambda t)^\theta} = 1 - e^{-t^2}$. The other parameters are (Wang 1997)

$$c_f = \$1 \qquad c_v = \$0.06 \qquad c_p = \$4 \qquad c_{fr} = \$12$$
$$\alpha = 0.95 \qquad \beta = 1.05 \qquad \eta = 0.03$$

Then the mean life of this unit is

$$\mu = \frac{1}{\lambda}\Gamma(\frac{1}{\theta}+1) = \Gamma(1.5) = 0.88623 \quad \text{time unit}$$

Substituting the above parameters into Equation (4.18) we obtain

$$L(T,k;\alpha,\beta) = \frac{(k-1)+\frac{(k-1)(k-2)}{2}\cdot 0.06 - 8\cdot\exp\left[-(0.95^{1-k}\,T)^2\right]+12}{\frac{0.88623\cdot(1-0.95^{k-1})}{1-0.95}+\frac{0.03\cdot(1.05^{k-1}-1)}{1.05-1}+\int_0^T\exp\left[-(0.95^{1-k}\,x)^2\right]dx}$$

Using nonlinear integer programming software, an optimum maintenance policy to minimize the maintenance cost rate can be found to be

$$k^* = 9 \qquad\qquad T^* = 0.0599$$

and the corresponding optimal maintenance cost rate is

$$L(0.0599,9;0.95,1.05) = \$2.178 \quad \text{per unit time}$$

This result indicates that the optimal maintenance policy for Model 8 is that the first eight failures of the unit will be imperfectly repaired at low costs, and after the eighth repair at failures the unit will be either preventively replaced at the age of 0.0599 time units at a cost of \$4 or replaced at next failure at a cost of \$12, whichever occurs first.

Table 4.1. Optimal maintenance policies for Model 8

Parameter(s) Changed		$L(k^*,T^*)$	k^*	T^*
$\beta = 1.1$	$c_{fr} = 8$	2.148	9	0.1182
$c_v = 0.1$	$c_{fr} = 8$	2.311	7	0.1561
$c_p = 6$	$c_{fr} = 8$	2.450	10	0.2433
$c_p = 2$	$c_{fr} = 8$	1.801	6	0.0899

If c_p is changed to \$8 from \$4 and the other parameters are kept unchanged, then the optimal solution is

$$L(k^*,T^*) = 2.742 \qquad k^* = 11 \qquad T^* = 0.1229$$

Similarly, we can change other parameters and leave the remaining unchanged, and compute the corresponding optimal solutions as shown in Table 4.1. Note that the failure rate of the unit $r_1(t) = \theta\lambda(\lambda t)^{\theta-1} = 2t$ is continuous and strictly increasing to infinity. From Table 4.1 we can see that for these situations the optimum solutions always exist and they are finite and unique.

4.4.2 Model 9: Imperfect CM and Imperfect PM

This model is exactly like Model 8 in Section 4.4.1 except that since the $(k-1)^{\text{th}}$ repair at failure the unit will be imperfectly maintained at age T at a cost c_p or perfectly repaired at next failure at a cost c_{fr}, whichever occurs first. The imperfect PM is treated by the (p,q) rule. In practice, after the $(k-1)$ minor repairs, although a PM is expected to be perfect it turns out to be not perfect due to maintenance cost and maintenance performers, *etc.* Hence, PMs are imperfect generally. Note that perfect PM is an extreme type of imperfect PM as discussed in Chapter 2.

Considering T and k as decision variables, α, β and p as parameters in this section, we have the following proposition:

PROPOSITION 4.12 *The long-run expected maintenance cost per unit time is*

$$L(T,k;\alpha,\beta,p) =$$

$$\frac{(k-1)c_f + \frac{(k-1)(k-2)}{2}c_v + c_p \sum_{i=1}^{\infty} q^{i-1}s_1(i\alpha^{1-k}T) + c_{fr}\left[1 - p\sum_{j=1}^{\infty} q^{j-1}s_1(j\alpha^{1-k}T)\right]}{\dfrac{\mu(1-\alpha^{k-1})}{1-\alpha} + \dfrac{\eta(1-\beta^{k-1})}{1-\beta} + \sum_{i=1}^{\infty} q^{i-1}\displaystyle\int_{(i-1)T}^{iT} s_1(\alpha^{1-k}x)dx}$$

Proof. See Wang and Pham (1996a) for proof. The key point of the proof is that $\sum_{i=1}^{\infty} q^{i-1}s_1(i\alpha^{1-k}T)$ and $[1 - p\sum_{j=1}^{\infty} q^{j-1}s_1(j\alpha^{1-k}T)]$ are the probabilities that PM and CM (repair) occur, respectively, in a renewal cycle (note that a renewal cycle may end with a perfect PM or a CM). For example, $q \cdot s_1(2\alpha^{1-k}T)$ represents the probability that the unit has never failed in the interval $(0, 2T)$ and the first PM turns out to be not perfect (with probability q). ◆

When $p = 1$, *i.e.*, PM is perfect, Proposition 4.12 becomes identical to Proposition 4.10.

It is important to determine k and T which minimize the expected maintenance cost rate. The optimal values for k and T can be obtained by differentiating $L(T,k;\alpha,\beta,p)$ with respect to T and k and setting them equal to zero respectively if they exist. Wang and Pham (1996a) prove that the optimum solution (T, k) satisfies the following two simultaneous equations:

$$\left\{(-c_p + pc_{fr})\alpha^{1-k}\sum_{i=1}^{\infty} iq^{i-1}f_1\left(\frac{i}{\alpha^{k-1}}T\right)\right\}$$

$$\times \left[\frac{\mu(1-\alpha^{k-1})}{1-\alpha} + \frac{\eta(1-\beta^{k-1})}{1-\beta} + \sum_{i=1}^{\infty} q^{i-1}\int_{(i-1)T}^{iT} s_1\left(\frac{x}{\alpha^{k-1}}\right)dx\right]$$

$$-\left\{(k-1)\left(c_f + \frac{(k-2)}{2}c_v\right) + c_p\sum_{i=1}^{\infty} q^{i-1}s_1\left(\frac{iT}{\alpha^{k-1}}\right) + c_{fr}\left[1 - p\sum_{j=1}^{\infty} q^{j-1}s_1\left(\frac{jT}{\alpha^{k-1}}\right)\right]\right\}$$

$$\times \left[\sum_{i=1}^{\infty} q^{i-1} \Big[i \cdot s_1(\alpha^{1-k}iT) - (i-1) \cdot s_1(\tfrac{i-1}{\alpha^{k-1}}T) \Big] \right] = 0$$

and

$$[c_f + \tfrac{2k-3}{2}c_v + (c_p - pc_{fr})\alpha^{1-k}T \ln \alpha \sum_{i=1}^{\infty} iq^{i-1} f_1\Big(\tfrac{iT}{\alpha^{k-1}}\Big)]$$

$$\times \left[\frac{\mu(1-\alpha^{k-1})}{1-\alpha} + \frac{\eta(1-\beta^{k-1})}{1-\beta} + \sum_{i=1}^{\infty} q^{i-1} \int_{(i-1)T}^{iT} s_1\Big(\tfrac{1}{\alpha^{k-1}}x\Big)dx \right] -$$

$$\left\{ (k-1)c_f + \frac{(k-1)(k-2)}{2}c_v + c_p \sum_{i=1}^{\infty} q^{i-1} s_1\Big(i\tfrac{1}{\alpha^{k-1}}T\Big) + c_{fr}[1-p\sum_{i=1}^{\infty} q^{i-1}s_1\Big(i\tfrac{1}{\alpha^{k-1}}T\Big)] \right\}$$

$$\times \left[-\frac{\mu\alpha^{k-1}\ln\alpha}{1-\alpha} - \frac{\eta\beta^{k-1}\ln\beta}{1-\beta} + \alpha^{1-k}\ln\alpha \sum_{i=1}^{\infty} q^{i-1} \int_{(i-1)T}^{iT} xf_1\Big(\tfrac{1}{\alpha^{k-1}}x\Big)dx \right] = 0$$

One can determine the optimal values for T and k from the above two equations by numerical computation methods and nonlinear programming software. Note that k is taken as a real number in the above equations but the final optimal k value should be rounded to an integer.

4.4.3 Model 10: Two Imperfect Repairs

This model is the same as Model 8 in Section 4.4.1 except that since the $(k-1)^{th}$ repair at failure the unit will be perfectly maintained at age T at a cost c_p, or imperfectly repaired at next failure at a cost c_{fr}, whichever occurs first. The imperfect repair is modeled by the (p,q) rule. If the repair is perfect, the next PM with the same cost c_p will be rescheduled at a time T since this perfect repair. If the repair is minimal, the unit is put back into operation and continues to operate until receiving a perfect maintenance, corrective or preventive.

We consider T and k as decision variables, α, β and p as parameters in this section.

PROPOSITION 4.13 *The long-run expected maintenance cost per unit time is*

$$L(T,k;\alpha,\beta,p) =$$

$$\frac{(k-1)c_f + \dfrac{(k-1)(k-2)}{2}c_v + c_p \cdot \big[s_1(\alpha^{1-k}T)\big]^p + c_{fr} \cdot \Big\{1 - \big[s_1(\alpha^{1-k}T)\big]^p\Big\}/p}{\dfrac{\mu(1-\alpha^{k-1})}{1-\alpha} + \dfrac{\eta(1-\beta^{k-1})}{1-\beta} + \displaystyle\int_0^T \big[s_1(\alpha^{1-k}t)\big]^p \, dt}$$

Proof. From renewal reward theory,

$$L(T,k;\alpha,\beta,p) = \frac{C(T,k;\alpha,\beta,p)}{D(T,k;\alpha,\beta,p)}$$

where $C(T,k;\alpha,\beta,p)$ is expected total maintenance cost per renewal cycle (until a perfect repair or perfect PM) and $D(T,k;\alpha,\beta,p)$ is the mean length of a renewal cycle. As mentioned in Chapter 2, Brown and Proschan (1983) prove that without PM the survival function of the time between successive perfect repairs of a unit is

$$\overline{F}_p(t) = \left[s_1(\alpha^{1-k}t)\right]^p .$$

if its *pdf* is $F_1(\alpha^{1-k}t)$ and imperfect repair is modeled by the (p,q) rule. Hence

$$D(T,k;\alpha,\beta,p) = \mu + \eta + \alpha\mu + \beta\eta + \cdots + \alpha^{k-2}\mu + \beta^{k-2}\eta + \int_0^T \overline{F}_p(t)dt$$

$$= \frac{\mu(1-\alpha^{k-1})}{1-\alpha} + \frac{\eta(1-\beta^{k-1})}{1-\beta} + \int_0^T \left[s_1(\alpha^{1-k}t)\right]^p dt$$

The expected cost per renewal cycle is

$$C(T,k;\alpha,\beta,p) = c_f + (c_f + c_v) + \cdots + [c_f + (k-2)c_v] + C_1$$

Using Lemma 2.1 of Fontenot and Proschan (1984), it follows that

$$C_1 = c_{fr}\int_0^T \left[1 + qR(t)\right]dF_p(t) + \left[c_{fr}qR(T) + c_p\right]\overline{F}_p(T)$$

where $R(t) = \int_0^t \alpha^{1-k}r_1(\alpha^{1-k}x)dx = -\ln\overline{F}_p(t)/p$ and $\overline{F}_p(t) = 1 - F_p(t)$.

Substituting C_1 into the $C(T,k;\alpha,\beta,p)$ expression it follows that

$$C(T,k;\alpha,p) = (k-1)c_f + \frac{(k-1)(k-2)}{2}c_v$$

$$+ c_{fr}\int_0^T \left[1 - q\ln\overline{F}_p(t)/p\right]dF_p(t) + \left[c_{fr}qR(T) + c_p\right]\overline{F}_p(T)$$

$$= (k-1)c_f + \frac{(k-1)(k-2)}{2}c_v + c_p \cdot \left[s_1(\alpha^{1-k}T)\right]^p + c_{fr}\cdot\left\{1 - \left[s_1(\alpha^{1-k}T)\right]^p\right\}/p$$

From which we obtain the expression for $L(T,k;\alpha,\beta,p)$.   ◆

PROPOSITION 4.14 *For a fixed k value, the following conclusions hold regarding the optimum solution T^*, which minimizes the maintenance cost rate:*

(a) *The optimal age T^* is infinite if $c_{fr} < pc_p$*

(b) *An optimal age T^* exists if $c_{fr} > pc_p$ and*

$$\left[\frac{\mu(1-\alpha^{k-1})}{1-\alpha}+\frac{\eta(1-\beta^{k-1})}{1-\beta}\right]\left[r_1(0)\alpha^{1-k}\right]\left(c_{fr}-pc_p\right)<\left(c_p+(k-1)c_f+\frac{(k-1)(k-2)}{2}c_v\right)$$

$$(4.23)$$

(c) *The equation* $\dfrac{\partial L(T,k;\alpha,p)}{\partial T}=0$ *has at most one solution. If a solution exists it must be an optimal solution.*

(d) *There is a finite unique optimal solution if* $c_{fr}>pc_p$ *and the condition (4.23) is satisfied as well as*

$$r_1(\infty)>\frac{c_{fr}+p(k-1)c_f+\dfrac{(k-1)(k-2)}{2}pc_v}{p(c_{fr}-pc_p)\alpha^{1-k}\left[\displaystyle\int_0^\infty s_1^p(\alpha^{1-k}t)dt+\dfrac{\mu(1-\alpha^{k-1})}{(1-\alpha)}+\dfrac{\eta(1-\beta^{k-1})}{(1-\beta)}\right]}$$

Proof. Differentiating $L(T,k;\alpha,\beta,p)$ with respect to T yields

$$\frac{\partial L}{\partial T}=\left(\frac{\mu(1-\alpha^{k-1})}{1-\alpha}+\frac{\eta(1-\beta^{k-1})}{1-\beta}+\int_0^T\left[s_1(\alpha^{1-k}t)\right]^p dt\right)^{-2}$$

$$\times\left\{\left\{\left[\frac{\mu(1-\alpha^{k-1})}{1-\alpha}+\frac{\eta(1-\beta^{k-1})}{1-\beta}+\int_0^T\left(s_1(\alpha^{1-k}t)\right)^p dt\right]\right.\right.$$

$$\times\left[r_1(\alpha^{1-k}T)\alpha^{1-k}\right]+\frac{\left[s_1(\alpha^{1-k}T)\right]^p}{p}\right\}\left(c_{fr}-pc_p\right)-\frac{c_{fr}}{p}-(k-1)c_f-\frac{(k-1)(k-2)}{2}c_v\right\}$$

$$\times\left[s_1(\alpha^{1-k}T)\right]^p$$

from which using the methods similar to Fontenot and Proschan (1984) and Section 4.4.1 we can see that for fixed k:

(a) If $c_{fr}<pc_p$ then $\dfrac{\partial L(T,k;\alpha,\beta,p)}{\partial T}<0$, i.e., the maintenance cost rate function L is decreasing. Hence, the optimal age $T^*=\infty$.

(b) Let

$$E=\left\{\left[\frac{\mu(1-\alpha^{k-1})}{1-\alpha}+\frac{\eta(1-\beta^{k-1})}{1-\beta}+\int_0^T\left(s_1(\alpha^{1-k}t)\right)^p dt\right]\left[r_1(\alpha^{1-k}T)\alpha^{1-k}\right]+\frac{\left[s_1(\alpha^{1-k}T)\right]^p}{p}\right\}$$

$$\times(c_{fr}-pc_p)-\frac{c_{fr}}{p}-(k-1)c_f-\frac{(k-1)(k-2)}{2}c_v$$

It is easy to verify that:

$$\frac{dE}{dT} = (c_{fr} - pc_p)\Big[r_1'(\alpha^{1-k}T)\alpha^{2-2k}\Big]\Big[\int_0^T \big(s_1(\alpha^{1-k}t)\big)^p \, dt + \frac{\mu(1-\alpha^{k-1})}{1-\alpha} + \frac{\eta(1-\beta^{k-1})}{1-\beta}\Big] > 0$$

Note that $\dfrac{\partial L(0,k;\alpha,\beta,p)}{\partial T} < 0$ when the condition (4.23) is satisfied. Therefore, if

$\dfrac{\partial L(T,k;\alpha,p)}{\partial T} = 0$ has a solution T^* then $\dfrac{\partial L(T,k;\alpha,p)}{\partial T} > 0$ in the interval (T^*,∞)

and there is an optimal age T^*. If $\dfrac{\partial L(T,k;\alpha,p)}{\partial T} = 0$ has no solution then

$\dfrac{\partial L(T,k;\alpha,p)}{\partial T} < 0$ and the optimal age T^* is ∞.

(c) From the proof of (b) it is easy to draw such a conclusion.

(d) Note that at this time there is a finite unique T^*, which makes

$$\frac{\partial L(T,k;\alpha,p)}{\partial T} < 0 \quad \text{when } 0 < T < T^* \qquad \text{and}$$

$$\frac{\partial L(T,k;\alpha,p)}{\partial T} > 0 \quad \text{when } T^* < T < \infty$$

Then the conclusion follows. ◆

If k is also regarded as a decision variable, differentiating $L(T,k;\alpha,p)$ with respect to k results in

$$\frac{\partial L}{\partial k} = \Big[\frac{\mu(1-\alpha^{k-1})}{1-\alpha} + \frac{\eta(1-\beta^{k-1})}{1-\beta} + \int_0^T \big[s_1(\alpha^{1-k}t)\big]^p dt\Big]^{-2}\Big\{\Big[c_f + (k-\tfrac{3}{2})c_v + r_1(\alpha^{1-k}T)$$

$$\cdot(pc_p - c_{fr})T\big[s_1(\alpha^{1-k}T)\big]^p \frac{1}{\alpha^{k-1}}\ln\alpha\Big]\Big[\frac{\mu(1-\alpha^{k-1})}{1-\alpha} + \frac{\eta(1-\beta^{k-1})}{1-\beta} + \int_0^T \big[s_1(\alpha^{1-k}t)\big]^p dt\Big]$$

$$-\Big\{(k-1)c_f + \frac{(k-1)(k-2)}{2}c_v + c_{fr}/p + (c_p - c_{fr}/p)\cdot\big[s_1(\alpha^{1-k}T)\big]^p\Big\}$$

$$\times\Big[\frac{-\mu\alpha^{k-1}\ln\alpha}{1-\alpha} + \frac{-\eta\beta^{k-1}\ln\beta}{1-\beta} + p\alpha^{1-k}\ln\alpha\int_0^T \big[s_1(\alpha^{1-k}t)\big]^{p-1} f_1(\alpha^{1-k}t)t\,dt\Big]\Big\}$$

The optimal solution in terms of k and T which minimizes $L(T,k;\alpha,p)$ satisfies the following simultaneous equations if it exists:

$$\frac{\partial L(T,k;\alpha,p)}{\partial T} = 0 \quad \text{and} \quad \frac{\partial L(T,k;\alpha,p)}{\partial k} = 0$$

4.4.4 Model 10a: Two Imperfect Repairs Considering Repair Time

Assume that in the maintenance Model 10 in Section 4.4.3 the PM time duration is a random variable W with mean w, and the duration of perfect repair time at failure is a random variable V with mean v. The duration of minimal repair time at failure is negligible because it is smaller than perfect repair duration generally.

4.4.4.1 Cost Rate and Availability

PROPOSITION 4.15 *The maintenance cost rate is given by*

$$L(T,k;\alpha,\beta,p) =$$

$$\frac{(k-1)c_f + \dfrac{(k-1)(k-2)}{2}c_v + c_p \cdot \left[s_1(\alpha^{1-k}T)\right]^p + c_{fr} \cdot \left\{1 - \left[s_1(\alpha^{1-k}T)\right]^p\right\}/p}{\dfrac{\mu(1-\alpha^{k-1})}{1-\alpha} + \dfrac{\eta(1-\beta^{k-1})}{1-\beta} + \int_0^T \left[s_1(\alpha^{1-k}t)\right]^p dt + (w-v)\left[s_1(\alpha^{1-k}T)\right]^p + v} \tag{4.24}$$

and the asymptotic average availability is

$$A(T,k;\alpha,\beta,p) =$$

$$\frac{\dfrac{\mu(1-\alpha^{k-1})}{1-\alpha} + \int_0^T \left[s_1(\alpha^{1-k}t)\right]^p dt}{\dfrac{\mu(1-\alpha^{k-1})}{1-\alpha} + \dfrac{\eta(1-\beta^{k-1})}{1-\beta} + \int_0^T \left[s_1(\alpha^{1-k}t)\right]^p dt + (w-v)\left[s_1(\alpha^{1-k}T)\right]^p + v} \tag{4.25}$$

Proof. From renewal reward theory,

$$L(T,k;\alpha,\beta,p) = \frac{C(T,k;\alpha,\beta,p)}{D(T,k;\alpha,\beta,p)}$$

$$A(T,k;\alpha,\beta,p) = \frac{U(T,k;\alpha,\beta,p)}{D(T,k;\alpha,\beta,p)}$$

It is easy to verify that

$$D(T,k;\alpha,\beta,p) = \mu + \eta + \alpha\mu + \beta\eta + \cdots + \alpha^{k-2}\mu + \beta\eta^{k-2}$$

$$+ \int_0^T \overline{F}_p(t)dt + w\overline{F}_p(T) + v\left[1 - \overline{F}_p(T)\right]$$

$$= \frac{\mu(1-\alpha^{k-1})}{1-\alpha} + \frac{\eta(1-\beta^{k-1})}{1-\beta}$$

$$+ \int_0^T \left[s_1(\alpha^{1-k}t)\right]^p dt + w\left[s_1(\alpha^{1-k}T)\right]^p + v\left\{1 - \left[s_1(\alpha^{1-k}T)\right]^p\right\}$$

$$C(T,k;\alpha,\beta,p) = (k-1)c_f + \frac{(k-1)(k-2)}{2}c_v + c_p \cdot \left[s_1(\alpha^{1-k}T)\right]^p$$

$$+ c_{fr} \cdot \left\{1 - \left[s_1(\alpha^{1-k}T)\right]^p\right\}/p$$

$$U(T,k;\alpha,\beta,p) = \mu + \alpha\mu + \cdots + \alpha^{k-2}\mu + \int_0^T \overline{F}_p(t)dt$$

$$= \frac{\mu(1-\alpha^{k-1})}{1-\alpha} + \int_0^T \left[s_1(\alpha^{1-k}t)\right]^p dt$$

From them the expressions for cost rate and the availability follow.   ◆

4.4.4.2 Optimal Maintenance Policies

As noted in Chapter 3, sometimes it may be required that when some availability requirements are satisfied the optimum maintenance policy is attained or when some maintenance cost requirements are satisfied the optimum reliability measures are attained. For the maintenance model in Section 4.4.4.1, the following optimization problem can be formulated:

Minimize
$$L(T,k;\alpha,\beta,p) =$$

$$\frac{(k-1)c_f + \frac{(k-1)(k-2)}{2}c_v + c_p \cdot \left[s_1(\alpha^{1-k}T)\right]^p + c_{fr} \cdot \left\{1 - \left[s_1(\alpha^{1-k}T)\right]^p\right\}/p}{\frac{\mu(1-\alpha^{k-1})}{1-\alpha} + \frac{\eta(1-\beta^{k-1})}{1-\beta} + \int_0^T \left[s_1(\alpha^{1-k}t)\right]^p dt + (w-v)\left[s_1(\alpha^{1-k}T)\right]^p + v}$$

Subject to
$$A(T,k;\alpha,\beta,p) =$$

$$\frac{\frac{\mu(1-\alpha^{k-1})}{1-\alpha} + \int_0^T \left[s_1(\alpha^{1-k}t)\right]^p dt}{\frac{\mu(1-\alpha^{k-1})}{1-\alpha} + \frac{\eta(1-\beta^{k-1})}{1-\beta} + \int_0^T \left[s_1(\alpha^{1-k}t)\right]^p dt + (w-v)\left[s_1(\alpha^{1-k}T)\right]^p + v} \geq A_0$$

where constant A_0 is the specified availability requirements, $T > 0$, and $k = 2,3,\ldots$

4.4.5 Model 11: Imperfect Repair and Perfect PM

For periodic maintenance Model 6 in Section 4.2.6 we now assume that after the first $(k-1)$ imperfect repairs the system will be subject to a perfect maintenance at age T $(T > 0)$ or at the first Type 1 failure, whichever occurs first. This process continues in infinite time horizon. That is, after the $(k-1)^{\text{th}}$ imperfect repair the system is subject to a perfect PM whenever it reaches age T. Otherwise, there are no PMs and after Type 1 failure the system age is set to zero and is counted again. When a failure after the first $(k-1)$ imperfect repairs occurs it is a Type 1 failure with probability $p(t)$ and a Type 2 failure with probability $q(t) = 1 - p(t)$. Type 2 failure is subject to minimal repair. Suppose that the minimal repair time is negligible, perfect maintenance (corrective or preventive) time is a random variable Q with mean τ.

We consider T and k as decision variables, α, β and p as parameters in this section. For this maintenance model, the times between consecutive perfect maintenance, corrective or preventive, constitute a renewal cycle. The long-run expected maintenance cost per system time, or cost rate, is

$$L(T,k) = \frac{C(T,k)}{D(T,k)} \tag{4.26}$$

Wang and Pham (1999) obtain

$$D(T,k) = \frac{\mu(1-\alpha^{k-1})}{1-\alpha} + \frac{\eta(1-\beta^{k-1})}{1-\beta} + \int_0^T \overline{S}(t)dt + \tau \tag{4.27}$$

Note the derivation of Equation.(4.15). It follows that

$$
\begin{aligned}
C(T,k) &= (k-1)c_f + \frac{(k-1)(k-2)}{2}c_v \\
&\quad + \left[c_2 E(Z_T \mid Y_p < T) + c_1\right]S(T) + \left[c_2 E(Z_T \mid Y_p \geq T) + c_p\right]\overline{S}(T) \\
&= (k-1)c_f + \frac{(k-1)(k-2)}{2}c_v + c_1 S(T) \\
&\quad + c_2\left[E(Z_T \mid Y_p < T)S(T) + E(Z_T \mid Y_p \geq T)\overline{S}(T)\right] + c_p\overline{S}(T) \\
&= (k-1)c_f + \frac{(k-1)(k-2)}{2}c_v \\
&\quad + c_2\left[\alpha^{1-k}\int_0^T \overline{S}(x)r(\alpha^{1-k}x)dx - S(T)\right] + c_1 S(T) + c_p\overline{S}(T) \\
&= (k-1)c_f + \frac{(k-1)(k-2)}{2}c_v \\
&\quad + c_2\left[\alpha^{1-k}\int_0^T \overline{S}(x)r(\alpha^{1-k}x)dx\right] + (c_1 - c_2)S(T) + c_p\overline{S}(T)
\end{aligned}
\tag{4.28}
$$

Substituting Equations (4.27) and (4.28) into Equation (4.26), we have:

PROPOSITION 4.16 *The long-run expected maintenance cost rate is given by*

$$L(T,k) =$$

$$\frac{(k-1)c_f + \frac{(k-1)(k-2)}{2}c_v + c_2\left[\alpha^{1-k}\int_0^T \overline{S}(x)r(\alpha^{1-k}x)dx\right] + (c_1 - c_2)S(T) + c_p\overline{S}(T)}{\frac{\mu(1-\alpha^{k-1})}{1-\alpha} + \frac{\eta(1-\beta^{k-1})}{1-\beta} + \int_0^T \overline{S}(t)dt + \tau} \tag{4.29}$$

The optimal maintenance policy (T^*, k^*) to minimize the expected cost rate can also be determined from Equation (4.29) using a nonlinear integer programming software.

4.5 Concluding Discussions

All imperfect maintenance models in this chapter are based on the quasi-renewal process: successive operating times of a system are independent and decreasing by a fraction $(1-\alpha)$, and successive maintenance times are independent and increasing by a fraction $(\beta-1)$, or alternatively, upon each repair the time to failure will be reduced to a fraction α of the immediately previous one and be independent of all previous ones and the repair time will be increased to a multiple β of the immediately previous one and be independent of all the previous ones. In other words, the successive times to failure form a decreasing quasi-renewal process with parameter α, and the successive maintenance times constitute an increasing quasi-renewal process with parameter β. One can see that most results obtained in this chapter are in closed forms. Based on these results the optimal maintenance policies can be easily obtained by using any nonlinear programming software. Therefore, the quasi-renewal process is effective to treat hardware imperfect maintenance.

For most technical systems, such as cars or refrigerators, when a new one is put into use, it will be in a good operating state and may not need major repairs at the beginning period. After a period of operating, to say, after a certain number of imperfect (usually minor) repairs at failure the system will be in a worse operating condition and then a better or perfect maintenance (preventive or unplanned, especially preventive) is necessary at a higher cost to bring the system to a better operating condition with higher reliability. Therefore, the imperfect maintenance models discussed in this chapter will be practical in reliability and maintenance practice.

Appendix

Assume that a unit has a distribution function $F(t)$ and survival function $s(t) = 1 - F(t)$ with mean μ. We have the following definitions (For details see Barlow and Proschan 1965):

- s is NBU if $s(x+y) \leq s(x) \cdot s(y)$ for all $x, y \geq 0$.

  s is NWU if $s(x+y) \geq s(x) \cdot s(y)$ for all $x, y \geq 0$.

- s is NBUE if $\int_t^\infty s(x)dx \leq \mu s(t)$ for all $t \geq 0$.

  s is NWUE if $\int_t^\infty s(x)dx \geq \mu s(t)$ for all $t \geq 0$.

- s is IFRA if $[s(x)]^{1/x}$ is decreasing in x for $x \geq 0$.

  s is DFRA if $[s(x)]^{1/x}$ is increasing in x for $x \geq 0$.

- s is IFR if and only if $[F(t+x)-F(t)]/s(t)$ is increasing in t for $x \geq 0$.

  s is DFR if and only if $[F(t+x)-F(t)]/s(t)$ is decreasing in t for $x \geq 0$.

- Minimal repair: restore the system to its condition just prior to failure, $i.e.$, if a system fails at age t and undergoes minimal repair, then the repaired system has survival function $s'(x) = s(x+t)/s(t)$.

5

Reliability and Optimal Maintenance of Series Systems with Imperfect Repair and Dependence

The series system is one of the most important and common systems in reliability theory and applications. This chapter discusses availability, maintenance cost, and optimal maintenance policies of the series system with n constituting components under the general assumption that each component is subject to correlated failure and repair, imperfect repair, and arbitrary distributions of times to failure and to repair under shut-off rules. Imperfect repair is modeled through quasi-renewal processes: the successive times to failure form a decreasing quasi-renewal process and the successive repair times constitute an increasing quasi-renewal process. System availability, mean time between system failures, mean time between system repairs, asymptotic fractional down time of the system, *etc.*, are derived, and a numerical example is presented to compare with the models by Barlow and Proschan (1975). Then two classes of maintenance cost models are proposed, and system maintenance cost rates are modeled. Finally, properties of system availability and maintenance cost rates are studied. Optimization models to optimize system availability and/or system maintenance costs are furnished, and optimum system maintenance policies are discussed through a numerical example.

5.1 Introduction

In reliability engineering the series system is an important system as most systems in practice can be regarded or simplified as series systems. Its reliability, availability and maintenance have been studied in the reliability literature. Barlow and Proschan (1975) study the availability of the series system assuming that the repair at failure is perfect. Zhao (1994) extends their availability results to a general repair case. Schneeweiss (2005) investigates the availability of series systems without aging during repairs, proving the steady-state availability of a series-system with no aging of components during the repair of another one can be interpreted as a conditional probability of the well-known s-independent case conditioned on allowing for at most one failed component at any time. Trivially, this conditional probability is larger than the standard probability, *i.e.*, the product of all components' probabilities (Pham 2003a). Blumenthal *et al.* (1976) discuss the transient reliability behavior of series system. Khalil (1985) investigates some shut-off rules for the series system. Wang and Pham (2006) study availability measures, maintenance cost modeling and optimal maintenance policies of series systems whose components are subject to imperfect repair as well as correlated failure and repair. This chapter introduces their work in which imperfect CM is modeled in a way that after repair the lifetime of each component will decrease to a fraction of its preceding one, and the repair time will increase to a multiple of its preceding one, given all lifetimes are independent and so are all repair times. In other words, the successive times to failure constitute a decreasing quasi-renewal process with parameter α and the successive repair times form an increasing quasi-renewal process with parameter β. The correlated failure and repair for each component are modeled by arbitrary bivariate distributions.

The rest of this chapter is organized as follows. Section 5.2 explores system reliability measures: asymptotic average system availability, mean time between system failures, mean time between system repairs, *etc.* Section 5.3 investigates system maintenance costs. Section 5.4 discusses the optimal system maintenance policies combining both system availability and maintenance cost rate. Several numerical examples are presented to demonstrate models derived.

The following notations will be used throughout this chapter:

NOTATION

POS	Period of Service of a component. It begins when the component is new and ends when it is replaced by a new one (perfect repair)
n	Number of components in the system
i	Index of component position or component position identification, $\forall i, i = 1,2,...,n$
k	Component identification number for each component position, or number of times which a new component has occupied this component position, $k = 1,2,....$
$k_i - 1$	Maximum number of (imperfect) repairs on any component at component position i where $k_i \geq 1$
j	Index : number of distinct contiguous periods the component has been operating in a single POS for any component position; $(j - 1)$ is the number of failures/imperfect repairs; $\forall j, j = 1,2,...,k_i$ for component position i
Z_{ik}	POS of component k at component position i
X_{ijk}	Time to failure of component k in component position i which has been repaired $(j - 1)$ times.
X_{i1k}	Time to the first failure of component k in component position i
Y_{ijk}	Time of the jth imperfect repair of the kth component in component position i where $j = 1,2,...,k_i - 1$

Y_{ik_jk}	Time of the perfect repair of the kth component in component position i
μ_{ij}	The expected value of X_{ijk} : $E[X_{ijk}]$
η_{ij}	The expected value of Y_{ijk} : $E[Y_{ijk}]$
μ_i	$\sum_{j=1}^{k_i}\mu_{ij}$, total operating time of a component in component position i in a POS
η_i	$\sum_{j=1}^{k_i}\eta_{ij}$, total repair (perfect and imperfect) time of a component in component position i in a POS
$U(t)$	System uptime accumulated during $[0, t]$
$D(t)$	System downtime accumulated during $[0, t]$
$U_i(t)$	Operating-time in component position i during $[0, t)$
$D_i(t)$	Downtime in component position i resulting from failures there in $[0, t]$
$\widetilde{N}_i(t)$	Number of failures in component position i during $[0, t]$
$\widetilde{N}(t)$	Total number of system failures during $[0, t]$
$\widetilde{N}_{ij}(t)$	Number of failures of components at component position i for which each has been repaired $(j-1)$ times during $[0, t]$
$U_{ij}(t)$	Accumulated operating time of the component in position i which has been repaired $(j-1)$ times during $[0, t]$
$a.s.$	Almost surely: a statement is true with probability one
α_i	Mean reduction factor of time to failure for components at component position i
β_i	Mean repair time growth factor for components at component position i

States, time to failure, time to repair and their relationship of components at component position i are shown in Figure 5.1 (Wang and Pham 2006). The vertical axis represents component state: up or down, and the horizontal axis is system operating time.

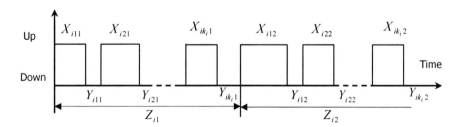

Figure 5.1. States of components in component position i

This chapter assumes:

i)   Each of the n components in the series system is new and starts to operate at time 0.

ii) The component in component position i is repaired at the jth failure if and only if $j < k_i$. The repair is imperfect and modeled by the quasi-renewal process.

iii) The component in component position i is replaced at the k_i^{th} failure by a new one where k_i is an integer. This is a perfect repair.

iv) The time to failure and corresponding time to repair of each component in the system are correlated.

v)   The times to failure or times to repair of the components in the same component position between the POSs are stochastically independent and have the same distributions if the corresponding components have the same history of repair.

vi) The steady-state availability exists.

vii) Two or more components cannot fail at the same time.

In addition, we assume that the series system is subject to the same shut-off rule as the one described in Barlow and Proschan (1975), and studied in Khalil (1985): while a failed component is in repair, all other components remain in "suspended animation". After the repair is completed, the system is returned to operation. At that instant, the components in "suspended animation" are as good as they were when the system stopped operating. In other words, the failure of any one component shuts off all other components in this series system.

Generally, when a new component is put into operation, the first $(k-1)$ repairs at failures will be performed at a low cost. This is because the system is young at that time and these repairs turn out to be imperfect. Usually, these repairs may be minor repairs because the component is in good operating condition. After the $(k-1)^{th}$ imperfect repair, the component may eventually be in a deteriorating operating condition and then a perfect (major) repair may be necessary.

5.2 System Availability Indices Modeling

Note that we consider the case that the time to failure of component k at position i which has been repaired for $(j-1)$ times is correlated with the jth repair time. The dependence of the time to failure X_{ijk} and the time to repair Y_{ijk} is modeled by the joint distribution density $f_{ij}(x, y)$ in this chapter. Assume that the marginal *pdf*s of X_{ijk} and Y_{ijk} are $f_{ij}(x)$ and $g_{ij}(y)$, respectively, and

$$E(X_{ijk}) = \int_0^\infty xf_{ij}(x)dx = \mu_{ij} \quad \text{and} \quad E(Y_{ijk}) = \int_0^\infty yg_{ij}(y)dy = \eta_{ij}.$$

Next we derive the asymptotic availability of the system. Note that $U_i(t)$ is the operating time in the component position i and the component in component position i will not operate during repair of components in other positions per the shut-off rule. Note also that a series system functions if and only if all its constituting components function. Therefore, $U(t) = U_i(t)$. Since $U(t) + D(t) = t$ and $U(t) = U_i(t)$ we have

$$\sum_{i=1}^{n}\sum_{j=1}^{k_i}\sum_{k=1}^{\tilde{N}_{ij}(t)-1} Y_{ijk} \leq D(t) \leq \sum_{i=1}^{n}\sum_{j=1}^{k_i}\sum_{k=1}^{\tilde{N}_{ij}(t)} Y_{ijk}$$

$$\frac{U(t)}{t} = \left[1 + \frac{D(t)}{U(t)}\right]^{-1} \geq \left[1 + \sum_{i=1}^{n}\sum_{j=1}^{k_i}\frac{1}{\tilde{N}_{ij}(t)}\sum_{k=1}^{\tilde{N}_{ij}(t)} Y_{ijk}\frac{N_{ij}\left(U_{ij}(t)\right)U_{ij}(t)}{U_{ij}(t)}\frac{U_{ij}(t)}{U_i(t)}\right]^{-1}$$

where $N_{ij}(U_{ij}(t)) = \tilde{N}_{ij}(t)$ and $\{N_{ij}(t), t \geq 0\}$ is the renewal counting process associated with $\{X_{ijk}, k \geq 1\}$.

By the strong law of large numbers it follows that

$$\lim_{t\to\infty}\frac{1}{\tilde{N}_{ij}(t)}\sum_{k=1}^{\tilde{N}_{ij}(t)} Y_{ijk} \overset{a.s.}{=} \eta_{ij}$$

Next we first prove

$$\lim_{t\to\infty}\frac{U_{ij}(t)}{U_i(t)} \overset{a.s.}{=} \frac{\mu_{ij}}{\mu_i} \tag{5.1}$$

Let

$$X_{ik} = \sum_{j=1}^{k_i} X_{ijk}$$

Then

$$\frac{\sum\limits_{k=1}^{N_i(t)-1} X_{ijk}}{\sum\limits_{k=1}^{N_i(t)+1} X_{ik}} \leq \frac{U_{ij}(t)}{U_i(t)} \leq \frac{\sum\limits_{k=1}^{N_i(t)+1} X_{ijk}}{\sum\limits_{k=1}^{N_i(t)-1} X_{ik}}$$

where $\{N_i(t), t \geq 0\}$ is the renewal counting process associated with $\{X_{ik}, k \geq 1\}$.

Note that

$$\frac{\sum\limits_{k=1}^{N_i(t)+1} X_{ijk}}{\sum\limits_{k=1}^{N_i(t)-1} X_{ik}} = \frac{\sum\limits_{k=1}^{N_i(t)+1} X_{ijk}}{N_i(t)+1}\frac{N_i(t)-1}{\sum\limits_{k=1}^{N_i(t)-1} X_{ik}}\frac{N_i(t)+1}{N_i(t)-1}$$

Because $N_i(t) \overset{a.s.}{\to} \infty$ as $t \to \infty$, we have by the strong law of large numbers,

$$\lim_{t \to \infty} \frac{U_{ij}(t)}{U_i(t)} \le \lim_{t \to \infty} \frac{E[X_{ij1}]}{E[X_{i1}]} \overset{a.s.}{\to} \frac{\mu_{ij}}{\mu_i}$$

The reverse inequality can be shown similarly. Thus, the above Limit (5.1) holds.

From the elementary renewal theorem it follows that

$$\lim_{t \to \infty} \frac{N_{ij}(t)}{t} \overset{a.s.}{=} \frac{1}{\mu_{ij}} \qquad \text{or}$$

$$\lim_{t \to \infty} \frac{N_{ij}(U_{ij}(t))}{U_{ij}(t)} \overset{a.s.}{=} \frac{1}{\mu_{ij}}$$

Using this result and Equation (5.1) we have

$$\lim_{t \to \infty} \frac{U(t)}{t} \overset{a.s.}{\ge} \left[1 + \sum_{i=1}^{n} \sum_{j=1}^{k_i} \eta_{ij} \cdot \frac{1}{\mu_{ij}} \cdot \frac{\mu_{ij}}{\mu_i}\right]^{-1} = \left[1 + \sum_{i=1}^{n} \frac{\eta_i}{\mu_i}\right]^{-1}$$

The reverse inequality can be proved similarly. Therefore, we have

$$\lim_{t \to \infty} \frac{U(t)}{t} \overset{a.s.}{=} \left[1 + \sum_{i=1}^{n} \frac{\eta_i}{\mu_i}\right]^{-1} \tag{5.2}$$

Since $0 \le U(t)/t \le 1$ for all $t > 0$, and the above asymptotic expression exists (the steady-state availability exists as per the assumptions in Section 5.1), it follows from the Lebesgue dominated convergence theorem that

$$A_{av} = \lim_{t \to \infty} \frac{E[U(t)]}{t} = \left[1 + \sum_{i=1}^{n} \frac{\eta_i}{\mu_i}\right]^{-1} \tag{5.3}$$

Now we derive the limiting MTBSF $\bar{\mu}$ and MTBSR $\bar{\eta}$. The average of system up times in $[0, t]$ will be approximately

$$\frac{U(t)}{\tilde{N}(t)} = \frac{U(t)/t}{\sum_{i=1}^{n} \tilde{N}_i(t)/t}$$

The difference between the above expression and the average of system up times is a term that converges to 0 with probability 1 as $t \to \infty$. Noting that the asymptotic number of failures in component position i per unit of time is

$$\bar{N}_i = \lim_{t \to \infty} \frac{\tilde{N}_i(t)}{t} = \lim_{t \to \infty} \frac{N_i(U(t))/U(t)}{t/U(t)} \overset{a.s.}{=} \frac{k_i}{\mu_i} A_{av} \tag{5.4}$$

and Equation (5.2), we have

$$\bar{\mu} = \lim_{t \to \infty} \frac{U(t)}{\tilde{N}(t)} \overset{a.s.}{=} \frac{A_{av}}{\sum\limits_{i=1}^{n} \frac{k_i}{\mu_i} A_{av}} = \left[\sum_{i=1}^{n} \frac{k_i}{\mu_i} \right]^{-1} \tag{5.5}$$

The average system downtime, $\bar{\eta}$, during $[0, t]$ will be approximately

$$\frac{1}{\tilde{N}(t)} \sum_{i=1}^{n} \sum_{j=1}^{k_i} \sum_{k=1}^{\tilde{N}_{ij}(t)} Y_{ijk}$$

The difference between the above expression and the average system downtime is a term that converges to 0 with probability 1 as $t \to \infty$. Wang and Pham (2006) prove the following results:

$$\bar{\eta} \overset{a.s.}{=} \bar{\mu} \sum_{i=1}^{n} \frac{\eta_i}{\mu_i} \tag{5.6}$$

$$D_{av,i} = \lim_{t \to \infty} \frac{D_i(t)}{t} \overset{a.s.}{=} \frac{\eta_i}{\mu_i} A_{av} \tag{5.7}$$

$$D_{av} = \lim_{t \to \infty} \frac{D(t)}{t} \overset{a.s.}{=} A_{av} \sum_{i=1}^{n} \frac{\eta_i}{\mu_i} \tag{5.8}$$

Note that in the above expressions, the dependence of the time to failure X_{ijk} and the time to repair Y_{ijk} is allowed.

Suppose that in this series system of n components upon each repair the time to failure of component i will decrease to a fraction α_i of its immediately previous one and the repair time will increase to a multiple β_i of its immediately previous one where successive times to failure are independent and so are successive repair times, $\forall i, i = 1,2,...,n$. In other words, the successive times to failure compose a decreasing quasi-renewal process with parameter α_i and the successive repair times form an increasing quasi-renewal process with parameter β_i for component i. Accordingly, upon each repair the expected time to failure of the component will decrease to a fraction α_i of its immediately previous value and the expected repair time will increase to a multiple β_i of its previous value, *i.e.*,

$$\mu_{i(j+1)} = \alpha_i \mu_{ij} \qquad\qquad \eta_{i(j+1)} = \beta_i \eta_{ij} \tag{5.9}$$

where $1 \le j \le k_i - 1$ and $k_i - 1 \ge 1$ for component position i. Note that $(k_i - 1)$ is the maximum number of imperfect repairs at component position i in a POS and that $k_i = 1$ indicates perfect repairs only for component position i. Obviously, that $\alpha_i \equiv 1$ also corresponds to a perfect repair only for the component in component position i.

Generally in practice $0 < \alpha_i \leq 1$ and $1 \leq \beta_i$. However, this chapter considers imperfect repair and assumes that the factors $\alpha_i, \beta_i \neq 1$ for $1 \leq i \leq n$ later in this chapter.

From the above relationship (5.9), it follows that the average accumulating operating time in a POS for component position i resulting from failure there is given by

$$\mu_i = \sum_{j=1}^{k_i} \mu_{ij} = \mu_{i1} + \alpha_i \mu_{i1} + \alpha_i^2 \mu_{i1} + \alpha_i^3 \mu_{i1} + \cdots + \alpha_i^{k_i-1} \mu_{i1} = \frac{\mu_{i1}(1-\alpha_i^{k_i})}{1-\alpha_i} \qquad (5.10)$$

Similarly, the average accumulating down time (including perfect repair time) in a POS for component position i resulting from failure there is

$$\eta_i = \sum_{j=1}^{k_i} \eta_{ij} = \eta_{i1} + \beta_i \eta_{i1} + \beta_i^2 \eta_{i1} + \beta_i^3 \eta_{i1} + \cdots + \beta_i^{k_i-1} \eta_{i1} = \frac{\eta_{i1}(1-\beta_i^{k_i})}{1-\beta_i} \qquad (5.11)$$

Alternatively, assume that the perfect repair time is τ_i, which is independent of other imperfect repair times. Then the average accumulating down time in a POS for component position i resulting from failure there is

$$\eta_i = \sum_{j=1}^{k_i} \eta_{ij} = \frac{\eta_{i1}(1-\beta_i^{k_i-1})}{1-\beta_i} + \tau_i \qquad (5.11a)$$

where $k_i \geq 1$.

Upon substituting Equations (5.10) and (5.11) into Equations (5.1)–(5.3) and simplifying, the following proposition follows:

PROPOSITION 5.1 *The mean asymptotic availability of the series system defined in Section 5.1 is*

$$A_{av} = \lim_{t \to \infty} \frac{E[U(t)]}{t} = \left[1 + \sum_{i=1}^{n} \frac{\eta_{i1}(1-\beta_i^{k_i})(1-\alpha_i)}{\mu_{i1}(1-\alpha_i^{k_i})(1-\beta_i)} \right]^{-1} \qquad (5.12a)$$

The limiting average functioning time between system failures (MTBSF) is

$$\bar{\mu} = \lim_{t \to \infty} \frac{U(t)}{\tilde{N}(t)} \overset{a.s.}{=} \left[\sum_{i=1}^{n} \frac{k_i(1-\alpha_i)}{\mu_{i1}(1-\alpha_i^{k_i})} \right]^{-1} \qquad (5.12b)$$

The limiting mean time between system repairs (MTBSR) is

$$\bar{\eta} = \lim_{t \to \infty} \frac{D(t)}{\tilde{N}(t)} \overset{a.s.}{=} \bar{\mu} \sum_{i=1}^{n} \frac{\eta_{i1}(1-\beta_i^{k_i})(1-\alpha_i)}{\mu_{i1}(1-\alpha_i^{k_i})(1-\beta_i)} \qquad (5.12c)$$

The asymptotic fractional down time due to the failure from component position i is

$$D_{av,i} = \lim_{t \to \infty} \frac{D_i(t)}{t}$$

$$\overset{a.s.}{=} \left[1 + \sum_{i=1}^{n} \frac{\eta_{i1}(1-\beta_i^{k_i})(1-\alpha_i)}{\mu_{i1}(1-\alpha_i^{k_i})(1-\beta_i)} \right]^{-1} \frac{\eta_{i1}(1-\beta_i^{k_i})(1-\alpha_i)}{\mu_{i1}(1-\alpha_i^{k_i})(1-\beta_i)} \tag{5.12d}$$

The asymptotic fractional down time of the system is

$$D_{av} = \lim_{t \to \infty} \frac{D(t)}{t} \overset{a.s.}{=} \left\{ 1 + \left[\sum_{i=1}^{n} \frac{\eta_{i1}(1-\beta_i^{k_i})(1-\alpha_i)}{\mu_{i1}(1-\alpha_i^{k_i})(1-\beta_i)} \right]^{-1} \right\}^{-1} \tag{5.12e}$$

The asymptotic number of failures in component position i per unit of time is

$$\overline{N}_i = \lim_{t \to \infty} \frac{\widetilde{N}_i(t)}{t} \overset{a.s.}{=} \frac{k_i(1-\alpha_i)}{\mu_{i1}(1-\alpha_i^{k_i})} A_{av} \tag{5.12f}$$

The asymptotic number of failures of the system per unit of time is

$$\overline{N} = \lim_{t \to \infty} \frac{\widetilde{N}(t)}{t} = \lim_{t \to \infty} \frac{\sum^{n} \widetilde{N}_i(t)}{t} \overset{a.s.}{=} A_{av} \left[\sum_{i=1}^{n} \frac{k_i(1-\alpha_i)}{\mu_{i1}(1-\alpha_i^{k_i})} \right] \tag{5.12g}$$

We can see that the above results $(5.12\,a-g)$ depend only on the expected time to failure μ_{i1} and the expected repair time η_{i1} of each new component (not yet repaired) in the system as well as the maximum number of imperfect repair $(k_i - 1)$, factors α_i and β_i but not on distributions of actual times to failure and repair of the components. Thus, these asymptotic results hold for arbitrary distributions of times to failure and repair, and are mathematically very simple in spite of the fact that the age distribution of components in the system quickly becomes stochastically very complicated.

It follows that that from Equations (5.10) and (5.11),

$$\lim_{\alpha_i \to 1} \mu_i = k_i \mu_{i1}$$

$$\lim_{\beta_i \to 1} \eta_i = k_i \eta_{i1}$$

which are identical to total operating time and down time for component position i in a POS for the perfect repair case respectively. Note $\alpha_i = 1$ corresponds to a perfect repair for the component in component position i. Therefore, the limiting values from the above results $(5.12\,a-g)$ for the imperfect repair case when $\alpha_i \to 1$ and $\beta_i \to 1$ can be used for the perfect repair case.

EXAMPLE 5.1. Barlow and Proschan (1975) study a system of four components in series with expected times to failure and repair given in Table 5.1. Given the time to failure and the time to repair of each component are independent they compute the availability measures for the perfect repair case. For this example, Wang and Pham (2006) consider imperfect repair case as described in the assumptions in Section 5.1, and assume that each component can be imperfectly repaired at most ($k_i - 1$) times. When each component fails at failure k_i, it undergoes a perfect repair (note that perfect repair times at different repair stages with different deteriorating component conditions may be different in practice). Correlated failure and repair are allowed. k_i, α_i and β_i values for each component are summarized in Table 5.1.

Table 5.1. System parameters

Component index	Component type	Mean time to failure μ_{i1} (hours)	Mean repair time η_{i1} (hours)	k_i	α_i	β_i
1	Power supply	50	0.1	6	0.90	1.05
2	Analog equipment	100	0.2	5	0.90	1.05
3	Digital equipment	1000	1.0	6	0.95	1.05
4	Mechanical	10000	20.0	7	0.92	1.05

Table 5.2. Comparison of numerical results

Availability measures	Perfect repair (Barlow-Proschan)	Imperfect repair (Wang-Pham)
Limiting availability	0.993	0.9903
Limiting average of system up times (hrs)	32.15	25.5890
Limiting average of system down times (hrs)	0.225	0.2516
Limiting number of system failures per hour		0.0387
$D_{av,1}$	0.002	0.0029
$D_{av,2}$	0.002	0.0027
$D_{av,3}$	0.001	0.0013
$D_{av,4}$	0.002	0.0029
$\overline{N}_1$	0.020	0.0254
$\overline{N}_2$	0.010	0.0121
$\overline{N}_3$	0.001	0.0011
$\overline{N}_4$	0.0001	0.0001

Using the formulae derived in this section, Wang and Pham (2006) evaluate availability measures for this system. The obtained numerical results are shown in Table 5.2 in which the numerical availability indices for the perfect repair case by Barlow and Proschan (1975) are also listed. Comparing these results with those by Barlow and Proschan, we can see that:

(a) The system availability and the limiting average of system up times for the imperfect repair case are significantly smaller than for the perfect repair case.

(b) The long-run fraction of the time that the system is down due to the failure of each of the four component types, and the long-run average number of failures per hour of each of the component types are larger for imperfect repair case. Obviously this is because the repair is imperfect, *i.e.*, repair doesn't make a component "good as new".

(c) If $k_i = 1$ for all four components, these results and those by Barlow and Proschan will be the same. Alternatively, the limiting results for the imperfect repair case when $\alpha_i \to 1$ and $\beta_i \to 1$ for all four components can expect to be the same as those by Barlow and Proschan (1975). This is because imperfect repair can include perfect repair as special case.

5.3 Modeling of Maintenance Costs

Section 5.2 has studied one system performance measure – system availability indices. For a repairable system the maintenance cost per unit of time, or maintenance cost rate is another interesting system performance measure. For example, a car owner may want to know how much to spend on his car maintenance every year. This section investigates the system maintenance cost rate. Two maintenance cost models will be presented, following Wang and Pham (2006). Note assumption (vii) in Section 5.1: no more than one failure occurs at the same time.

5.3.1 Cost Model 1

Assume that it costs d_i dollars per unit of down time for component i to perform repair. d_i consists of two parts: d_0 and d_{im}, where d_0 is the loss cost per unit of system down time because the system is not available (loss from service interruption) and the same for all component positions, and d_{im} is the repair cost per unit of time for the component in component position i where $1 \le i \le n$. Obviously, $d_i = d_0 + d_{im}$. Repair cost rate d_{im} may include two parts: materials and labor costs, and can be estimated from historical repair data, material cost, labor rate, *etc.* It follows that the total cost of the series system maintenance and service interruption accrued during the time interval $[0, t]$ is

$$C_1(t) = \sum_{i=1}^{n} d_i D_i(t)$$

The system maintenance cost per unit of time, or system maintenance cost rate is, in the limit,

$$\overline{C}_1 = \lim_{t \to \infty} \frac{C_1(t)}{t} = \lim_{t \to \infty} \sum_{i=1}^{n} d_i \frac{D_i(t)}{t}$$

Per Section 5.2 and Zhao (1994),

$$\lim_{t \to \infty} \frac{D_i(t)}{t} = D_{av,i}$$

Using the quasi-renewal process in Chapter 4, Wang and Pham (2006) show

$$\overline{C}_1 = \lim_{t \to \infty} \frac{C_1(t)}{t} = A_{av} \left[\sum_{i=1}^{n} d_i \frac{\eta_{i1}(1 - \beta_i^{k_i})}{1 - \beta_i} \frac{1 - \alpha_i}{\mu_{i1}(1 - \alpha_i^{k_i})} \right]$$

or

$$\overline{C}_1 = \frac{\sum_{i=1}^{n} d_i \dfrac{\eta_{i1}(1 - \alpha_i)(1 - \beta_i^{k_i})}{\mu_{i1}(1 - \alpha_i^{k_i})(1 - \beta_i)}}{1 + \sum_{i=1}^{n} \dfrac{\eta_{i1}(1 - \alpha_i)(1 - \beta_i^{k_i})}{\mu_{i1}(1 - \alpha_i^{k_i})(1 - \beta_i)}} \qquad (5.13)$$

Next we consider another type of repair cost for each component in the system: repair cost is in a lump sum instead of cost per unit of time.

5.3.2 Cost Model 2

Suppose that component position i costs d_{i1} dollars to imperfect repair and service interruption each time it fails and d_{i2} dollars to perfect repair and service interruption every k_i repairs, regardless of the time to complete repair. Assume further that $d_{i1} < d_{i2}$. d_{i1} and d_{i2} may include system loss cost, materials cost, and labor cost, and can be estimated from service importance, historical repair data, material cost, labor rate, *etc.* Note that service interruption cost may tend to be related to time to repair in many situations, as in Cost Model 1. Therefore, Cost Model 1 and Model 2 should be chosen per actual application. Note that a lot of research in the maintenance and reliability field considers repair/maintenance cost only and ignores system loss cost. Then the total cost of system maintenance and service interruption accrued during $[0, t]$ is

$$C_2(t) = \sum_{i=1}^{n} \left\{ d_{i1} \left[\tilde{N}_i(t) - \left\lfloor \tilde{N}_i(t)/k_i \right\rfloor \right] + d_{i2} \left\lfloor \tilde{N}_i(t)/k_i \right\rfloor \right\}$$

where $\lfloor x \rfloor$ means the largest integer less than or equal to x.

The limiting system maintenance cost per unit of time, or maintenance cost rate is

$$\overline{C}_2 = \lim_{t \to \infty} \frac{C_2(t)}{t} = \lim_{t \to \infty} \sum_{i=1}^{n} \left\{ d_{i1} \cdot \tilde{N}_i(t)/t + (d_{i2} - d_{i1}) \lfloor \tilde{N}_i(t)/k_i \rfloor / t \right\}$$

Following Section 5.2, we have that

$$\lim_{t \to \infty} \tilde{N}_i(t)/t = \overline{N}_i = \lim_{t \to \infty} \left[\tilde{N}_i(t) - k_i \right]/t$$

and

$$\left[\left(\tilde{N}_i(t) - k_i \right)/k_i \right]/t \le \lfloor \tilde{N}_i(t)/k_i \rfloor / t \le \left[\tilde{N}_i(t)/k_i \right]/t$$

Thus,

$$\lim_{t \to \infty} \lfloor \tilde{N}_i(t)/k_i \rfloor / t = \overline{N}_i / k_i$$

and

$$\overline{C}_2 = \lim_{t \to \infty} \frac{C_2(t)}{t} = A_{av} \sum_{i=1}^{n} \mu_i^{-1} \left[(k_i - 1)d_{i1} + d_{i2} \right]$$

Finally, we have

$$\overline{C}_2 = \lim_{t \to \infty} \frac{C_2(t)}{t} = A_{av} \sum_{i=1}^{n} \left[(k_i - 1)d_{i1} + d_{i2} \right] \frac{1 - \alpha_i}{\mu_{i1}(1 - \alpha_i^{k_i})}$$

or

$$\overline{C}_2 = \left[1 + \sum_{i=1}^{n} \frac{\eta_{i1}(1 - \beta_i^{k_i})(1 - \alpha_i)}{\mu_{i1}(1 - \alpha_i^{k_i})(1 - \beta_i)} \right]^{-1} \sum_{i=1}^{n} \frac{(1 - \alpha_i)\left[(k_i - 1)d_{i1} + d_{i2}\right]}{\mu_{i1}(1 - \alpha_i^{k_i})} \tag{5.14}$$

It is easy to verify that $\overline{C}_1 = \overline{C}_2$ when $d_i \eta_i = \left[(k_i - 1)d_{i1} + d_{i2}\right]$, *i.e.*, when the total repair cost in a POS per Cost Model 1 is equal to the total repair cost in a POS per Cost Model 2.

5.4 Optimal System Maintenance Policies

5.4.1 Optimality of Availability and Maintenance Cost Rates

In Sections 5.2 and 5.3 we have derived systems availability measures and maintenance cost measures respectively. Next, we discuss how to determine the optimal number of repairs for each component position in a POS that maximizes the limiting system availability and/or minimizes the limiting system maintenance cost rate. Note that the number of all repairs on a component in component position i in a POS is k_i where $k_i \ge 1$. We have the following propositions from Wang and Pham (2006):

PROPOSITION 5.2 *The optimal value for the vector of repair numbers* $(k_1, k_2, ..., k_n)$ *in a POS is* $(1,1,...,1)$, *given any of the following criteria: maximization of the limiting system availability* A_{av}, *or MTBSF, or minimization of MTBSR, or fractional system down time* D_{av} .

Proof. Consider A_{av} first. Let

$$A_{av}(k_1, k_2, ..., k_n) = A_{av} = \left[1 + \sum_{i=1}^{n} \frac{\eta_{i1}(1 - \beta_i^{k_i})(1 - \alpha_i)}{\mu_{i1}(1 - \beta_i)(1 - \alpha_i^{k_i})} \right]^{-1}$$

Then

$$\Delta A_{av}(j) = A_{av}(k_1, k_2, ..., k_j + 1, ..., k_n) - A_{av}(k_1, k_2, ..., k_j, ..., k_n)$$

$$= \left[1 + \sum_{i=1}^{n} \frac{\eta_{i1}(1 - \alpha_i)(1 - \beta_i^{k_i})}{\mu_{i1}(1 - \beta_i)(1 - \alpha_i^{k_i})} \right]^{-1}$$

$$\times \left[1 + \sum_{\substack{i=1 \\ i \neq j}}^{n} \frac{\eta_{i1}(1 - \alpha_i)(1 - \beta_i^{k_i})}{\mu_{i1}(1 - \beta_i)(1 - \alpha_i^{k_i})} + \frac{\eta_{j1}(1 - \alpha_j)(1 - \beta_j^{k_j+1})}{\mu_{j1}(1 - \beta_j)(1 - \alpha_j^{k_j+1})} \right]^{-1}$$

$$\times \frac{\eta_{j1}(1 - \alpha_j)}{\mu_{j1}(\beta_j - 1)} \left(\frac{\beta_j^{k_j} - 1}{1 - \alpha_j^{k_j}} - \frac{\beta_j^{k_j+1} - 1}{1 - \alpha_j^{k_j+1}} \right)$$

Let $G(x) = \dfrac{\beta_j^{x} - 1}{1 - \alpha_j^{x}}$ where x is a real number.

Then

$$\frac{dG(x)}{dx} = \frac{\beta_j^{x}\left(1 - \alpha_j^{x}\right)\ln \beta_j + \alpha_j^{x}\left(\beta_j^{x} - 1\right)\ln \alpha_j}{\left(1 - \alpha_j^{x}\right)^2}$$

Now let

$$W(x) = \beta_j^{x}\left(1 - \alpha_j^{x}\right)\ln \beta_j + \alpha_j^{x}\left(\beta_j^{x} - 1\right)\ln \alpha_j$$

Noting that $W(0) = 0$, and

$$\frac{dW(x)}{dx} = \beta_j^{x}\left(1 - \alpha_j^{x}\right)\left(\ln \beta_j\right)^2 + \alpha_j^{x}\left(\beta_j^{x} - 1\right)\left(\ln \alpha_j\right)^2 > 0$$

it follows that $W(x) > 0$ for $x > 0$ and then $G'(x) > 0$ for $x > 0$.

Therefore, we have that

$$\frac{\beta_j^{k_j} - 1}{1 - \alpha_j^{k_j}} - \frac{\beta_j^{k_j+1} - 1}{1 - \alpha_j^{k_j+1}} < 0 \quad \text{for} \quad j = 1, 2, \ldots, n$$

It follows that $\Delta A_{av}(j) < 0$ for $j = 1, 2, \ldots, n$, and then the maximum system asymptotic availability is obtained when $k_1 = k_2 = \cdots = k_n = 1$.

Next we consider the asymptotic mean functioning time between system failures $\bar{\mu}$. Let

$$\bar{\mu}(k_1, k_2, \ldots, k_n) = \bar{\mu} = \left[\sum_{i=1}^{n} k_i \frac{1 - \alpha_i}{\mu_{i1}(1 - \alpha_i^{k_i})} \right]^{-1}$$

Then

$$\Delta \bar{\mu}_j = \bar{\mu}(k_1, \ldots, k_j + 1, \ldots, k_n) - \bar{\mu}(k_1, \ldots, k_j, \ldots, k_n)$$

$$= \left[\sum_{i=1}^{n} \frac{k_i(1 - \alpha_i)}{\mu_{i1}(1 - \alpha_i^{k_i})} \right]^{-1} \left[\sum_{\substack{i=1 \\ i \neq j}}^{n} \frac{k_i(1 - \alpha_i)}{\mu_{i1}(1 - \alpha_i^{k_i})} + \frac{(k_j + 1)(1 - \alpha_j)}{\mu_{j1}(1 - \alpha_j^{k_j+1})} \right]^{-1}$$

$$\times \frac{1 - \alpha_j}{\mu_{j1}} \frac{-1 + \alpha_j^{k_j} + \alpha_j^{k_j} k_j(1 - \alpha_j)}{(1 - \alpha_j^{k_j})(1 - \alpha_j^{k_j+1})}$$

Let

$$G(k_j) = -1 + \alpha_j^{k_j} + \alpha_j^{k_j} k_j (1 - \alpha_j)$$

Then

$$\Delta G_j = G(k_j + 1) - G(k_j) = -\alpha_j^{k_j}(k_j + 1)(1 - \alpha_j)^2 < 0$$

Note that

$$G(1) = -(1 - \alpha_j)^2 < 0$$

From them it follows that $G(k_j) < 0$ and then $\Delta \mu_j < 0$ for $j = 1, 2, \ldots, n$.
For D_{av}, let

$$D_{av}(k_1, k_2, \ldots, k_n) = D_{av} = \left\{ 1 + \left[\sum_{i=1}^{n} \frac{\eta_{i1}(1 - \beta_i^{k_i})}{1 - \beta_i} \middle/ \frac{\mu_{i1}(1 - \alpha_i^{k_i})}{1 - \alpha_i} \right]^{-1} \right\}^{-1}$$

It is easy to obtain that

$$\Delta D_{av}(j) = D_{av}(k_1, k_2, \ldots, k_j + 1, \ldots, k_n) - D_{av}(k_1, k_2, \ldots, k_j, \ldots, k_n)$$

$$= \left\{ 1 + \left[\sum_{i=1}^{n} \frac{\eta_{i1}(1-\alpha_i)(1-\beta_i^{k_i})}{\mu_{i1}(1-\beta_i)(1-\alpha_i^{k_i})} \right] \right\}^{-1}$$

$$\times \left\{ 1 + \left[\frac{\eta_{j1}(1-\alpha_j)(1-\beta_j^{k_j})}{\mu_{j1}(1-\beta_{ij})(1-\alpha_j^{k_j})} + \sum_{\substack{i=1 \\ i \neq j}}^{n} \frac{\eta_{i1}(1-\alpha_i)(1-\beta_i^{k_i})}{\mu_{i1}(1-\beta_i)(1-\alpha_i^{k_i})} \right] \right\}^{-1}$$

$$\times \frac{\eta_{j1}(1-\alpha_j)}{\mu_{j1}(\beta_j-1)} \left(\frac{\beta_j^{k_j+1}-1}{1-\alpha_j^{k_j+1}} - \frac{\beta_j^{k_j}-1}{1-\alpha_j^{k_j}} \right) > 0 \qquad \text{for} \quad j = 1,2,...,n \; .$$

From them Proposition 5.2 follows.   ◆

The above proposition is intuitively obvious. $k_1, k_2, k_3, k_4 \geq 1$ and they are all integers. Note that A_{av} is limiting *average* availability over infinite POSs which may be insensitive to changes of k_i values.

PROPOSITION 5.3 *The necessary condition for minimizing* $\overline{C}_1$ *is to do perfect repair only on the component in the component position which has the largest maintenance cost per unit of time.*

Proof. See Wang and Pham (2006). An interpretation to this proposition is given after the following proposition.   ◆

PROPOSITION 5.4 *For the component position with the smallest maintenance cost per unit of down time, say, position j, there may exist a finite unique number of all repairs in a POS, say,* k_j^*, *for all components in this position if*

$$\sum_{i=1}^{n} \frac{\eta_{i1}}{\mu_{i1}}(d_i - d_j) < d_j$$

and it is the optimal solution for component position j to minimize $\overline{C}_1$. *Otherwise, the minimum system maintenance cost is obtained when* $k_j \to \infty$.

Proof. Following Wang and Pham (2006), let

$$\Delta_j = d_j - \sum_{i=1}^{n} \frac{\eta_{i1}(1-\alpha_i)(1-\beta_i^{k_i})}{\mu_{i1}(1-\beta_i)(1-\alpha_i^{k_i})}(d_i - d_j)$$

and
$$d_j = \min\{d_1, d_2, ..., d_n\}$$

If at least one $k_i \to \infty$ for $i = 1, 2, ..., j-1, j+1, ..., n$, then $\Delta_j \to -\infty < 0$ and $\dfrac{\partial \overline{C}_1}{\partial k_j} < 0$. Note that

$$\frac{\eta_{i1}(1-\alpha_i)(1-\beta_i^{k_i})}{\mu_{i1}(1-\beta_i)(1-\alpha_i^{k_i})}$$

is increasing in k_i and its minimum value is obtained when $k_i = 1$ for $i = 1, 2, ..., n$. When $k_1 = k_2 = \cdots = k_{j-1} = k_{j+1} = \cdots = k_n = 1$, Δ_j has the maximum value

$$\Delta_{j,\max} = d_j - \sum_{i=1}^{n} \frac{\eta_{i1}}{\mu_{i1}}(d_i - d_j)$$

if $\displaystyle\sum_{i=1}^{n} \frac{\eta_{i1}}{\mu_{i1}}(d_i - d_j) < d_j$ then $\Delta_j > 0$ and thus $\dfrac{\partial \overline{C}_1}{\partial k_j} > 0$.

Therefore, there may exist an optimal k_j which minimizes $\overline{C}_1$ when the above condition is satisfied.

On the other hand, if $\displaystyle\sum_{i=1}^{n} \frac{\eta_{i1}}{\mu_{i1}}(d_i - d_j) > d_j$, then $\Delta_j \le 0$ and thus $\dfrac{\partial \overline{C}_1}{\partial k_j} \le 0$ for any $k_j \ge 1$. Hence, the minimum $\overline{C}_1$ is obtained when $k_j \to \infty$.   ◆

Propositions 5.3 and 5.4 imply that each component may not be treated as a single-unit system individually and local optimal maintenance policies for individual components may not result in the global optimal maintenance policy for the whole system. Now let's see a numerical example. Assume that in Example 5.1 in Section 5.2, $d_1 = 20$, $d_2 = 80$, $d_3 = 18$, and $d_4 = 0.2$ dollars per hour. Note

$$\sum_{i=1}^{n} \frac{\eta_{i1}}{\mu_{i1}}(d_i - d_j) = 0.2170 > d_4 = 0.2 \quad \text{at this time.}$$

Using nonlinear integer programming software, and noting that $k_1, k_2, k_3, k_4 \ge 1$ and they are all integers, we obtain that if there is no constraint on k_4, the minimum system maintenance cost per hour $\overline{C}_1$ is \$0.2 which is reached when $k_2 = 1$ together with, $k_1 = 1$, $k_3 = 1$, and $k_4 \to +\infty$. One explanation may be that it is most expensive to restore the component at component position 2 once it fails. Note that the minimum system maintenance cost per hour of \$0.2 is equal to d_4. This is because the component at component position 2 is always repaired if $k_4 \to +\infty$. d_4 is so low as compared with repair cost d_1, d_2, and d_3 that to always repair the component at component position 4 is of low cost for the entire system. However, this special situation may rarely happen in practice since loss cost from service interruption needs to be very low at this time.

If $d_4 = 4.4$ then

$$\sum_{i=1}^{n} \frac{\eta_{i1}}{\mu_{i1}}(d_i - d_j) = 0.1960 < d_4 = 4.4 \quad \text{at this time}$$

and the optimal system maintenance cost rate is attained when repair numbers $k_1 = k_2 = k_3 = k_4 = 1$. Note this k_4 is finite. The corresponding $A_{av} = 0.993049$.

Propositions 5.3 and 5.4 can be explained as follows. When the component in component position j with the smallest maintenance cost d_j is in repair, the components in the other component positions which have larger maintenance costs per unit of time will be suspended (neither age nor experience repair) and thus result in no additional repair cost. If the component in position j fails frequently, the other components with larger repair cost per unit time will be subject to less repair costs. Hence, there exists a trade-off between the increase of the maintenance cost resulting from frequent repair of the component in position j and less maintenance costs resulting from the suspended animation of the other components during the repair of the component at position j.

Similarly, we can consider minimizing system maintenance cost rate $\overline{C}_2$. Existence of the optimal solution to minimize $\overline{C}_2$ can also be explained from the physical meaning. According to the meaning of d_i, when the number of imperfect repairs in a POS becomes larger the repair time and repair frequency will become larger and thus the maintenance cost per unit of time will increase. On the other hand, if only perfect repair is performed, the maintenance cost rate may be big because $d_{i1} < d_{i2}$. Hence there may exist a trade-off between them, and so an optimal solution.

EXAMPLE 5.2. Suppose that in Example 5.1, repair costs for four components are $d_{11} = 1$, $d_{21} = 10$, $d_{31} = 15$, $d_{41} = 10$, $d_{12} = 2$ $d_{22} = 16$, $d_{32} = 18$, $d_{42} = 14$ in dollars. Determine the minimum system maintenance cost per hour using cost model 2 in Section 5.3.2.

Note that $k_1, k_2, k_3, k_4 \geq 1$ and they are all integers. Using nonlinear integer programming software, Wang and Pham (2006) obtain the minimum system maintenance cost per hour:

$$\overline{C}_{2\min} = \$0.178596 \quad \text{when} \quad k_1 = 4, \ k_2 = 3, \ k_3 = 2, \ k_4 = 5$$

and the corresponding limiting system availability:

$$A_{av} = 0.991594.$$

5.4.2 Optimal Repair Policy

From Section 5.4.1, we can see that when the minimum system maintenance rate $\overline{C}_1$ is obtained, the corresponding system availability may be so low that it may not be acceptable in practice. Therefore, both the system availability and maintenance cost rate must be considered together to obtain the optimal system maintenance policy. For example, in view of the maintenance cost constraints, one may determine the optimal number of repairs in a POS for all component positions to maximize the system availability. This class of problems can be formulated as:

Optimization Problem 1

Maximize
$$A_{av} = \left[1 + \sum_{i=1}^{n} \frac{\eta_{i1}(1-\beta_i^{k_i})(1-\alpha_i)}{\mu_{i1}(1-\alpha_i^{k_i})(1-\beta_i)} \right]^{-1}$$

Subject to
$$\begin{cases} \overline{C}_2 \leq C_{20} \\ k_1, k_2, ..., k_n \geq 1 \\ k_1, k_2, ..., k_n = \text{integer} \end{cases}$$

where constant C_{20} is the predetermined requirement for system maintenance cost per unit of time.

Optimization Problem 2

Minimize
$$\overline{C}_1 = \frac{\sum_{i=1}^{n} d_i \frac{\eta_{i1}(1-\beta_i^{k_i})}{1-\beta_i} \frac{1-\alpha_i}{\mu_{i1}(1-\alpha_i^{k_i})}}{1 + \sum_{i=1}^{n} \frac{\eta_{i1}(1-\beta_i^{k_i})}{1-\beta_i} \frac{1-\alpha_i}{\mu_{i1}(1-\alpha_i^{k_i})}}$$

Subject to
$$\begin{cases} \overline{N} = A_{av} \left[\sum_{i=1}^{n} \frac{k_i(1-\alpha_i)}{\mu_{i1}(1-\alpha_i^{k_i})} \right] \leq N_0 \\ k_1, k_2, ..., k_n \geq 1 \\ k_i \leq k_i^0, \quad \forall i = 1, ..., n \\ k_1, k_2, ..., k_n = \text{integer} \end{cases}$$

where constant N_0 is the pre-specified requirement for the number of failures of the system per unit of time, and k_i^0 is the upper bound for the $k_i, i = 1, ..., n$.

Note Proposition 5.4. Thus, a upper bound k_i^0 for the k_i where $i = 1, ..., n$ is set for Optimization Problem 2.

Similarly, according to different requirements and circumstances we can devise different optimization models based on the system availability and maintenance cost rates derived in Sections 5.2 and 5.3. These models can be solved using any nonlinear programming software. Now let's see an example.

EXAMPLE 5.3 Assume that in Example 5.2, $C_{20} = \$0.18$ per hour of down time. Determine the maximum system availability subject to this cost constraint.

Using optimization model 1, the optimal solution can be found to be

$$A_{av,max} = 0.992354 \qquad \text{when} \quad k_1 = 3 \quad k_2 = 3 \quad k_3 = 2 \quad k_4 = 1$$

If $C_{20} = \$0.20$ per hour,

$$A_{av,max} = 0.992893 \qquad \text{when} \quad k_1 = 1 \quad k_2 = 2 \quad k_3 = 1 \quad k_4 = 1$$

If $C_{20} = \$5$ per hour,

$$A_{av,max} = 0.993049 \qquad \text{when} \quad k_1 = 1 \quad k_2 = 1 \quad k_3 = 1 \quad k_4 = 1$$

At this point, the system availability is equal to the one for the case of perfect repair only given by Barlow and Proschan (1975) because the maintenance cost constraint is not tight.

5.5 Concluding Discussions

This chapter models imperfect repair using the quasi-renewal process: upon repair the time to failure of a unit will decrease to some proportion of its immediately previous one and be independent of all previous ones, and the repair time will increase to a multiple of its immediately previous one and be independent of all previous ones, *i.e.*, the successive times to failure follow a decreasing quasi-renewal process and the successive repair times form an increasing quasi-renewal process. One can see that the quasi-renewal process makes availability and maintenance cost modeling of the series system mathematically tractable given repair is imperfect. Most results obtained in this chapter are in closed forms. The optimization problems proposed in this chapter and optimal maintenance policies can be easily solved using mathematical programming software.

Availability indices derived in this chapter are compared with the well-known results by Barlow and Proschan (1975) which assume perfect repair through an example of a four-components series system. The comparison reveals: (a) The system availability and the limiting average of system up times for the imperfect repair case are smaller than for the perfect repair case; (b) The long-run fraction of the time that the system is down due to the failure of each of the four components, and the long-run average number of failures per hour of each of the four components are larger for imperfect repair case. If $k_i = 1$ for all four components, results in this chapter and those by Barlow and Proschan (1975) will be the same. Alternatively, the limiting results for the imperfect repair case when $\alpha_i \to 1$ and $\beta_i \to 1$ for all four components can expect to be the same as those by Barlow and Proschan.

Many systems are the series system in practice, and their maintenance and availability are subject to various shut-off rules (Khalil 1985). Further work on optimum maintenance and availability modeling of the series system can be performed for other shut-off rules.

6

Opportunistic Maintenance of Multi-unit Systems

Block, age and sequential PM policies have been studied extensively in the literature, as shown in Chapter 3. However, these PM policies are designed for a system composed of a single stochastically deteriorating subsystem (McCall 1965). A natural generalization of the underlying maintenance model is to consider a system with multi-subsystems. Optimal maintenance policies for such systems reduce to those for systems with a single subsystem only if there exists no economic dependence, failure dependence and structural dependence. In this case, maintenance decisions are also independent and the optimal policy is to employ an optimal block, age, failure limit, or sequential PM policy for each separate subsystem. However, if there is economic dependence, then the optimal maintenance policy is not one of considering each subsystem separately and maintenance decisions will not be independent, as mentioned in Chapter 1. Obviously, the optimal maintenance action for a given subsystem at any point of time depends on the states of all subsystems, *i.e.*, maintenance is opportunistic. Radner and Jorgenson (1963), Cho and Parlar (1991), Dekker and Smeitink (1991), Zheng (1995), Jesen (1996), Wang and Pham (1996), and Wang (2002) summarize existing work on multi-component systems and opportunistic maintenance.

Chapters 1 and 3 indicate that most existing maintenance models in the reliability and maintenance literature have been developed for one-unit systems, and maintenance models for multi-subsystem systems are only a small proportion. It should also be noted that economic dependence is ignored in some previous maintenance models of multi-unit systems. Some models consider economic dependence but they suppose that all maintenance is perfect. This chapter, based on Wang (1997), investigates the optimal PM policy for a system with economic dependence and imperfect repair.

We assume that this system consists of $n+1$ subsystems and all of them are monitored. We suppose also that in this system one subsystem has increasing failure rate and the remaining n subsystems have constant failure rates. Next, the subsystem with increasing failure rate is denoted by subsystem 0 while the remaining subsystems are labeled by subsystem 1, subsystem 2,..., subsystem n. The failure rate function for each subsystem is given by $\lambda_i(t)$, $\forall i, i = 0,1,...,n$ where

$$\lambda_i(t) = \lambda_i, \quad \forall i, i = 1,...,n \quad \text{and}$$

$$\lambda_0'(t) > 0$$

Since subsystem 1, subsystem 2,..., and subsystem n fail exponentially, they will never be replaced before failure, hence no PM will be performed on them.

The following notation will be used in this chapter:

NOTATION

T	Critical age at which a PM is performed on subsystem 0
t_i	Critical age of subsystem i for $i = 1,2,...,n$
λ_i	Failure rate of subsystem i for $i = 1,2,...,n$
n	Number of subsystems with constant failure rates
C_0, C_{00}	PM cost of subsystem 0 at T and repair cost at its failure
w_0, w_{00}	PM time of subsystem 0 and perfect repair time at its failure
C_i, w_i	Cost and time to replace subsystem i for $i = 1,2,...,n$
C_{0i}, w_{0i}	Cost and time to maintenance subsystem 0 and i together, $\forall i, i = 1,2,...,n$
p	Probability that repair is perfect in section 6.1
q	Probability that repair is minimal, $p + q = 1$ in Section 6.1
$p(t)$	Probability that maintenance or repair is perfect in Section 6.2
$q(t)$	Probability that maintenance or repair is minimal, $p(t) + q(t) = 1$ in Section 6.2
q_{0i}	Probability that the renewal cycle ends with a replacement of subsystem i and PM of subsystem 0 together
q_{00}	Probability that the cycle ends with a perfect repair of subsystem 0
d_i	Probability that the renewal cycle ends on the interval $[t_i, t_{i+1}]$
L	Asymptotic system maintenance cost per unit of time
A	Asymptotic average system availability
B	Random variable: the renewal cycle duration
D	Expected duration of a renewal cycle
C	Expected system maintenance cost per renewal cycle
U	Expected accumulating system failure-free time per renewal cycle
S_i	Time spent on replacing subsystem i alone, $\forall i, i = 1,2,...,n$
S	$\sum_{i=1}^{n} S_i$
Y	Age of subsystem 0 when perfectly repaired or preventively maintained, whichever occurs first
Z	Time spent on performing perfect repair or PM on subsystem 0, possibly with other subsystems (at end of cycle). The minimal repair time is ignored
V_i	Duration of the interval over which subsystem i alone would be replaced
R	Expected system maintenance (down) time per renewal cycle

$\varphi(x)$ Probability density function of Y

$\lambda_0(t), f_0(t)$ Failure rate and probability density function of the life of subsystem 0

$F_0(t), \overline{F}_0(t)$ Cumulative failure distribution and survival function of Subsystem 0

Since this chapter assumes that there exists economic dependence, it spends less cost and time to repair subsystem 0 and any other subsystem together than to repair each subsystem separately, and the optimal action for subsystem 0 depends on the state of the other (exponentially failing) subsystems. Throughout this chapter, we will make such an assumption, *i.e.*,

$$C_0, C_i < C_{0i} < C_0 + C_i \text{ and } w_0, w_i < w_{0i} < w_0 + w_i \qquad (6.1)$$

Radner and Jorgenson (1963) devise an opportunistic maintenance policy. Using a dynamic programming formulation, Radner and Jorgenson show that the optimum replacement policy is what they call a (t_i, T) type of policy. Barlow and Proschan (1975), and Khalil (1985) propose and study such a shut-off rule for system maintenance, which is called shut-off rule 1 in Chapter 1:

While a failed subsystem is in replacement or maintenance, all other subsystems remain in "suspended animation (do not age or fail)". After the repair is completed, the system is returned to operation. At that instant the subsystems in "suspended animation" are as good as they were when the system stopped operating.

The maintenance policy by Radner and Jorgenson (1963) and the shut-off by Barlow and Proschan (1975) and Khalil (1985) are practical in many engineering applications. Let x be the age of subsystem 0 since last replacement of subsystem 0 and T be some fixed time – a decision variable. This chapter considers the following opportunistic maintenance policy, based on the maintenance policy by Radner and Jorgenson (1963), the above shut-off rule used by Barlow and Proschan (1975) and Khalil (1985), and the imperfect repair concepts discussed in Chapter 2:

(i)   If subsystem 0 fails at any time before T, perform imperfect repair on it at a cost C_{00}.

(ii)   If subsystem i fails when the age of subsystem 0 is in the time interval $0 \le x < t_i$, replace subsystem i alone at a cost C_i and at a time w_i, $\forall i, i = 1, 2, ..., n$.

(iii) If subsystem i fails when the age of subsystem 0 is in the time interval $t_i \le x < T$, replace subsystem i and do perfect PM on subsystem 0, $\forall i, i = 1, 2, ..., n$. The total maintenance cost is C_{0i} and total maintenance time is w_{0i}.

(iv) If subsystem 0 survives until $x = T$, perform PM on subsystem 0 at a cost C_0 and at a maintenance time w_0.

Note that in the above maintenance policy the repair before T is imperfect, and PM at T, if any, is perfect. The reason for such a policy is that before T the system is young and in a good operating condition and thus no major repair is needed. Therefore, after repair the subsystem is not as good as new. However, when subsystem 0 survives until $x = T$ it is in a bad operating condition and a perfect PM is necessary. The treatment methods for imperfect repair discussed in Chapter 2 will be used in this chapter.

The optimal maintenance policy for this opportunistic maintenance model of multi-component systems is characterized by $n+1$ decision variables, and is obtained by determining the optimal $(t_1, t_2, ..., t_n, T)$ to maximize the system availability or minimize the system maintenance cost rate or optimize one when the requirements for the other are satisfied. It is worth noting that to achieve good operating characteristics of systems, we might take into account system availability because while the system cost rate is minimized the system availability, however, may not achieve an acceptable level, as demonstrated in Chapter 5.

The above maintenance policy is plausible. Since we assume that it spends less cost and time to perform maintenance on subsystem 0 and any other subsystem together than on each subsystem separately, the optimal action for subsystem 0 will depend on the state of other subsystems. If an exponentially failing subsystem, say subsystem i, is good at some time, two actions are possible for subsystem 0: perform maintenance on it or do nothing. If subsystem i has failed, then there are again two possible actions: perform maintenance on the exponentially failing subsystem only or on both subsystems.

From Equation (6.1) we can see that for multi-component systems this opportunistic maintenance policy may result in higher system availability as compared with the case that each subsystem is separately maintained. This is because while any subsystem fails and is under maintenance the whole system is down, and it will save time to do PMs on unfailed subsystems during this down period and thus reduce the system downtime. Therefore, the optimal maintenance model in this chapter can be expected to be effective to approximate any type of multi-component systems.

6.1 Optimal Maintenance Policies by the (p, q) Rule

Suppose that the imperfect repair of subsystem 0 at failure is treated by the (p, q) rule. Given that the perfect repair time is w_{00}, note that C_{00} is the imperfect repair cost of subsystem 0 in this case. The PM time at T and the perfect repair time at failure are assumed to be different. In this section, PM of subsystem 0 at T or PM of subsystem 0 together with another subsystem before T are assumed to be perfect. Next we will first derive the long-run system maintenance cost per unit of time, or system maintenance cost rate, and the asymptotic average system availability, and then investigate other system operating characteristics and optimization problems.

6.1.1 Modeling of Availability and Cost Rate

Given the above opportunistic PM policy, the times between consecutive perfect repairs or PMs of subsystem 0 constitute a renewal cycle. From the renewal reward theory we have

$$L(T,t_1,t_2,...,t_n;p) = \frac{C(T,t_1,t_2,...,t_n;p)}{D(T,t_1,t_2,...,t_n;p)} \qquad (6.2)$$

$$A(T,t_1,t_2,...,t_n;p) = \frac{U(T,t_1,t_2,...,t_n;p)}{U(T,t_1,t_2,...,t_n;p) + R(T,t_1,t_2,...,t_n;p)} \qquad (6.3)$$

Note that

$$D(T,t_1,t_2,...,t_n;p) = U(T,t_1,t_2,...,t_n;p) + R(T,t_1,t_2,...,t_n;p)$$

Next, without loss of the generality, we assume that $t_1 \le t_2 \le \cdots \le t_n$. Let us first study the expected duration $D(T,t_1,t_2,...,t_n;p)$. The renewal cycle duration B is the sum of three random variables. The duration and expected duration are respectively

$$B = S + Y + Z$$

$$D(T,t_1,t_2,...,t_n;p) = E(B) = E(S) + E(Y) + E(Z) \qquad (6.4)$$

We evaluate the probability density and mean of Y first. Let U_i be the time to failure of subsystem i after t_i, given that subsystem i is good at t_i ($i = 1,2,...,n$), and U_0 is time to the first perfect repair of subsystem 0 since time 0. Let $t_0 = 0$. Noting shut-off rule 1, then,

$$Y = \min(U_0 + t_0, t_1 + U_1,..., t_n + U_n, T) \qquad (6.5)$$

The random variables U_i ($i = 1,2,...,n$) are statistically independent. For $i \ne 0$, U_i has an exponential distribution with failure rate λ_i. Let us denote the cumulative distribution of new subsystem 0 by F_0. Let $\overline{F}_0 = 1 - F_0$. We assume that F_0 is absolutely continuous with density f_0 and that $F_0(0) = 0$. The failure rate of subsystem 0 is supposed to be continuous and increasing. The distribution function and failure rate of the time between successive perfect repairs at failure, will be denoted by $H(t)$ and $r_H(t)$ respectively. We shall use the relationships, proven by Brown and Proschan (1983), that $\overline{H}(t) = \overline{F}_0^{\,p}(t)$ and $r_H(t) = p\lambda_0(t)$ where $\overline{H}(t) = 1 - H(t)$ if there is no PM (see Chapter 2). The density of $H(t)$ is herein denoted by $h(t)$. The cumulative distribution function of Y for $Y < T$ is

$$1 - \Pr(Y > y) = 1 - \Pr(t_i + U_i > y, \ \forall i, i = 0,1,2,...,n)$$

$$= 1 - \prod_{i=0}^{n} \Pr(U_i > y - t_i)$$

$$= 1 - \Pr(U_0 > y) \prod_{i=1}^{n} \Pr(U_i > y - t_i)$$

$$= 1 - \left[(1 - F_0(y))^p\right] \exp\left[-\sum_{\substack{i=1 \\ y - t_i \geq 0}}^{n} \lambda_i (y - t_i)\right]$$

and for $Y = T$

$$\Pr(Y = T) = \left[(1 - F_0(T))^p\right] \exp\left[-\sum_{i=1}^{n} \lambda_i (T - t_i)\right]$$

Next we investigate the probability density of Y. For $i = 1, 2, ..., n$, let

$$M_i = \left(\sum_{j=1}^{i} \lambda_j\right)$$

$$D_i = \exp\left(\sum_{j=1}^{i} \lambda_j t_j\right)$$

$$f_i(y) = \left(\sum_{j=1}^{i} \lambda_j\right) \exp\left[-\sum_{j=1}^{i} \lambda_j (y - t_j)\right]$$

$$= D_i M_i \exp(-M_i y)$$

The distribution of Y has probability density

$$\varphi(y) = \begin{cases} g_0(y), & 0 \leq y < t_1 \\ g_i(y), & t_i \leq y < t_{i+1} \\ g_n(y), & t_n \leq y < T \end{cases} \qquad \forall i, i = 1, 2, ..., n-1 \qquad (6.6)$$

with probability mass at $Y = T$

$$G_0(T) = \left[1 - F_0(T)\right]^p \exp[-\sum_{j=1}^{n} \lambda_j (T - t_j)] = \overline{F}_0^p(T) D_n \exp(-M_n T) \qquad (6.7)$$

where

$$g_i(y) = \begin{cases} h(y) & i = 0; \\ (1 - F_0(y))^p f_1(y) + h(y) \exp\left[-\lambda_1 (y - t_1)\right] & i = 1; \\ \quad \vdots \\ (1 - F_0(y))^p f_i(y) + h(y) \exp\left[-\sum_{j=1}^{i} \lambda_j (y - t_j)\right] & i = 2, 3, ..., n-1 \\ \quad \vdots \\ (1 - F_0(y))^p f_n(y) + h(y) \exp\left[-\sum_{j=1}^{n} \lambda_j (y - t_j)\right] & i = n \end{cases}$$

Therefore, the expected value of Y is given by

$$E(Y) = \sum_{i=0}^{n} \int_{t_i}^{t_{i+1}} y g_i(y) dy + T G_0(T) \tag{6.8}$$

where $t_{n+1} = T$.

Second, we derive the expected value of S. According to the previous definition, V_i is the duration of the interval over which subsystem i alone would be replaced if it were to fail ($i = 1,2,...n$). Then

$$V_i = \min(Y, t_i) \qquad \forall i, i = 1,2,...n$$
$$E(S_i) = \lambda_i E(V_i) w_i \qquad \forall i, i = 1,2,...n$$

Note that V_i has a probability density equal to that of Y for $Y < t_i$, and probability mass $\Pr(Y = t_i)$ concentrated at t_i. Therefore,

$$E(V_i) = \sum_{j=0}^{i-1} \int_{t_j}^{t_{j+1}} y g_i(y) dy + t_i \left[1 - \sum_{j=0}^{i-1} \int_{t_j}^{t_{j+1}} g_i(y) dy \right] \tag{6.9}$$

Recall that $S = \sum_{i=1}^{n} S_i$

Then

$$E(S) = \sum_{i=1}^{n} E(S_i)$$
$$= \sum_{i=1}^{n} \lambda_i w_i E(V_i)$$
$$= \sum_{i=1}^{n} \lambda_i w_i \left[\sum_{j=0}^{i-1} \int_{t_j}^{t_{j+1}} y g_i(y) dy + t_i \left(1 - \sum_{j=0}^{i-1} \int_{t_j}^{t_{j+1}} g_i(y) dy \right) \right] \tag{6.10}$$

Finally, let us derive an expression for $E(Z)$. Denote by d_i the probability that the renewal cycle ends on the interval $[t_i, t_{i+1}]$:

$$d_i = \Pr(t_i \le Y \le t_{i+1}), \qquad \forall, i = 0,1,2,...,n, \quad t_0 = 0, \quad t_{n+1} = T$$

Then

$$d_0 = \int_0^{t_1} g_0(y) dy$$

$$d_1 = (1 - d_0) \int_{t_1}^{t_2} \left[(1 - F_0(y))^p f_1(y) + h(y) \exp[-\lambda_1(y - t_1)] \right] dy$$

$$d_2 = (1 - d_0)(1 - \alpha_1) \int_{t_2}^{t_3} \left[(1 - F_0(y))^p f_2(y) + h(y) \exp\left(-\sum_{j=1}^{2} \lambda_j(y - t_j)\right) \right] dy$$

$$\vdots$$

$$d_i = \prod_{j=0}^{i-1} (1 - \alpha_j) \alpha_j$$

$$\vdots$$

$$d_{n+1} = G_0(T)$$

where $\alpha_0 = d_0$ and

$$\alpha_j = \int_{t_j}^{t_{j+1}} \left[(1 - F_0(y))^p f_j(y) + h(y) \exp\left(-\sum_{i=1}^{j} \lambda_i(y - t_i)\right) \right] dy$$

Let q_{0i} represents the probability that the renewal cycle ends with a replacement of subsystem i and subsystem 0 together. Then

$$q_{0i} = \sum_{j=i}^{n} (1 - d_{j-1}) \int_{t_j}^{t_{j+1}} (1 - F_0(y))^p f_j(y) \, dy \qquad \forall i, i = 1,2,...,n \qquad (6.11)$$

and q_{00}, the probability that the cycle ends with a perfect repair of subsystem 0, is

$$q_{00} = d_0 + \sum_{j=i}^{n} (1 - d_{j-1}) \int_{t_j}^{t_{j+1}} h(y) \exp\left(-\sum_{i=1}^{j} \lambda_i(y - t_i)\right) dy \qquad (6.12)$$

It follows that the third term in Equation (6.4) is given by

$$E(Z) = \sum_{j=i}^{n} q_{0i} w_{0i} + d_{n+1} w_0 \qquad (6.13)$$

This completes the derivation of the expected duration $D(T, t_1, t_2, ..., t_n; p)$ in Equation (6.4).

Next we derive the expected system maintenance cost rate. Noting that the expected number of replacements of subsystem i alone is $\lambda_i E(V_i)$ we obtain the corresponding expected replacement cost for subsystem i in one renewal cycle as $\lambda_i E(V_i) C_i$ where $i = 1,2,...,n$. The probability of a replacement of subsystem 0 and i together multiplied by the corresponding cost is $q_{0i} C_{0i}$, $\forall i, i = 1,...,n$. Using Lemma 2.1 of Fontenot and Proschan (1984), the probability that subsystem 0 is subject to PM at T, multiplied by the sum of the costs of PM at T and minimal repair of subsystem 0 during $[0, T]$, results in

$$d_{n+1}\left[C_0 + C_{00} \, q \int_0^T \lambda_0(t) \, dt \right] \quad \text{or} \quad G_0(T)\left[C_0 + C_{00} \, q \int_0^T \lambda_0(t) \, dt \right].$$

The probability that subsystem 0 is subject to perfect repair at its failure in the time interval $[0, T)$, multiplied by the cost of repair at its failure is $q_{00} C_{00}$. Due to the same lemma of Fontenot and Proschan (1984), the expected minimal repair cost of subsystem 0 during a single renewal cycle, if the renewal cycle ends with a perfect repair of subsystem 0, alone or together with another subsystem, is given by

$$C_{mini} = C_{00} \int_0^T \left[q \int_0^y \lambda_0(t)dt \right] \varphi(y)dy$$

$$= C_{00} \int_0^{t_1} \left[\int_0^y q\lambda_0(t)dt \right] h(y)dy$$

$$+ C_{00} \int_{t_1}^{t_2} \left[\int_0^y q\lambda_0(t)dt \right] \left[(1 - F_0(y))^p f_1(y) + h(y)\exp[-\lambda_1(y - t_1)] \right] dy + \cdots$$

$$+ C_{00} \int_{t_n}^T \left[\int_0^y q\lambda_0(t)dt \right] \left[(1 - F_0(y))^p f_n(y) + h(y)\exp\left[-\sum_{j=1}^n \lambda_j(y - t_j)\right] \right] dy$$

It follows that the expected system maintenance cost during one renewal cycle

$$C(T, t_1, t_2, ..., t_n; p)$$

$$= \sum_{i=1}^n \lambda_i E(V_i)C_i + \sum_{i=0}^n q_{0i}C_{0i} + C_{00} \int_0^T \left[\int_0^y q\lambda_0(t)dt \right] \varphi(y)dy$$

$$+ d_{n+1} \left[C_0 + C_{00} \int_0^T q\lambda_0(t)dt \right]$$ (6.14)

Substituting the above results into Equations (6.2) and (6.3), and noting the relationships

$$E(Y) = U(T, t_1, t_2, ..., t_n; p) \text{ and } E(S + Z) = R(T, t_1, t_2, ..., t_n; p),$$

we have the following proposition:

PROPOSITION 6.1 *The long-run system maintenance cost per unit of time, or system maintenance cost rate, and the asymptotic average system availability are respectively:*

$$L(T, t_1, t_2, ..., t_n; p) =$$

$$\frac{\sum_{i=1}^n \lambda_i E(V_i)C_i + \sum_{i=0}^n q_{0i}C_{0i} + qC_{00} \int_0^T \left[\int_0^y \lambda_0(t)dt \right] \varphi(y)dy + d_{n+1} \left[C_0 + qC_{00} \int_0^T \lambda_0(t)dt \right]}{D(T, t_1, t_2, ..., t_n; p)}$$

(6.15)

$$A(T, t_1, t_2, ..., t_n; p) = \frac{E(Y)}{D(T, t_1, t_2, ..., t_n; p)}$$ (6.16)

From Proposition 6.1, the optimal opportunistic PM policy $(T^*, t_1^*, t_2^*, ..., t_n^*)$ that minimizes the system maintenance cost rate or that maximizes the system availability can be obtained by a nonlinear programming software. Obviously, it would be difficult to obtain the analytical optimal solution.

Next we discuss the other operating characteristics of this opportunistic PM model.

6.1.2 Other Operating Characteristics

To evaluate the performance of the imperfect opportunistic PM policy in this chapter, and to predict supply and maintenance requirements, let us investigate its other operating characteristics besides system maintenance cost rate and availability. First, we note that for this imperfect maintenance model, four different maintenance actions are observed:

(i)   Repair of a failed subsystem 0.
(ii) Replacement of a failed part with a constant failure rate by itself.
(iii) The joint opportunistic maintenance of a failed subsystem with a constant failure rate and subsystem 0 unfailed.
(iv) Perfect maintenance of subsystem 0 unfailed at age T.

In addition to the system availability and maintenance cost rate derived in the last section, other important operating characteristics of this maintenance policy are the expected number of each of these maintenance actions per unit time, and expected maintenance cost of each of these maintenance actions per unit time. Another characteristic of interest is the probability of at least m failures of one of the subsystems in the interval $(0, t)$ (see McCall 1963). In this section, the following operating characteristics will be investigated:

r_{00} Expected rate of perfect repair of subsystem 0

r_{0f} Expected rate of failure of subsystem 0

r_i Expected rate of failure of subsystem i, $\forall i, i = 1, 2, ..., n$

r_{0i} Expected rate of joint opportunistic replacement of failed subsystem i and unfailed subsystem 0, $\forall i, i = 1, 2, ..., n$

r_0 Expected rate of PM of subsystem 0 at age T

r_{0p} Expected rate of total perfect maintenance of subsystem 0

c_{00} Expected rate of expenditure on repair of failed subsystem 0

c_i Expected rate of expenditure on replacement of subsystem i

c_{0i} Expected rate of expenditure on joint replacement of subsystems 0 and i

c_0 Expected rate of expenditure on PM of subsystem 0 at age T

$P_i(m,t)$ Probability of at least m failures of subsystem i in the interval $(0, t)$, $\forall i, i = 0, 1, ..., n$

Let us consider the subsystems with constant failure rates first. Clearly,

$$r_i = \lambda_i \qquad\qquad \forall i, i = 1,2,...,n \qquad (6.17)$$

or including w_i, the time to replace subsystem i,

$$r_i = \frac{1/\lambda_i}{1/\lambda_i + w_i} = \frac{\lambda_i}{\lambda_i w_i + 1} \qquad\qquad \forall i, i = 1,2,...,n \qquad (6.18)$$

Therefore,

$$c_i = r_i C_i = \frac{\lambda_i C_i}{\lambda_i w_i + 1} \qquad\qquad \forall i, i = 1,2,...,n \qquad (6.19)$$

Using the elementary renewal theorem we obtain that the rate of perfect maintenance of subsystem 0, is asymptotically equal to the reciprocal of the expected value of Y, the time to the first perfect maintenance of subsystem 0, that is,

$$\lim_{t \to \infty} r_{0p}(t) = [E(Y)]^{-1}$$

Thus, for large value of t

$$r_{0p} \approx [E(Y)]^{-1} \qquad (6.20)$$

On the other hand, from the foregoing definitions for rates of maintenance,

$$r_{0p} = r_0 + r_{00} + \sum_{i=1}^{m} r_{0i} \qquad (6.21)$$

That is, this expected value, r_{0p}, can be partitioned into three parts: the expected rate of PM at age T, the expected rate of repair of subsystem 0 at failure, the expected rate of joint opportunistic replacement with another subsystem. We can also see this relationship from the derivation of Equation (6.8). By the law of large numbers, the fraction of the total number of replacements of subsystem 0 that are preventive is equal to p_1, the probability that, starting with a new subsystem 0, this subsystem will not be replaced in the interval (0, t). From Equation (6.7) of Section 6.1.1, we have

$$p_1 = [1 - F_0(T)]^p \exp[-\sum_{j=1}^{m} \lambda_j (T - t_j)] \qquad (6.22)$$

Hence, in the long run, the expected rate of PM of subsystem 0 is

$$r_0 \approx p_1 [E(Y)]^{-1} \qquad (6.23)$$

or, including replacement and maintenance time,

$$r_0 \approx \frac{p_1 [E(Y)]^{-1}}{p_1 [E(Y)]^{-1} [E(Z) + E(S)] + 1} = \frac{p_1}{p_1 [E(Z) + E(S)] + E(Y)} \qquad (6.24)$$

The expected expenditure on PM is given by

$$c_0 = r_0 C_0 = \frac{p_1 C_0}{p_1 [E(Z) + E(S)] + E(Y)} \tag{6.25}$$

Hence, in the long run, $\sum_0^n r_{0i}$, the expected rate of repair of subsystem 0 plus the expected rate of joint opportunistic replacement of subsystem 0 is given by

$$r_{fi} = \sum_0^n r_{0i} = (1 - p_1)[E(Y)]^{-1} \tag{6.26}$$

From Section 6.1.1,

$$1 - p_1 = \sum_0^n q_{0i}$$

where q_{0i} s are given by Equations (6.11) and (6.12).
Therefore,

$$r_{fi} = [E(Y)]^{-1} \sum_0^n q_{0i} \tag{6.27}$$

Since the probability that subsystem 0 will be replaced jointly with subsystem i is q_{0i}, the asymptotic expected rate of opportunistic replacement of subsystem 0 and i is given by

$$r_{0i} = q_{0i}[E(Y)]^{-1} \qquad \forall i, \ i = 1,2,...,n \tag{6.28}$$

Noting that q_{00} is the probability that subsystem 0 will be perfectly repaired at failure, we have

$$r_{00} = q_{00}[E(Y)]^{-1} \tag{6.29}$$

or including maintenance time,

$$r_{0i} = \frac{q_{0i}[E(Y)]^{-1}}{q_{0i}[E(Y)]^{-1}[E(Z) + E(S)] + 1} = \frac{q_{0i}}{q_{0i}[E(Z) + E(S)] + E(Y)} \qquad \forall i, \ i = 1,2,...,n \tag{6.30}$$

and

$$r_{00} = \frac{q_{00}[E(Y)]^{-1}}{q_{00}[E(Y)]^{-1}[E(Z) + E(S)] + 1} = \frac{q_{00}}{q_{00}[E(Z) + E(S)] + E(Y)} \tag{6.31}$$

Accordingly, the expected rate of expenditure on opportunistic replacements is

$$c_{0i} = r_{0i} C_{0i} = \frac{q_{0i} C_{0i}}{q_{0i}[E(Z) + E(S)] + E(Y)} \qquad \forall i, \ i = 1,2,...,n \tag{6.32}$$

Since the repair of subsystem 0 at failure is perfect with probability p, the rate of failure of subsystem 0 is given by

$$r_{0f} = r_{00}/p = \frac{q_{00}}{q_{00}[E(Z)+E(S)]+E(Y)}\frac{1}{p} \tag{6.33}$$

Accordingly, the expected rate of expenditure on repair of subsystem 0 at failure is

$$c_{00} = r_{0f}C_{00} = \frac{q_{00}C_{00}}{q_{00}[E(Z)+E(S)]+E(Y)}\frac{1}{p} \tag{6.34}$$

Because the lifetime of subsystem i follows the exponential distribution, the probability of at least m replacements of subsystem i in the interval $(0, t)$ is

$$P_i(m,t) = \sum_{j=m}^{\infty} \exp(-\lambda_i t)\cdot\frac{(\lambda_i t)^j}{j!} \qquad \forall i, i = 1,...,n \tag{6.35}$$

6.1.3 Optimization Models

So far we have derived the system reliability measures: system availability, expected rate of failure of subsystem 0, expected rate of failure of subsystem i, etc., and system maintenance cost measures: system maintenance cost rate, expected rate of expenditure on repair of failed subsystem 0, expected rate of expenditure on replacement of subsystem i, etc. To obtain the optimal system maintenance policies the system reliability measures and system maintenance cost measures must be both acceptable. For example, if it is required that while the expected rate of failure of subsystem 0 is less than A_0 the system maintenance cost rate is minimized. For such a problem, we can formulate the following optimization model from Equations (6.15) and (6.33):

Minimize

$$L(T,t_1,t_2,...,t_n;p) =$$

$$\frac{\sum_{i=1}^{n}\lambda_i E(V_i)C_i + \sum_{i=0}^{n}q_{0i}C_{0i} + qC_{00}\int_0^T\left[\int_0^y\lambda_0(t)dt\right]\varphi(y)dy + d_{n+1}\left[C_0 + C_{00}\,q\int_0^T\lambda_0(t)dt\right]}{D(T,t_1,t_2,...,t_n;p)}$$

Subject to

$$\begin{cases} r_{0f} = \dfrac{q_{00}}{q_{00}[E(Z)+E(S)]+E(Y)}\dfrac{1}{p} \le r^+ \\[4mm] t_1,t_2,...,t_n,T \ge 0 \end{cases}$$

where r^+ is the predetermined requirement for expected rate of failure of subsystem 0.

This model can be solved by nonlinear programming software to obtain an optimal system maintenance policy $(t_1^*, t_2^*, ..., t_n^*, T^*)$.

6.2 Optimal Maintenance Policies by the $(p(t), q(t))$ Rule

Suppose that the imperfect repair of subsystem 0 at failure is treated by the $(p(t), q(t))$ rule introduced in Section 2.1.1. Note that C_{00} is the imperfect repair cost of subsystem 0 in this case, and the PM time at T and the perfect repair time at failure are assumed to be different. In this section, PM of subsystem 0 at T or PM of subsystem 0 together with another subsystem before T are assumed to be perfect. Next we will derive the long-run system maintenance cost per unit of time, or system maintenance cost rate, and the asymptotic average system availability.

6.2.1 Modeling of Availability and Cost Rate

Given that the above opportunistic PM policy, the times between consecutive perfect repairs or PMs of subsystem 0 constitute a renewal cycle. From the renewal reward theory,

$$L(T, t_1, t_2, ..., t_n; p) = \frac{C(T, t_1, t_2, ..., t_n; p)}{D(T, t_1, t_2, ..., t_n; p)} \tag{6.36}$$

$$A(T, t_1, t_2, ..., t_n; p) = \frac{U(T, t_1, t_2, ..., t_n; p)}{U(T, t_1, t_2, ..., t_n; p) + R(T, t_1, t_2, ..., t_n; p)} \tag{6.37}$$

Note that

$$D(T, t_1, t_2, ..., t_n; p) = U(T, t_1, t_2, ..., t_n; p) + R(T, t_1, t_2, ..., t_n; p)$$

Next without loss of the generality we assume that $t_1 \le t_2 \le \cdots \le t_n$. Let us first investigate the expected duration $D(T, t_1, t_2, ..., t_n; p)$. The renewal cycle duration B is the sum of three random variables. The duration and expected duration are respectively

$$B = S + Y + Z$$

$$\begin{aligned} D(T, t_1, t_2, ..., t_n; p) &= E(B) \\ &= E(S) + E(Y) + E(Z) \end{aligned} \tag{6.38}$$

For the three terms in Equation (6.38), we evaluate the probability density and mean of Y first. Let U_i be the time to failure of subsystem i after t_i , given that subsystem i is good at t_i ($i = 1, 2, ..., n$), and U_0 is time to the first perfect repair of subsystem 0 since time 0. Let $t_0 = 0$. Noting shut-off rule 1, then

$$Y = \min(U_0 + t_0, t_1 + U_1, ..., t_n + U_n, T)$$

The random variables U_i $(i = 1, 2, ..., n)$ are statistically independent. For $i \neq 0$, U_i has an exponential distribution with constant failure rate λ_i. Denote the *cdf* of new subsystem 0 by F_0. Let $\overline{F}_0 = 1 - F_0$. We assume that F_0 is absolutely continuous with density f_0 and that $F_0(0) = 0$. The failure rate, $\lambda_0(t)$, of subsystem 0 is supposed to be continuous and increasing. The *cdf* and failure rate of the time between successive perfect repairs of subsystem 0 at failures will be denoted by $H(t)$ and $r_H(t)$ respectively. We shall use the facts, proven by Block *et al.* (1985), that

$$\overline{H}(t) = \exp\left[-\int_0^t p(x)\lambda_0(x)dx\right] \text{ and } r_H(t) = p(t)\lambda_0(t)$$

where $\overline{H}(t) = 1 - H(t)$, given that there is no PM and $\int_0^\infty p(x)\overline{F}_0^{-1}(x)F_0(dx) = +\infty$ (see Chapter 2). The derivative of $H(t)$ is herein denoted by $h(t)$.

It is easy to verify that the cumulative distribution function of Y for $Y < T$ is given by

$$1 - \Pr(Y > y) = 1 - \Pr(t_i + U_i > y, \forall i, i = 0, 1, 2, ..., n)$$

$$= 1 - \prod_{i=0}^n \Pr(U_i > y - t_i)$$

$$= 1 - \Pr(U_0 > y)\prod_{i=1}^n \Pr(U_i > y - t_i)$$

$$= 1 - \exp\left[-\int_0^y p(x)\lambda_0(x)dx\right]\exp\left[-\sum_{\substack{i=1 \\ y \geq t_i}}^n \lambda_i(y - t_i)\right]$$

and for $Y = T$

$$\Pr(Y = T) = \exp\left[-\int_0^T p(x)\lambda_0(x)dx\right]\exp\left[-\sum_{i=1}^n \lambda_i(T - t_i)\right]$$

Now we evaluate the probability density function of Y. For $i = 1, 2, ..., n$, let

$$M_i = \left(\sum_{j=1}^i \lambda_j\right)$$

$$D_i = \exp\left(\sum_{j=1}^i \lambda_j t_j\right)$$

$$f_i(y) = \left(\sum_{j=1}^i \lambda_j\right)\exp\left[-\sum_{j=1}^i \lambda_j(y - t_j)\right]$$

$$= D_i M_i \exp(-M_i y)$$

Then the distribution of Y has probability density

$$\varphi(y) = \begin{cases} g_0(y), \ 0 \le y < t_1 \\ g_i(y), \ t_i \le y < t_{i+1} \\ g_n(y), \ t_n \le y < T \end{cases} \qquad \forall i, i = 1,2,...,n-1 \qquad (6.39)$$

with probability mass at $Y = T$

$$G_0(T) = \exp\left[-\int_0^T p(x)\lambda_0(x)dx\right]\exp\left[-\sum_{j=1}^m \lambda_j(T-t_j)\right]$$

$$= \exp\left[-\int_0^T p(x)\lambda_0(x)dx\right]D_n \exp(-M_n T) \qquad (6.40)$$

where

$$g_i(y) = \begin{cases} h(y) & i = 0; \\ \exp\left[-\int_0^y p(x)\lambda_0(x)dx\right]f_1(y) + h(y)\exp[-\lambda_1(y-t_1)] & i = 1; \\ \quad \vdots \\ \exp\left[-\int_0^y p(x)\lambda_0(x)dx\right]f_i(y) + h(y)\exp\left[-\sum_{j=1}^i \lambda_j(y-t_j)\right] \\ \qquad\qquad\qquad\qquad\qquad\qquad\qquad\qquad i = 2,3,...,n-1 \\ \quad \vdots \\ \exp\left[-\int_0^y p(x)\lambda_0(x)dx\right]f_n(y) + h(y)\exp\left[-\sum_{j=1}^m \lambda_j(y-t_j)\right] & i = n \end{cases}$$

Therefore, the expected value of Y is given by

$$E(Y) = \sum_{i=0}^n \int_{t_i}^{t_{i+1}} y g_i(y)dy + TG_0(T) \qquad (6.41)$$

where $t_{n+1} = T$.

Second, we explore the expected value of S. According to the previous definition, V_i is the duration of the interval over which subsystem i alone would be replaced if it were to fail ($i = 1,2,...n$). Then

$$V_i = \min(Y, t_i) \qquad \forall i, i = 1,2,...n$$
$$E(S_i) = \lambda_i E(V_i)w_i \qquad \forall i, i = 1,2,...n$$

Note that V_i has a probability density equal to that of Y for $Y < t_i$, and probability mass $\Pr(Y = t_i)$ concentrated at t_i. Therefore,

$$E(V_i) = \sum_{j=0}^{i-1} \int_{t_j}^{t_{j+1}} y g_i(y)dy + t_i\left[1 - \sum_{j=0}^{i-1} \int_{t_j}^{t_{j+1}} g_i(y)dy\right] \qquad (6.42)$$

Recall that $S = \sum_{i=1}^{n} S_i$

Then

$$E(S) = \sum_{i=1}^{n} E(S_i)$$

$$= \sum_{i=1}^{n} \lambda_i w_i E(V_i)$$

$$= \sum_{i=1}^{n} \lambda_i w_i \left[\sum_{j=0}^{i-1} \int_{t_j}^{t_{j+1}} y g_i(y) dy + t_i \left(1 - \sum_{j=0}^{i-1} \int_{t_j}^{t_{j+1}} g_i(y) dy \right) \right] \qquad (6.43)$$

Finally, let us derive an expression for $E(Z)$. Denote by d_i the probability that the renewal cycle ends on the interval $[t_i, t_{i+1}]$:

$$d_i = \Pr(t_i \leq Y \leq t_{i+1}), \qquad \forall, i = 0,1,2,...,n, \ t_0 = 0, \ t_{n+1} = T$$

Then

$$d_0 = \int_0^{t_1} g_0(y) dy$$

$$d_1 = (1 - d_0) \int_{t_1}^{t_2} \left\{ \exp\left[-\int_0^y p(x) \lambda_0(x) dx \right] f_1(y) + h(y) \exp[-\lambda_1(y - t_1)] \right\} dy$$

$$d_2 = (1 - d_0)(1 - \alpha_1)$$

$$\times \int_{t_2}^{t_3} \left\{ \exp\left[-\int_0^y p(x) \lambda_0(x) dx \right] f_2(y) + h(y) \exp\left(-\sum_{j=1}^{2} \lambda_j (y - t_j) \right) \right\} dy$$

$$\vdots$$

$$d_i = \prod_{j=0}^{i-1} (1 - \alpha_j) \alpha_j$$

$$\vdots$$

$$d_{n+1} = G_0(T)$$

where $\alpha_0 = d_0$ and

$$\alpha_j = \int_{t_j}^{t_{j+1}} \left\{ \exp\left[-\int_0^y p(x) \lambda_0(x) dx \right] f_j(y) + h(y) \exp\left(-\sum_{i=1}^{j} \lambda_i (y - t_i) \right) \right\} dy$$

Let q_{0i} represents the probability that the renewal cycle ends with a replacement of subsystem i and subsystem 0 together. Then,

$$q_{0i} = \sum_{j=i}^{n} \left(1 - d_{j-1}\right) \int_{t_j}^{t_{j+1}} f_j(y) \exp\left[-\int_0^y p(x)\lambda_0(x)dx\right]dy \qquad \forall i, i = 1,2,...,n \quad (6.44)$$

and q_{00}, the probability that the cycle ends with a perfect repair of subsystem 0, is given by

$$q_{00} = d_0 + \sum_{j=i}^{n} \left(1 - d_{j-1}\right) \int_{t_j}^{t_{j+1}} h(y) \exp\left(-\sum_{i=1}^{j} \lambda_i(y - t_i)\right)dy \qquad (6.45)$$

It follows that the third term in Equation (6.38) is given by

$$E(Z) = \sum_{j=i}^{n} q_{0i} w_{0i} + d_{n+1} w_0 \qquad (6.46)$$

This completes the derivation of the expected duration $D(T, t_1, t_2, ..., t_n; p)$.

Now we derive the expected system maintenance cost rate. Noting that the expected number of replacements of subsystem i alone is $\lambda_i E(V_i)$, it follows that the corresponding expected replacement cost for subsystem i in one renewal cycle is $\lambda_i E(V_i)C_i$, $i = 1,2,...,n$. The probability of a replacement of subsystem 0 and i together multiplied by the corresponding cost is $q_{0i}C_{0i}$, $\forall i, i = 1,2,...,n$. The probability that subsystem 0 alone is subject to perfect repair at its failure in the time interval $(0, T)$, multiplied by the cost of repair at its failure is $q_{00}C_{00}$.

Next we investigate the expected imperfect repair cost of subsystem 0 during a single renewal cycle. Consider the non-homogeneous Poisson process $\{N(t), t > 0\}$ with intensity function $\lambda_0(t)$ and successive arrival times $s_1, s_2, ...$ At time s_n we flip a coin. Designate the outcome by W_n which takes the value 1 (head) with probability $p(s_n)$ and the value 0 (tail) with probability $q(s_n)$. Let

$$I(t) = \sum_{n=1}^{N(t)} W_n$$

$$M(t) = N(t) - I(t)$$

According to Savits (1988), the processes $\{I(t), t \geq 0\}$ and $\{M(t), t \geq 0\}$ are independent non-homogeneous Poisson processes with respective intensities $p(t)\lambda_0(t)$ and $q(t)\lambda_0(t)$. Hence, the conditional probability that k minimal repairs occur given that no perfect repair in $[0, y]$ is given by

$$P\{M(y) = k \mid I(y) = 0\} = P\{M(y) = k\}$$

$$= \frac{\exp\left[-\int_0^y q(t)\lambda_0(t)dt\right] \cdot \left[\int_0^y q(t)\lambda_0(t)dt\right]^k}{k!}$$

with mean of $\int_0^y q(t)\lambda_0(t)dt$.

Using the foregoing result of Savits (1988), the probability that subsystem 0 is subject to PM at T, multiplied by the sum of the costs of PM at T and minimal repair of subsystem 0 during $[0,T]$, results in

$$d_{n+1}\left[C_0 + C_{00}\int_0^T q(t)\lambda_0(t)dt\right].$$

Using this result by Savits (1988) again, the expected minimal repair cost of subsystem 0 during a single renewal cycle, if the renewal cycle ends with a perfect repair of subsystem 0, alone or together with another subsystem, is given by

$$C_{mini} = C_{00}\int_0^T\left[\int_0^y q(t)\lambda_0(t)dt\right]\varphi(y)dy$$

$$= C_{00}\int_0^{t_1}\Psi(y)\cdot h(y)dy$$

$$+ C_{00}\int_{t_1}^{t_2}\Psi(y)\left\{\exp\left[-\int_0^y p(x)\lambda_0(x)dx\right]f_1(y) + h(y)\exp\left[-\lambda_1(y-t_1)\right]\right\}dy + \ldots$$

$$+ C_{00}\int_{t_n}^{T}\Psi(y)\left\{\exp\left[-\int_0^y p(x)\lambda_0(x)dx\right]f_n(y) + h(y)\exp\left[-\sum_{j=1}^{n}\lambda_j(y-t_j)\right]\right\}dy$$

$$(6.47)$$

where

$$\Psi(y) = \int_0^y q(t)\lambda_0(t)dt \qquad (6.47a)$$

From the above analysis it follows that the expected system maintenance cost during one renewal cycle is given by

$$C(T,t_1,t_2,\ldots,t_n;p) = \sum_{i=1}^{n}\lambda_i E(V_i)C_i$$

$$+ \sum_{i=0}^{n}q_{0i}C_{0i} + C_{00}\int_0^T\left[\int_0^y q(t)\lambda_0(t)dt\right]\varphi(y)dy$$

$$+ d_{n+1}\left[C_0 + C_{00}\int_0^T q(t)\lambda_0(t)dt\right]$$

$$= \sum_{i=1}^{n}\lambda_i E(V_i)C_i$$

$$+ \sum_{i=0}^{n}q_{0i}C_{0i} + C_{00}\int_0^T\Psi(y)\cdot\varphi(y)dy + d_{n+1}\left[C_0 + C_{00}\int_0^T q(t)\lambda_0(t)dt\right]$$

$$(6.48)$$

Substituting the above results into Equations (6.36) and (6.37), and noting that $E(Y) = U(T, t_1, t_2, ..., t_n; p)$ and $E(S + Z) = R(T, t_1, t_2, ..., t_n; p)$, the following result follows:

PROPOSITION 6.2 *The long-run system maintenance cost per unit of time, or system maintenance cost rate, and the asymptotic average system availability are, respectively,*

$$L(T, t_1, t_2, ..., t_n; p) =$$

$$\frac{\sum_{i=1}^{n} \lambda_i E(V_i) C_i + \sum_{i=0}^{n} q_{0i} C_{0i} + C_{00} \int_0^T \Psi(y) \cdot \varphi(y) dy + d_{n+1} \left[C_0 + C_{00} \int_0^T q(t) \lambda_0(t) dt \right]}{D(T, t_1, t_2, ..., t_n; p)}$$

(6.49)

$$A(T, t_1, t_2, ..., t_n; p) = \frac{E(Y)}{D(T, t_1, t_2, ..., t_n; p)}$$
(6.50)

From Proposition 6.2, the optimal opportunistic PM policy $(T^*, t_1^*, t_2^*, ..., t_n^*)$ to minimize the system maintenance cost rate or to maximize the system availability can be obtained by a nonlinear programming software. Next we discuss the other operating characteristics of this opportunistic PM model.

6.2.2 Other Performance Measures

As in Section 6.1.2, we can derive other important operating characteristics of this policy, such as the expected number of each of these maintenance actions per unit time, and expected maintenance cost of each of these maintenance actions per unit time, in addition to the system availability and maintenance cost rate derived in Section 6.2.1. For example, using the elementary renewal theorem, the rate of perfect maintenance of subsystem 0, is asymptotically equal to the reciprocal of the expected value of Y, the time to the first perfect maintenance of subsystem 0, *i.e.*,

$$\lim_{t \to \infty} r_{0p}(t) = [E(Y)]^{-1}$$

Thus, for large value of t,

$$r_{0p} \approx [E(Y)]^{-1}$$
(6.51)

On the other hand, from the definitions for rates of maintenance (replacement),

$$r_{0p} = r_0 + r_{00} + \sum_1^n r_{0i}$$
(6.52)

That is, this expected value, r_{0p}, can be partitioned into three parts: the expected rate of PM at age T, the expected rate of perfect repair of subsystem 0 at failure, and the expected rate of joint opportunistic replacement with another subsystem. We can also see this relationship from the derivation of Equation (6.41). By the law of large numbers, the fraction of the total number of perfect maintenance of subsystem 0 that is preventive equals p_1, the probability that, starting with a new subsystem 0, this subsystem will not be perfectly maintained in the interval (0, T). From Equation (6.40) of Section 6.2.1, we have

$$p_1 = \exp\left[-\int_0^T p(t)\lambda_0(t)dt \right]\exp\left[-\sum_{j=1}^n \lambda_j(T - t_j) \right] \tag{6.53}$$

where $p(t)$ is the probability that a repair of subsystem 0 at failure is perfect. Hence, in the long run, the expected rate of PM of subsystem 0 is

$$r_0 \approx p_1[E(Y)]^{-1} \tag{6.54}$$

or, including replacement and maintenance time,

$$r_0 \approx \frac{p_1[E(Y)]^{-1}}{p_1[E(Y)]^{-1}[E(Z) + E(S)] + 1} = \frac{p_1}{p_1[E(Z) + E(S)] + E(Y)} \tag{6.55}$$

Hence, in the long run, $\sum_0^n r_{0i}$, the expected rate of perfect repair of subsystem 0 plus the expected rate of joint opportunistic replacement of subsystem 0 is:

$$r_{fj} = \sum_0^n r_{0i} = (1 - p_1)[E(Y)]^{-1} \tag{6.56}$$

From Section 6.2.1,

$$1 - p_1 = \sum_0^n q_{0i}$$

where q_{0i}s are given by Equations (6.44) and (6.45).
Therefore,

$$r_{fj} = [E(Y)]^{-1}\sum_0^n q_{0i} \tag{6.57}$$

Other operating characteristics can be derived in a way similar to Section 6.1.2. Results can be found in Wang (1997).

6.2.3 Optimal Maintenance Policy

To obtain the optimal system maintenance policies, the system reliability measures and system maintenance cost measures must both be acceptable. Based on the obtained system reliability measures – system availability, expected rate of failure of subsystem 0, expected rate of failure of subsystem i, etc., and system maintenance cost measures – system maintenance cost rate, expected rate of expenditure on repair of failed subsystem 0, expected rate of expenditure on replacement of subsystem i, etc., we can formulate different optimization models. For example, based on Equations (6.49) and (6.50), we can have the following optimization model:

Maximize

$$A(T, t_1, t_2, ..., t_n; p) = \frac{E(Y)}{D(T, t_1, t_2, ..., t_n; p)}$$

Subject to

$$
\begin{cases}
L(T, t_1, t_2, ..., t_n; p) = \\[2mm]
\dfrac{\displaystyle\sum_{i=1}^{n} \lambda_i E(V_i) C_i + \sum_{i=0}^{n} q_{0i} C_{0i} + C_{00} \int_0^T \Psi(y)\varphi(y)dy + d_{n+1}[C_0 + C_{00} \int_0^T q(t)\lambda_0(t)dt]}{D(T, t_1, t_2, ..., t_n; p)} \leq \varpi \\[4mm]
t_1, t_2, ..., t_n, T \geq 0
\end{cases}
$$

where ϖ is the predetermined requirement for the rate of perfect maintenance of subsystem 0.

The above optimization model can be solved by nonlinear programming software to obtain the optimal system maintenance policy $(t_1^*, t_2^*, ..., t_n^*, T^*)$.

6.3 Concluding Remarks

The optimal maintenance of a system with $n+1$ subsystems is studied in this chapter. We assume that in this system there is economic dependency, *i.e.*, both maintenance costs and times are less for several subsystems simultaneously than for each subsystem taken separately. We also suppose that repair is imperfect and the imperfect repair is modeled by the (p, q) rule and $(p(t), q(t))$ rule. The realistic shut-off rule is used in this chapter. The system availability, system maintenance cost rate, and other operating characteristics of this multi-unit system are derived and the optimum system maintenance policies to optimize the system operating characteristics are proposed.

For multi-component systems the opportunistic maintenance policy in this chapter may result in higher system availability as compared with the case that each subsystem is separately maintained. This is because while any subsystem fails and is under maintenance the whole system is down, and it will save time to do PMs on unfailed subsystems during this down period and thus reduce the system downtime.

Noting the relationship between the (p, q) rule and $(p(t), q(t))$ rule, the results in Section 6.2 are general. Both the (p, q) rule and $(p(t), q(t))$ rule are convenient to model imperfect maintenance of multi-component systems.

Different from single component systems, one of the key problems for multi-component systems is economic dependence. Imperfectness of maintenance is another important factor. This chapter considers both factors which affect optimal maintenance policies of multi-unit systems and produces some results on this aspect. The maintenance policies are realistic and the results obtained in this chapter can be expected to be useful in practice. Further work includes extending this work to multi-unit systems with two or more IFR subsystems which are subject to imperfect maintenance and economic dependence or considering other shut-off rules. Various shut-off rules for system maintenance can be found in Khalil (1985).

7

Optimal Preparedness Maintenance of Multi-unit Systems with Imperfect Maintenance and Economic Dependence

A system is placed in storage and is called on to perform a given task only if a specific but unpredictable emergency occurs. Some maintenance actions may be taken while the system is in storage and the objective is to choose the sequence of maintenance actions resulting in the highest level of system "preparedness for field use". This maintenance policy is known as the preparedness maintenance policy (McCall 1963). In Menipaz (1978), various inspection and preparedness models are examined, which deal with stochastically failing systems, in which failure is detected by inspection only. He provides the analysis of various models and various objective functions, and the analysis of those models while the maintenance costs are changing over time. The preparedness maintenance of a multi-unit system with $n+1$ subsystems subject to imperfect maintenance and opportunistic maintenance is presented in this chapter, following Wang (1997) and Wang *et al.* (2001). It is assumed that in the system the total maintenance costs and times are less for several subsystems simultaneously than for each subsystem separately as they often do in practice, *i.e.*, economic dependence exists. An opportunistic maintenance policy is incorporated in this model. It is also assumed that PM is imperfect since in practice most maintenance actions tend to make systems not "as good as new" but younger, as discussed in Chapter 1. The system storage 'availability', system maintenance cost rate, and other operating characteristics of this multi-unit system are discussed. The optimum opportunistic preparedness maintenance policies to optimize the system operating performance and to provide the best operational readiness are then investigated.

7.1 Introduction

This chapter discusses an optimal preparedness maintenance policy for a system with $n+1$ subsystems considering imperfect maintenance and economic dependency. Assume that the system is placed in storage and is called upon to perform a given task only if a specific but unpredictable emergency occurs. Some maintenance actions may be taken while the system is in storage or long-term cold standby and the objective is to choose the maintenance action sequence providing the best level of preparedness. McCall (1963, 1965) applies this preparedness maintenance policy to ballistic missile maintenance and obtains an optimal preparedness maintenance policy. The ballistic missile studied was composed of one uninspected subsystem: the rocket engines, as well as three subsystems which are continuously inspected – the nozzle control units, the guidance and control system, and the re-entry vehicle. Obviously, to keep ballistic missiles at the highest level of operational readiness and thus to prevent them from failure in use they should be subject to frequent inspection and maintenance when in storage.

The main difference between the preparedness maintenance model for missiles, rockets, *etc.* and the other maintenance models for automobiles, aircraft, *etc.* lies in the way in which failures are detected. With the automobiles, aircraft, *etc.*, failure occurring while the system is not in operation will be detected whenever an attempt at operation is made. The state of the system is always known with certainty. In fact, continuous operation provides assurance that the state of the system is always known with certainty. However, in a missile system, such a failure will go undetected indefinitely; the state of the system (at least some of its subsystems) is not known with certainty unless some definite maintenance or inspection action is taken. The difference directly affects the design of optimal maintenance policies for each kind of system. Those for automobiles must be designed to overcome the effects of uncertainty about when failures will occur. The policies for missile systems must overcome the same uncertainty, plus another as well: uncertainty about the actual state of the system at any given time – that is, whether it is good or has failed (McCall 1965). If the actual state of the system is known with certainty, either through continuous inspection or continuous operation, the theory of maintenance for the preparedness model becomes the same as for the other maintenance models such as the age replacement policy and block replacement policy. In this sense, the theory of maintenance for the preparedness model is more general than the other maintenance models.

The preparedness model is characterized by three different uncertainties. First, it is impossible to predict the exact time of system failure. Second, the time of emergency use is also not susceptible to exact prediction. Finally, the state of the system is known only at the time of certain maintenance or inspection actions (Radner and Jorgenson 1963).

This chapter considers imperfectness of PM and economic dependence in this multi-unit system, *i.e.*, maintenance costs and times are less for several subsystems simultaneously than for each subsystem taken separately. We suppose that the times to failure of the subsystems in this system are stochastically independent. We also assume that one subsystem in this system has an increasing failure rate while the remaining n subsystems have constant failure rates. The subsystem with an increasing failure rate is a uninspected subsystem and the other n subsystems are inspected ones. The optimal policy for these kinds of systems possesses an opportunistic characteristic. For example, the failure of one subsystem results in a possible opportunity to perform PM on other subsystems.

NOTATION

λ_i	Failure rate of subsystem i, $\forall i$, $i = 1,2,...,n$
n	Number of subsystems with constant failure rates
T	Time interval at the end of which a PM is performed on subsystem 0
t_i	Critical age of subsystem i, $\forall i$, $i = 1,2,...,n$
C_0	PM cost of subsystem 0 at jT
w_0	PM time of subsystem 0
C_i, w_i	Cost and time to replace subsystem i, $\forall i$, $i = 1,2,...,n$
C_{0i}, w_{0i}	Cost and time to maintenance subsystem 0 and i together, $\forall i$, $i = 1,2,...,n$
p	Probability that PM is perfect
q	Probability that PM is minimal, $p + q = 1$
q_{0i}	Probability that the renewal cycle ends with a replacement of subsystem i and PM of subsystem 0 together
d_i	Probability that the renewal cycle ends on the interval $[t_i, t_{i+1})$
L	Asymptotic system maintenance cost per unit of time
A	Asymptotic average system (storage) 'availability'
D	Expected duration of a renewal cycle
C	Expected system maintenance cost per renewal cycle
U	Expected accumulating system failure-free time per renewal cycle
R	Expected system maintenance (down) time per renewal cycle
B	Random variable: the renewal cycle duration
S_i	Random variable: time spent on replacing subsystem i alone in one renewal cycle, $\forall i$, $i = 1,2,...,n$
Y	Random variable: age of subsystem 0 when perfectly preventively maintained
Z	Random variable: time spent on performing perfect repair or perfect PM on subsystem 0, possibly with other subsystems (at end of renewal cycle)
V_i	Random duration of the interval over which subsystem i alone would be replaced
$\varphi(x)$	Probability density function of Y
$\lambda_0(t), f_0(t)$	Failure rate and probability density function of the life of subsystem 0
$F_0(t), \overline{F}_0(t)$	Cumulative failure distribution and survival function of subsystem 0

The distinctive feature of preparedness models is that the state of the system is ascertained only at the time of inspection or maintenance. Next, the subsystem with increasing failure rate $\lambda_0(t)$ is denoted by subsystem 0 while the remaining subsystems are labeled by subsystem 1, subsystem 2,..., and subsystem n. The failure rate function for each remaining subsystem is denoted by $\lambda_i(t)$, $\forall i, i = 1,...,n$ where

$$\lambda_i(t) = \lambda_i, \qquad \forall i, i = 1,...,n \quad \text{and}$$

$$\lambda_0'(t) > 0$$

Since subsystems 1,2,..., n fail exponentially, they are never replaced before failure, that is, no PM will be performed on them.

As stated at the beginning of this chapter, we assume that it spends less cost and time to perform maintenance on subsystem 0 and any other subsystem together than to do maintenance on each subsystem separately, that is,

$$C_0, C_i < C_{0i} < C_0 + C_i \quad \text{and} \quad w_0, w_i < w_{0i} < w_0 + w_i \quad (7.1)$$

At any point in time the maintenance performer must choose among four alternatives: perform maintenance on the un-inspected subsystem; on an inspected subsystem; on the un-inspected subsystem and an inspected subsystem together; or do nothing (no maintenance). Using a dynamic programming formulation, Radner and Jorgenson (1963) show that the optimum maintenance policy is what they call a (t_i, T) type of policy and proposed an opportunistic maintenance policy. Note that shut-off rule 1 for maintenance by Barlow and Proschan (1975) and Khalil (1985) is realistic:

> While a failed subsystem is in repair or maintenance, all other subsystems remain in "suspended animation". After the repair or maintenance is completed, the system is returned to operation. At that instant the subsystems in "suspended animation" are as good as they were when the system stopped operating.

Let x be the age of subsystem 0 since last replacement of subsystem 0. This chapter investigates such an opportunistic preparedness maintenance policy, based on the preparedness maintenance model developed by Radner and Jorgenson (1963), the shut-off rule by Barlow and Proschan (1975) and Khalil (1985), as well as the method to model imperfect maintenance studied by Brown and Proschan (1983) and Fontenot and Proschan (1984):

(i)   If subsystem i fails when the age of subsystem 0 is in the time interval $[0, t_i)$, replace subsystem i alone at a cost of C_i and at a time of w_i $\forall i, i = 1,2,...,n$.

(ii) If subsystem i fails when the age of subsystem 0 is in the time interval $[t_i, T)$, replace subsystem i and do perfect PM on subsystem 0 $\forall i, i = 1,2,...,n$. The total maintenance cost is C_{0i} and total maintenance time is w_{0i}.

(iii) If subsystem 0 survives until its age $x = T$, perform PM on subsystem 0 alone at a cost of C_0 and at a maintenance time of w_0 at $x = T$. PM is imperfect.

(iv) If subsystem 0 has not received a perfect PM, perform PM on it alone at time jT ($j = 2,3,...$) until it receives a perfect PM. If subsystem 0 does not receive a perfect maintenance and subsystem i fails after some PM,

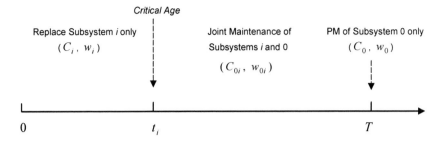

Figure 7.1. Opportunistic maintenance policy

replace subsystem i and do perfect PM on subsystem $0, \forall i, i = 1,2,...,n$, and the total maintenance cost is still C_{0i} and total maintenance time is w_{0i}. This process continues until subsystem 0 receives a perfect maintenance.

The optimal maintenance policy for this opportunistic preparedness maintenance model of multi-component systems is characterized by $(n+1)$ decision variables $(t_1,t_2,...,t_n,T)$, and is obtained by determining the optimal $(t_1,t_2,...,t_n,T)$ that maximizes the system availability, or minimizes the system maintenance cost rate, or optimizes one when the predetermined requirements for the other are satisfied. It is worth noting that to achieve good operating characteristics of systems, we might take into account system availability because while the system cost rate is minimized the system availability may sometimes not be maximized and is even very low, as shown in Chapter 5.

From Equation (7.1) we can see that for multi-component systems this opportunistic maintenance policy may result in higher system availability as compared with the case that each subsystem is separately maintained. This is because while any subsystem fails and is under maintenance the entire system is down, and it will save time to do PM on unfailed subsystems during this down period and thus reduce the system downtime. Note that shut-off rule 1 is plausible for the series system. The optimal maintenance policy discussed in this chapter can be expected to approximate any type of multi-component systems since maintenance time is short relative to operating time.

In this chapter, we suppose that imperfect PM of subsystem 0 is modeled by the (p,q) rule. Upon each PM there is a perfect inspection requiring negligible time and yielding perfect information as to whether PM is perfect or minimal. Assume further that PM of subsystem 0 together with another subsystem between jT s are assumed to be perfect, $\forall j, j = 1,2,3,...$ Next we will first derive the long-run expected system maintenance cost per unit of time, or system maintenance cost rate, the asymptotic average system (storage) 'availability', and then evaluate other system operating performance characteristics and investigate the optimal maintenance polices.

7.2 System Maintenance Cost Rate and 'Availability'

Given the above opportunistic preparedness maintenance policy, the times between consecutive perfect maintenance of subsystem 0 constitute a renewal cycle. From the renewal reward theory, the system maintenance cost rate is

$$L(T, t_1, t_2, ..., t_n; p) = \frac{C(T, t_1, t_2, ..., t_n; p)}{D(T, t_1, t_2, ..., t_n; p)}$$

(7.2)

Asymptotic average system storage 'availability' is defined as:

$$A(T, t_1, t_2, ..., t_n; p) = \frac{U(T, t_1, t_2, ..., t_n; p)}{U(T, t_1, t_2, ..., t_n; p) + R(T, t_1, t_2, ..., t_n; p)}$$

(7.3)

where $C(T, t_1, t_2, ..., t_n; p)$ is the expected system maintenance cost per renewal cycle, $D(T, t_1, t_2, ..., t_n; p)$ is the expected duration of a renewal cycle, and $U(T, t_1, t_2, ..., t_n; p)$ and $R(T, t_1, t_2, ..., t_n; p)$ are, respectively, the accumulating system storage time and the maintenance time of this system in one renewal cycle. Obviously,

$$D(T, t_1, t_2, ..., t_n; p) = U(T, t_1, t_2, ..., t_n; p) + R(T, t_1, t_2, ..., t_n; p)$$

Next, without loss of the generality, we assume that $t_1 \le t_2 \le \cdots \le t_n$. Let us first evaluate the expected duration $D(T, t_1, t_2, ..., t_n; p)$. The renewal cycle duration B is the sum of three random variables. The duration and expected duration are respectively

$$B = S + Y + Z$$

$$D(T, t_1, t_2, ..., t_n; p) = E(B) = E(S) + E(Y) + E(Z)$$

(7.4)

First, we investigate the cumulative distribution function (*cdf*), probability density function (*pdf*) and mean of Y. Let U_i be the time to failure of subsystem i after t_i, given subsystem i is good at t_i ($i = 1, 2, ..., n$). Let H denote the time until subsystem 0 alone is subject to a perfect PM. Noting shut-off rule 1 and the memoryless property (Ross 1983) of the exponential distribution, then the age of subsystem 0 is given by

$$Y = \min(t_1 + U_1, ..., t_n + U_n, H)$$

Random variables U_i ($i = 1, 2, ..., n$) are statistically independent and U_i has an exponential distribution with failure rate λ_i. Let us denote the *cdf* of new subsystem 0 by $F_0(t)$. Let $\overline{F}_0 = 1 - F_0$. We assume that F_0 is absolutely continuous with density $f_0(t)$ and that $F_0(0) = 0$. The failure rate of subsystem 0 is supposed to be increasing. It is easy to show that random variable H has a discrete distribution given by

$$\Pr(H = jT) = q^{j-1}p \qquad \forall j, j = 1,2,\ldots$$

and the *cdf* of Y for $y < T$ is as follows:

$$\Pr(Y \le y) = 1 - \Pr(Y > y)$$
$$= 1 - \Pr(t_i + U_i > y, \ \forall i, i = 1,2,\ldots,n)$$
$$= 1 - \prod_{i=1}^{n} \Pr(U_i > y - t_i)$$
$$= 1 - \exp\left[-\sum_{i=1}^{n} \lambda_i (y - t_i)\right]$$
$$\scriptstyle y-t_i \ge 0$$

and for $y = T$ we have

$$\Pr(Y = T) = p \cdot \exp\left[-\sum_{i=1}^{n} \lambda_i (T - t_i)\right]$$

The *cdf* of Y for $T < y < 2T$ is given by

$$\Pr(Y \le y) = 1 - \Pr(Y > y)$$
$$= 1 - \Pr(t_i + U_i \ge y, \ i = 1,2,\ldots,n; \ \text{1st PM} = \text{imperfect})$$
$$= 1 - q\prod_{i=1}^{n} \Pr(U_i > y - t_i)$$
$$= 1 - q\exp\left[-\sum_{i=1}^{n} \lambda_i (y - t_i)\right]$$

and for $y = 2T$

$$\Pr(Y = 2T) = qp \cdot \exp\left[-\sum_{i=1}^{n} \lambda_i (2T - t_i)\right]$$

Generally, for $(j-1)T < Y < jT$ where $j = 1,2,3,\ldots$ we have

$$\Pr(Y \le y) = 1 - \Pr(t_i + U_i \ge y, \ \forall i, i = 1,2,\ldots,n; \ 1^{\text{st}} \text{ perfect PM is the } j^{\text{th}} \text{ PM})$$
$$= 1 - q^{j-1}\prod_{i=1}^{n} \Pr(U_i > y - t_i)$$
$$= 1 - q^{j-1}\exp\left[-\sum_{i=1}^{n} \lambda_i (y - t_i)\right]$$
$$\scriptstyle y-t_i \ge 0$$

and for $y = jT$

$$\Pr(Y = jT) = q^{j-1}p \cdot \exp\left[-\sum_{i=1}^{n} \lambda_i (jT - t_i)\right]$$

Next we derive the *pdf* of Y. For $i = 1,2,...,n$, let

$$\begin{cases} M_i = \left(\sum_{j=1}^{i} \lambda_j\right) \\ D_i = \exp\left(\sum_{j=1}^{i} \lambda_j t_j\right) \\ f_i(y) = \left(\sum_{j=1}^{i} \lambda_j\right) \exp\left[-\sum_{j=1}^{i} \lambda_j(y - t_j)\right] \\ \quad\quad = D_i M_i \exp(-M_i y) \end{cases} \tag{7.5}$$

Then Y has probability density given by, for $y < T$

$$\varphi(y) = \begin{cases} 0, & 0 \le y < t_1 \\ f_i(y), & t_i \le y < t_{i+1} \\ f_n(y), & t_n \le y < T \end{cases} \quad\quad \forall i, i = 1,2,...,n-1 \tag{7.6}$$

with probability mass at $Y = T$

$$G_0(T) = p \cdot \exp\left[-\sum_{j=1}^{n} \lambda_j(T - t_j)\right] \tag{7.7}$$
$$= p \cdot D_n \exp(-M_n T)$$

For $(j-1)T < Y < jT$, $\forall j, j = 2,3,...$, Y has probability density as follows:

$$\varphi(y) = q^{j-1} f_n(y) \tag{7.7a}$$

with probability mass at $Y = jT$

$$G_0(jT) = \Pr(Y = jT)$$
$$= q^{j-1} p \cdot \exp\left[-\sum_{i=1}^{n} \lambda_i(jT - t_i)\right]$$
$$= q^{j-1} p D_n \exp\left[-jTM_n\right]$$

$$\tag{7.7b}$$

It follows that the expected value of Y is given by

$$E(Y) = \sum_{i=1}^{n} \int_{t_i}^{t_{i+1}} y f_i(y)\,dy + \sum_{j=1}^{\infty} \int_{jT}^{(j+1)T} q^{j-1} f_n(y)\,dy + \sum_{j=1}^{\infty} jT G_0(jT)$$

$$= \sum_{i=1}^{n} \int_{t_i}^{t_{i+1}} y f_i(y)\,dy + D_n M_n \sum_{j=1}^{\infty} q^{j-1} \int_{jT}^{(j+1)T} \exp(-M_n y)\,dy$$

$$+ \sum_{j=1}^{\infty} jT q^{j-1} p D_n \exp\left[-jTM_n\right]$$

$$= \sum_{i=1}^{n} \int_{t_i}^{t_{i+1}} y f_i(y) dy + D_n \sum_{j=1}^{\infty} q^{j-1} \{ \exp(-jM_nT) - \exp[-(j+1)M_nT] \}$$

$$+ D_n p T \sum_{j=1}^{\infty} j q^{j-1} \exp(-jTM_n) \qquad (7.8)$$

where $t_{n+1} = T$. For each of the integrals in the first sum we have (Radner and Jorgenson 1963; McCall 1965):

$$\int_{t_i}^{t_{i+1}} y f_i(y) dy = D_i \left[\left(t_i + M_i^{-1} \right) \exp(- M_i t_i) - \left(t_{i+1} + M_i^{-1} \right) \exp(- M_i t_{i+1}) \right]$$

Second, we derive the expected value of S. Recall that V_i is the duration of the interval over which subsystem i alone would be replaced if it were to fail, $\forall i, i = 1,2,...,n$. Then

$$V_i = \min(Y, t_i) \qquad \forall i, i = 1,2,...n$$
$$E(S_i) = \lambda_i E(V_i) w_i \qquad \forall i, i = 1,2,...n$$

Note that V_i has *pdf* equal to that of Y for $Y < t_i$, and probability mass $\Pr(Y \geq t_i)$ concentrated at t_i. Therefore,

$$E(V_i) = \sum_{j=1}^{i-1} \int_{t_j}^{t_{j+1}} y f_j(y) dy + t_i \left[1 - \sum_{j=1}^{i-1} \int_{t_j}^{t_{j+1}} f_j(y) dy \right] \qquad (7.9)$$

Recall that $S = \sum_{i=1}^{n} S_i$, then

$$E(S) = \sum_{i=1}^{n} E(S_i)$$

$$= \sum_{i=1}^{n} \lambda_i w_i E(V_i)$$

$$= \sum_{i=1}^{n} \lambda_i w_i \left[\sum_{j=0}^{i-1} \int_{t_j}^{t_{j+1}} y f_j(y) dy + t_i \left(1 - \sum_{j=0}^{i-1} \int_{t_j}^{t_{j+1}} f_j(y) dy \right) \right] \qquad (7.10)$$

Finally, let us derive an expression for $E(Z)$. Denote by d_i the probability that the renewal cycle ends on the interval $[t_i, t_{i+1})$, i.e.,

$$d_i = \Pr(t_i \leq Y < t_{i+1}), \qquad \forall i, i = 1,2,...,n, \text{ and } t_{n+1} = T$$

Then

$$\begin{cases} d_1 = 1 - \exp\left[- M_1(t_2 - t_1)\right] \\ d_2 = (1 - \alpha_1)\{1 - \exp\left[- M_2(t_3 - t_2)\right]\} \\ \quad\vdots \\ d_i = \prod_{j=1}^{i-1}(1 - \alpha_j)\alpha_i \\ \quad\vdots \\ d_n = \prod_{j=1}^{n-1}(1 - \alpha_j)\alpha_i \\ d_{n+1} = \prod_{j=1}^{n}(1 - \alpha_j) = p \cdot D_n \exp\left[- M_n T\right] \end{cases}$$

where $\alpha_j = 1 - \exp[-M_j(t_{j+1} - t_j)]$ and d_{n+1} is the probability that the renewal cycle ends at T.

It is easy to verify that the probability that the renewal cycle ends on the interval $((j-1)T, jT)$ and at jT, $\forall j, j = 2,3,\ldots$ respectively, from Equations (7.7a) and (7.7b):

$$d_{1j} = \Pr\{(j-1)T < Y < jT\}$$

$$= \int_{(j-1)T}^{jT} q^{j-1} f_n(y)dy$$

$$= q^{j-1} \int_{(j-1)T}^{jT} f_n(y)dy$$

$$d_{2j} = \Pr(Y = jT)$$

$$= q^{j-1} pD_n \exp\left[- jTM_n\right]$$

Denote by q_{0i} the probability that the renewal cycle ends with a replacement of subsystem i and subsystem 0 together. Noting that for two independent exponential random variables Z_1 and Z_2 with rate η_1 and η_2 respectively, there exists the following equation (Ross 1993):

$$\Pr(Z_1 < Z_2) = \frac{\eta_1}{\eta_1 + \eta_2}$$

Then, we have

$$q_{0i} = \sum_{j=i}^{n} d_j \frac{\lambda_i}{M_j} + \sum_{j=1}^{n} d_{1j} \frac{\lambda_i}{M_n}$$

$$= \lambda_i \sum_{j=i}^{n} \frac{d_j}{M_j} + \frac{\lambda_i}{M_n} \sum_{j=1}^{\infty} d_{1j} \qquad \forall i, i = 1,2,\ldots,n \qquad (7.11)$$

The probability that the renewal cycle ends with a replacement of subsystem 0 alone is given by

$$d_{0p} = d_{n+1} + \sum_{j=2}^{\infty} d_{2j} = pD_n \sum_{j=1}^{\infty} q^{j-1} \exp\left[-jTM_n\right] \tag{7.11a}$$

Therefore, it follows that the third term in Equation (7.4) is given by

$$E(Z) = \sum_{i=1}^{n} q_{0i} w_{0i} + d_{0p} w_0 \tag{7.12}$$

Recall that

$$D(T, t_1, t_2, ..., t_n; p) = E(B) = E(S) + E(Y) + E(Z)$$

Thus,

$$D(T, t_1, t_2, ..., t_n; p) = \sum_{i=1}^{n} \lambda_i w_i \left[\sum_{j=0}^{i-1} \int_{t_j}^{t_{j+1}} y f_j(y) dy + t_i \left(1 - \sum_{j=0}^{i-1} \int_{t_j}^{t_{j+1}} f_j(y) dy \right) \right]$$

$$+ \sum_{i=1}^{n} \int_{t_i}^{t_{i+1}} y f_i(y) dy + D_n \sum_{j=1}^{\infty} q^{j-1} \{ \exp(-jM_n T) - \exp[-(j+1)M_n T] \}$$

$$+ D_n T \sum_{j=1}^{\infty} j q^{j-1} \exp\left[-jTM_n\right] + \sum_{i=1}^{n} q_{0i} w_{0i} + d_{0p} w_0 \tag{7.13}$$

This completes the derivation of the expected duration of a renewal cycle $D(T, t_1, t_2, ..., t_n; p)$.

Wang et al. (2001) investigate the expected system maintenance cost over a single renewal cycle, and here we use their result without proof:

$$D(T, t_1, t_2, ..., t_n; p) = \sum_{i=1}^{n} \lambda_i C_i \left[\sum_{j=0}^{i-1} \int_{t_j}^{t_{j+1}} y f_j(y) dy + t_i \left(1 - \sum_{j=0}^{i-1} \int_{t_j}^{t_{j+1}} f_j(y) dy \right) \right]$$

$$+ \sum_{i=1}^{n} q_{0i} C_{0i} + d_{0p} C_0 \tag{7.14}$$

Substituting the above results into Equations (7.2) and (7.3), and noting that

$$E(Y) = U(T, t_1, t_2, ..., t_n; p) \text{ and}$$

$$E(S + Z) = R(T, t_1, t_2, ..., t_n; p),$$

we obtain the following proposition:

PROPOSITION 7.1 *The long-run expected system maintenance cost rate, and the asymptotic average system (storage) 'availability' are respectively:*

$$L(T,t_1,t_2,...,t_n;p) =$$

$$\frac{\sum_{i=1}^{n}\lambda_i C_i\left[\sum_{j=0}^{i-1}\int_{t_j}^{t_{j+1}}yf_j(y)dy +t_i\left(1-\sum_{j=0}^{i-1}\int_{t_j}^{t_{j+1}}f_j(y)dy\right)\right]+\sum_{i=1}^{n}q_{0i}C_{0i}+d_{0p}C_0}{D(T,t_1,t_2,...,t_n;p)}$$

(7.15)

$$A(T,t_1,t_2,...,t_n;p) = \frac{E(Y)}{D(T,t_1,t_2,...,t_n;p)}$$

(7.16)

From Proposition 7.1, the optimal opportunistic preparedness maintenance policy $(T^*,t_1^*,t_2^*,...,t_n^*)$ that minimizes the system maintenance cost rate or maximizes the asymptotic average system storage 'availability' can be obtained by using nonlinear programming software.

Next we discuss the other operating performance characteristics of this opportunistic preparedness maintenance model.

7.3 Other Operating Characteristics

To learn more about this imperfect opportunistic preparedness maintenance and to predict supply and maintenance requirements, let us investigate its other operating characteristics. First, we note that for this imperfect preparedness maintenance model, three different maintenance actions are observed:

(i)   Replacement (perfect repair) of a failed subsystem with a constant failure rate by itself.

(ii) Joint opportunistic maintenance of a failed subsystem with a constant failure rate and subsystem 0 unfailed.

(iii) PM of unfailed subsystem 0 at some time jT where j is a natural number.

Besides the system storage availability and maintenance cost rate derived in the last section, other important operating characteristics of this preparedness maintenance policy are the expected number of each of these maintenance actions per unit time, and expected maintenance cost of each of these maintenance actions per unit time. Another characteristic of interest is the probability of at least m failures of one of the subsystems in the interval $(0, t)$ (McCall 1963, 1965; Radner and Jorgenson 1963). Overall, the following operating characteristics will be investigated in this section:

r_i Expected rate of replacement of subsystem i, $\forall i$, $i = 1,2,...,n$

r_{0i} Expected rate of joint opportunistic replacement of failed subsystem i and unfailed subsystem 0, $\forall i$, $i = 1,2,...,n$

r_{00} Expected rate of planned maintenance of subsystem 0 at times jT

r_{0p} Expected rate of total perfect maintenance (alone and joint) of subsystem 0

r_{fj} Expected rate of joint opportunistic maintenance of subsystem 0 with another subsystem

c_{00} Expected rate of expenditure on PM of subsystem 0 at times jT

c_i Expected rate of expenditure on replacement of subsystem i

c_{0i} Expected rate of expenditure on joint replacement of subsystem 0 and subsystem i

$P_i(m,t)$ Probability of at least m failures of subsystem i in the interval $(0,\ t)$ where t is constant and $\forall i,\ i = 0,1,...,n$

Let us consider the subsystems with constant failure rates first. The time to failure for each of the inspected subsystems is an exponential random variable with rate λ_i. Obviously,

$$r_i = \lambda_i \qquad\qquad \forall i, i = 1,2,...,n \qquad (7.17)$$

or including w_i, the time to replace subsystem i,

$$r_i = \frac{1/\lambda_i}{1/\lambda_i + w_i} = \frac{\lambda_i}{\lambda_i w_i + 1} \qquad\qquad \forall i, i = 1,2,...,n \qquad (7.18)$$

Therefore,

$$c_i = r_i C_i = \frac{\lambda_i C_i}{\lambda_i w_i + 1} \qquad\qquad \forall i, i = 1,2,...,n \qquad (7.19)$$

Using the elementary renewal theorem we know that the rate of perfect maintenance of subsystem 0, is asymptotically equal to the reciprocal of the expected value of Y, the time to the first perfect maintenance of subsystem 0, that is,

$$\lim_{t \to \infty} r_{0p}(t) = [E(Y)]^{-1}$$

where $E(Y)$ is given in Equation (7.8). Thus, for large value of t,

$$r_{0p} \approx [E(Y)]^{-1} \qquad (7.20)$$

On the other hand, from the foregoing definitions for rates of maintenance we have

$$r_{0p} = r_{00}' + \sum_1^n r_{0i} \qquad (7.21)$$

That is, this expected rate, r_{0p}, can be partitioned into two parts: the expected rate of perfect PM at some time jT and the expected rate of joint opportunistic replacement with another subsystem. We can also see this relationship from the derivation of Equation (7.8). By the Law of Large Numbers, the fraction of the total number of perfect PM of subsystem 0 is equal to d_{0p} given in Equation

(7.11a) for large t. Hence, in the long run, the expected rate of perfect PM of subsystem 0 is given by

$$r_{00}' \approx d_{0p}\left[E(Y)\right]^{-1}$$

or, including maintenance time,

$$r_{00}' \approx \frac{d_{0p}\left[E(Y)\right]^{-1}}{d_{0p}\left[E(Y)\right]^{-1}\left[E(Z)+E(S)\right]+1} = \frac{d_{0p}}{d_{0p}\left[E(Z)+E(S)\right]+E(Y)}$$

It follows that the expected rate of PM of subsystem 0 is

$$r_{00} \approx r_{00}'/p = \frac{d_{0p}}{d_{0p}\left[E(Z)+E(S)\right]p+E(Y)p} \tag{7.22}$$

The expected expenditure on planned maintenance is given by

$$c_{00} = r_{00}C_0 = \frac{d_{0p}C_0}{d_{0p}\left[E(Z)+E(S)\right]p+E(Y)p} \tag{7.23}$$

It is easy to see that in the long run, $\sum_1^n r_{0i}$, the expected rate of joint opportunistic maintenance of subsystem 0 is

$$r_{fj} = \sum_1^n r_{0i} = (1-d_{0p})\left[E(Y)\right]^{-1} \tag{7.24}$$

From Section 7.2 ,

$$1-d_{0p} = \sum_1^n q_{0i}$$

where q_{0i} s are given by Equation (7.11).
Therefore,

$$r_{fj} = \left[E(Y)\right]^{-1}\sum_1^n q_{0i} \tag{7.25}$$

Since the probability that subsystem 0 will be replaced jointly with subsystem i is q_{0i}, the asymptotic expected rate of opportunistic replacement of subsystem 0 and i is given by

$$r_{0i} = q_{0i}\left[E(Y)\right]^{-1} \qquad \forall i,i=1,2,...,n \tag{7.26}$$

or including maintenance time,

$$r_{0i} = \frac{q_{0i}\left[E(Y)\right]^{-1}}{q_{0i}\left[E(Y)\right]^{-1}\left[E(Z)+E(S)\right]+1} = \frac{q_{0i}}{q_{0i}\left[E(Z)+E(S)\right]+E(Y)} \qquad \forall i,i=1,2,...,n$$

$$\tag{7.27}$$

Accordingly, the expected rate of expenditure on opportunistic maintenance is

$$c_{0i} = r_{0i}C_{0i} = \frac{q_{0i}C_{0i}}{q_{0i}[E(Z) + E(S)] + E(Y)} \qquad \forall i, i = 1,2,...,n \qquad (7.28)$$

Because the lifetime of subsystem i follows the exponential distribution, the probability of at least m replacements of subsystem i in the interval $(0, t)$ is given by

$$P_i(m,t) = \sum_{j=m}^{\infty} \exp(-\lambda_i t) \cdot \frac{(\lambda_i t)^j}{j!} \qquad \forall i, i = 1,...,n \qquad (7.29)$$

7.4 Optimization Models

So far we have derived the system reliability measures – system storage availability, probability of at least m failures of subsystem i in the interval $(0, t)$, expected rate of failure of subsystem i, etc., and system maintenance cost measures – system maintenance cost rate, expected rate of expenditure on planned maintenance of subsystem 0, expected rate of expenditure on replacement of subsystem i, etc. To obtain the optimal system maintenance policies the system reliability measures and system maintenance cost measures must both be acceptable. For example, it may be required that the system maintenance cost rate is minimized while the system availability is not less than some predetermined requirement A_0. For such a problem, we can formulate the following optimization model from Equations (7.15) and (7.16):

Maximize

$$A(T, t_1, t_2, ..., t_n; p) = \frac{E(Y)}{D(T, t_1, t_2, ..., t_n; p)}$$

Subject to

$$\begin{cases} L(T, t_1, t_2, ..., t_n; p) = \\ \dfrac{\displaystyle\sum_{i=1}^{n} \lambda_i C_i \left[\sum_{j=0}^{i-1} \int_{t_j}^{t_{j+1}} y f_j(y) dy + t_i \left(1 - \sum_{j=0}^{i-1} \int_{t_j}^{t_{j+1}} f_j(y) dy \right) \right] + \sum_{i=1}^{n} q_{0i} C_{0i} + d_{0p} C_0}{D(T, t_1, t_2, ..., t_n; p)} \leq L_0 \\ t_1, t_2, ..., t_n, T \geq 0 \end{cases}$$

where L_0 is the pre-determined requirement for system maintenance cost rate.

This model can be solved by nonlinear programming software to obtain an optimal system preparedness maintenance policy $(t_1^*, t_2^*, ..., t_n^*, T^*)$. Similarly, based on other operating characteristics derived we can formulate other optimization models as needed.

7.5 Concluding Discussions

Different from single component systems, one of the key problems for multi-component systems in modern maintenance practice is economic dependence (Wang *et al.* 2001). Besides, maintenance is often imperfect. This chapter has considered these two factors which greatly impact on optimal maintenance policies in multi-unit systems and presents some results on this aspect. Moreover, maintenance time is not ignored in this work. Both system reliability and maintenance cost measures are incorporated in the optimal opportunistic maintenance models in this chapter so that the optimal maintenance policies obtained may be optimal not only in terms of maintenance costs but also in terms of reliability measures. Therefore, the opportunistic maintenance model of the multi-component system with $(n+1)$ decision variables $(t_1, t_2, ..., t_n, T)$ introduced in this study is more realistic and the results obtained in this chapter expect to be effective in practice.

If the actual state of the system is known with certainty, either through continuous inspection or continuous operation, the theory of maintenance for the preparedness model becomes the same as for the regular maintenance models. In this sense, the theory of maintenance for the preparedness model is more general than the regular maintenance models.

This chapter has discussed optimal preparedness maintenance policy of multi-unit systems with one IFR subsystems given economic dependence and imperfectness of maintenance. It can be extended to multi-unit systems with two or more IFR subsystems which are subject to imperfect maintenance and economic dependence. Another extension is to use other shut-off rules.

Jia and Christer (2002) consider the periodic testing of a preparedness system where, in addition to working and failed state recognition, a working but defective state also exists, and demonstrate their availability models in the context of a missile buffer system. The possible extension is to consider a working but defective state for subsystem 0.

The imperfect PM is modeled by the (p, q) rule in this chapter. Similarly, one can consider other imperfect maintenance modeling methods discussed in Chapter 2, for example, the $(p(t), q(t))$ rule.

8

Optimal Opportunistic Maintenance Policies of *k*-out-of-*n* Systems

A *k*-out-of-*n*:G system is an important system in reliability engineering and could include series and parallel systems as special cases. This chapter introduces opportunistic maintenance of a *k*-out-of-*n*:G system with imperfect PM and economic dependence, studied by Pham and Wang (2000). Two new (τ, T) opportunistic maintenance models with consideration of reliability requirements and allowing partial failure are presented. In these two models, only minimal repairs are performed on failed components before a fixed time τ and CM of all failed components are combined with PM of all functioning but deteriorated components after τ and number of failed components triggering maintenance can be specified in advance or considered as a decision variable; If the system survives to another fixed time T without perfect maintenance it will be subject to PM at time T. τ and T are decision variables. System cost rate and availability are investigated for nonegligible maintenance time. The results, including 13 maintenance models as special cases, generalize and unify some previous work in this area.

8.1 Introduction

In this chapter, a *k*-out-of-*n* system is defined to be a complex coherent system with *n* independent components such that the system operates if and only if at least *k* of these components function successfully. For a complex and expensive system, it may not be advisable to replace the entire system just because of the failure of one component, especially for a *k*-out-of-*n* system . In fact, the system comes back into operation on repair or replacement of the failed component by a new one or by a used but operative one. Such maintenance actions do not renew the system completely but enable the system to continue to operate (Kapur *et al.* 1989). However, the system is usually deteriorating with usage and time. At some point of time or usage it may be in a poor operating condition and a perfect maintenance is necessary. Based on this situation, we formulate the following maintenance policy for a *k*-out-of-*n* system.

The new system starts to operate at time 0. Each failure of a component of this system during the time interval $(0, \tau)$ is immediately removed by a minimal repair. Components which fail in the time interval (τ, T) can be left idle. A CM on the failed components together with PM on all unfailed but deteriorating ones is performed at a cost of c_f once exactly m components are idle, or PM on the whole system is carried out at a cost of c_p once the total operating time reaches T, whichever occurs first. That is, if m components fail in the time interval (τ, T), CM combined with PM is performed; if less than m components fail in the time interval (τ, T), then PM is carried out at time T. After a perfect maintenance, either a CM combined with PM or a PM at T, the process repeats.

Figure 8.1. The (τ, T) opportunistic maintenance policy of k-out-of-n systems

This maintenance policy is shown in Figure 8.1. In this policy τ and T are decision variables. We assume that m is a pre-determined positive integer, $1 \leq m \leq n - k + 1$, and $\tau < T$. According to different reliability and cost requirements, m may take different values. Obviously, $m = 1$ means that the system is subject to maintenance whenever one component fails after τ. For a series system (n-out-of-n system) or a system with critical applications m may basically be required to be 1. If m is chosen as $n - k + 1$ then k-out-of-n system is maintained once the system fails. In most cases, the whole system is subject to a perfect CM together with a PM upon a system failure ($m = n - k + 1$) or partial failure, that is, some components may fail but the system still functions. However, if inspection is not continuous and the system operating condition can be known only through inspection, m can be a number greater than ($n - k + 1$). We assume that if CM together with PM is carried out both are perfect, and that CM combined with PM takes w_1 time units and PM at time T takes w_2 time units.

The justification for this policy is that before τ every component in the system is young and no major repair is necessary and only minimal repairs, which may not take much time and cost, are performed. The component is deteriorating as time passes and after τ time has elapsed the component may be in a weak operating condition and has a larger failure rate, and consequently a major or perfect repair is needed. Because there is economic dependence and availability requirements (less frequent shut-offs for maintenance), however, we may not replace it immediately but start CM until the number of failed components reaches some pre-specified number m. In fact, when the number of failed components reaches m, the remaining $(n - m)$ operating components may degrade to a worse operating condition and also need PM. Note that as long as m is less than $(n - k + 1)$ the system will not fail and will continue to operate.

As pointed in Chapter 1, economic dependence means that it takes less cost and time to perform maintenance on several components jointly than on each component separately. For a multi-component system, if there is strong economic dependency joint maintenance should be considered. The optimal maintenance policy for this kind of systems possesses an opportunistic characteristic, that is, the optimal maintenance actions for one component depends on the states of the other components (Zheng 1995). Obviously, the maintenance policy proposed above is an opportunistic one.

In this chapter, the following assumptions are made:

i) All failure events are s-independent.

ii) Each component has increasing failure rate (IFR).

iii) Minimal repair takes negligible time since minimal repair time is small relatively to perfect maintenance time.

iv) Minimal repair costs are random variables which depend on age and number of minimal repairs.

v) The planning horizon is infinite.

vi) k-out-of-n system consists of n i.i.d. components.

We assume that for each component in the system the cost of the ith minimal repair at age t consists of two parts: the deterministic part $c_1(t,i)$ which depends on the age of this component and the number of minimal repairs i, and the age-dependent random part $c_2(t)$. This general cost structure was used by Sheu (1991) in study of an age replacement model.

It is well-known that for a single-unit system PM is justified only if it has IFR. The above assumption that failure rate of each component has IFR is still necessary for the k-out-of-n system. This is because the system may be subject to a PM at time T. This requires that the system is IFR after τ. The following proposition states the relationship between component and system failure rates:

PROPOSITION 8.1 *If a k-out-of-n system is composed of independent, identical, IFR components, the system also has IFR.*

Proof. Assume that reliability of each component at some time is p. The survival function of a k-out-of-n system is given by

$$r(p) = \sum_{i=k}^{n} \binom{n}{i} p^i (1 - p)^{n-i}$$

Using binomial theorem it follows that

$$r(p) = \frac{n!}{(k-1)!(n-k)!} \int_0^p x^{k-1} (1 - x)^{n-k} \, dx \qquad (8.1)$$

It is easy to prove that

$$\frac{pr'(p)}{r(p)} = \left[\frac{r(p)}{pr'(p)}\right]^{-1}$$

$$= \left[\frac{1}{p}\int_0^p \left(\frac{x}{p}\right)^{k-1}\left(\frac{1-x}{1-p}\right)^{n-k}dx\right]^{-1}$$

$$= \left[\int_0^1 y^{k-1}\left(\frac{1-yp}{1-p}\right)^{n-k}dy\right]^{-1}$$

where $y = x/p$.

Since $[(1-yp)/(1-p)]$ is increasing in p, it is easy to see that $pr'(p)/r(p)$ is decreasing in p. Similar arguments are also found in Barlow and Proschan (1975), and Ross (1983). Note that the failure rate of a k-out-of-n system is given by

$$h_s(t) = \frac{-dr[\overline{F}(t),...,\overline{F}(t)]}{dt}\Big/ r[\overline{F}(t),...,\overline{F}(t)]$$

$$= \frac{-dr[\overline{F}(t),...,\overline{F}(t)]}{d\overline{F}}\frac{d\overline{F}}{dt}\frac{\overline{F}}{\overline{F}}\Big/ r[\overline{F}(t),...,\overline{F}(t)]$$

$$= q(t)\frac{pr'(p)}{r(p)}\Big|_{p=\overline{F}(t)}$$

where $r(p) = r(p,...,p)$.

It follows from the above equation that $h_s(t)$ has IFR noting that $q(t)$ has IFR, $\overline{F}(t)$ is a decreasing function of t, and $pr'(p)/r(p)$ is decreasing in p.   ◆

The following notation will be used throughout this chapter:

NOTATION

c_f	Cost of CM together with PM of a system
c_p	Cost of PM alone of a system
$g(c_1(t,i),c_2(t))$	Cost of the ith minimal repair at age t, where g is a positive, non-decreasing and continuous function
$c_1(t,i)$	Deterministic part of cost of the ith minimal repair at age t, which depends on the age and the number of minimal repairs
$c_2(t)$	Random part of cost of the ith minimal repair at age t
$V_t(x)$	Cumulative distribution function of $c_2(t)$
$v_t(x)$	Probability density function of $c_2(t)$
τ, T	Two decision variables constituting the (τ, T) policy,
$f(t)$	Probability density function of a component
$F(t)$	Cumulative density function of a component

$\overline{F}(t)$	Survival function of a component, $\overline{F}(t) - 1 - F(t)$
$\overline{G}(t)$	Residual survival function of a component
$\overline{F}_{n-k+1}(y)$	Survival function of the time to failure of a *k*-out-of-*n* system
$q(t)$	Failure rate of a component
$Q(t)$	Cumulative failure rate of a component, $Q(t) = \int_0^t q(x)dx$
n	Number of components in a system
k	Minimum number of operating components to make a system function
m	Minimum number of failed components needed to start maintenance.
w_1	Time to perform CM together with PM
w_2	Time to perform PM alone
p	Probability that PM is perfect
q	$1 - p$ in Section 8.3
$N(t)$	Number of minimal repairs during time interval $(0, t)$
$M(t)$	Expected number of minimal repairs during time interval $(0, t)$
$L(\tau,T)$	Long-run expected system maintenance cost per unit time, or cost rate

8.2 Perfect PM

We shall now characterize the classes of possible maintenance actions. Note that at any instant of time, the following alternative maintenance actions for a *k*-out-of-*n* system are to be performed, per the maintenance policies described in Section 8.1:

   i)   Keep the present system and no maintenance actions are given.

   ii) Performed minimal repair on a component of the system (before time τ).

  iii) Performed perfect repair on all failed components together with PM on all unfailed but deteriorating components (after time τ).

  iv) Performed PM on the system at time T.

    Assume that each component in the *k*-out-of-*n* system has cumulative distribution function (*cdf*) $F(x)$, and probability density function (*pdf*) $f(x)$. Then their failure rates (or the hazard rates) are $q(x) = f(x)/\overline{F}(x)$ and cumulative failure rates are $Q(x) = \int_0^x q(t)dt$ which have a relationship with their survival functions $\overline{F}(x) = \exp\{-Q(x)\}$, where $\overline{F}(x) = 1 - F(x)$. It is further assumed that the failure rate is differentiable, monotonely increasing, and remains undisturbed by minimal repair.

If there is no PM, the residual survival function of each component is given by

$$\overline{G}(y) = P\{Y \geq \tau + y \mid Y > \tau\}$$
$$= \int_{\tau+y}^{+\infty} f(t)dt \Big/ \int_{\tau}^{+\infty} f(t)dt$$
$$= \overline{F}(\tau + y)/\overline{F}(\tau)$$
$$= e^{-Q(\tau+y)+Q(\tau)}$$

where $y \geq 0$

Let $Y_1, Y_2, ... Y_n$ be $i.i.d.$ random variables with survival distribution $\overline{G}(y)$, and $Y_{(1)} \leq Y_{(2)} \leq \cdots \leq Y_{(n)}$ be the corresponding order statistics. Note that the order statistics may be interpreted as successive times of failures of components in the systems, and the $(n-k+1)^{th}$ order statistic is just the time to failure of the k-out-of-n system. The order statistic $Y_{(j)}$ has survival distribution, $\forall j, j = 1, 2, ..., n$,

$$\overline{F}_j(y) = \sum_{i=0}^{j-1} \binom{n}{i} [G(y)]^i [\overline{G}(y)]^{n-i}$$
$$= \sum_{i=0}^{j-1} \binom{n}{i} \left[1 - e^{-Q(\tau+y)+Q(\tau)}\right]^i e^{-(n-i)Q(\tau+y)+(n-i)Q(\tau)} \qquad (8.2)$$

In this section we assume that PM at time T is perfect. According to renewal theory, the times between consecutive perfect maintenance, preventive or corrective, constitute a renewal cycle. From the classical renewal reward theory, the long-run expected system maintenance cost per unit time, or cost rate, is

$$L(\tau, T) = \frac{C(\tau, T)}{D(\tau, T)}$$

where $C(\tau, T)$ is the expected system maintenance cost per renewal cycle and $D(\tau, T)$ is the expected duration of a renewal cycle.

Let $Z_1, Z_2, ...$ be $i.i.d.$ random variables with distribution function $F_m(y)$, and $Z_i^* = \min(Z_i, T)$, $\forall i, i = 1, 2, ...$ Then a renewal cycle consists of maintenance time and Z_i^* duration. It is easy to verify

$$D(\tau, T) = E[Z_i^*] + I_{\{Y_m < T-\tau\}} w_1 + I_{\{Y_m \geq T-\tau\}} w_2$$
$$= \tau + \int_0^{T-\tau} \overline{F}_m(t)dt + I_{\{Y_m < T-\tau\}} w_1 + I_{\{Y_m \geq T-\tau\}} w_2$$
$$= \tau + \int_0^{T-\tau} \overline{F}_m(t)dt + F_m(T-\tau)(w_1 - w_2) + w_2 \qquad (8.3)$$

Next we evaluate expected system maintenance cost per renewal cycle $C(\tau, T)$. Note that $C(\tau, T)$ consists of three parts: minimal repair cost, cost of CM combined with PM, and cost of PM at time T. For each component the failures between $(0, \tau)$ occur in accordance with a non-homogeneous Poisson process of rate $q(t)$. The cost of the ith minimal repair at age t is $g(c_1(t,i), c_2(t))$, where g is a positive, non-decreasing and continuous function of t, and is a positive, non-decreasing function of i. Suppose that the random part $c_2(t)$ at age t has distribution function $V_t(x)$, density function $v_t(x)$ and finite mean $E[c_2(t)]$. The total minimal repair cost for a *k*-out-of-*n* system in one cycle is given by

$$C_{smr} = nE\left[\sum_{i=1}^{N(\tau)} g(c_1(S_i,i), c_2(S_i))\right]$$

where $N(\tau)$ is number of minimal repairs during time interval $(0, \tau)$.

The further derivation of this cost expression needs a proposition from Sheu (1991) and we now state it without proof:

PROPOSITION 8.2 *Let* $\{N(t), t \geq 0\}$ *be a non-homogeneous Poisson process with intensity* $q(t)$ *and* $M(t) = E[N(t)] = \int_0^t q(u)du$. *Denote the successive arrival times of this process by* $S_1, S_2, \ldots$ *Assume that at time* S_i *a cost of* $g(c_1(S_i,i), c_2(S_i))$ *is incurred. Suppose that* $c_2(y)$ *at age* y *is a random variable with finite mean* $E[c_2(t)]$ *and g is a positive, non-decreasing and continuous function. If* $A(t)$ *is the total cost incurred over* $[0,t]$, *then*

$$E[A(t)] = \int_0^t \mu(y)q(y)dy$$

where $\mu(y) = E_{N(y)}[E_{C_2(y)}[g(c_1(y, N(y)+1), c_2(y))]]$ *which is the expectation with respect to random variables* $N(y)$ *and* $c_2(y)$.

According to the above proposition the total minimal repair cost in one cycle is given by

$$C_{smr} = n\int_0^\tau \mu(y)q(y)dy \tag{8.4}$$

The total cost of PM at time T and CM combined with PM is given by

$$C_{pf} = I_{\{Y_m < T-\tau\}}c_f + I_{\{Y_m \geq T-\tau\}}c_p$$
$$= F_m(T-\tau)(c_f - c_p) + c_p$$

Thus,

$$C(\tau, T) = C_{smr} + C_{pf}$$
$$= n\int_0^\tau \mu(y)q(y)dy + F_m(T-\tau)(c_f - c_p) + c_p \tag{8.5}$$

From Equations (8.3) and (8.5) the following proposition follows:

PROPOSITION 8.3 *If PM is always perfect, then the long-run expected system maintenance cost per unit time, or cost rate, for a k-out-of-n system is given by*

$$L(\tau,T) = \frac{n\int_0^\tau \mu(y)q(y)dy + F_m(T-\tau)(c_f - c_p) + c_p}{\tau + \int_0^{T-\tau} \overline{F}_m(t)dt + F_m(T-\tau)(w_1 - w_2) + w_2} \tag{8.6}$$

and the limiting average system availability is

$$A(\tau,T) = \frac{\tau + \int_0^{T-\tau} \overline{F}_m(t)dt}{\tau + \int_0^{T-\tau} \overline{F}_m(t)dt + F_m(T-\tau)(w_1 - w_2) + w_2} \tag{8.6a}$$

In what follows, we shall attempt to minimize $L(\tau,T)$ with respect to τ and T. Differentiating $L(\tau,T)$ with respect to T and τ, respectively, we have

$$\frac{\partial L(\tau,T)}{\partial T} = -\frac{\overline{F}_m(T-\tau)\left[n\int_0^\tau \mu(y)q(y)dy + (c_f - c_p)\cdot F_m(T-\tau) + c_p\right]}{\left[\tau + \int_0^{T-\tau} \overline{F}_m(t)dt + F_m(T-\tau)(w_1 - w_2) + w_2\right]^2} +$$

$$\frac{\left[\tau(c_f - c_p) + c_f w_2 - c_p w_1 + (c_f - c_p)\int_0^{T-\tau} \overline{F}_m(t)dt - n(w_1 - w_2)\int_0^\tau \mu(y)q(y)dy\right]}{\left[\tau + \int_0^{T-\tau} \overline{F}_m(t)dt + F_m(T-\tau)(w_1 - w_2) + w_2\right]^2}$$

$$\times \frac{\partial F_m(T-\tau)}{\partial T}$$

$$\frac{\partial L(\tau,T)}{\partial \tau} = \frac{\left[n\mu(\tau)q(\tau) + \frac{\partial F_m(T-\tau)}{\partial \tau}(c_f - c_p)\right]}{\left[\tau + \int_0^{T-\tau} \overline{F}_m(t)dt + F_m(T-\tau)(w_1 - w_2) + w_2\right]^2} \times$$

$$\left[\tau + \int_0^{T-\tau} \overline{F}_m(t)dt + F_m(T-\tau)(w_1 - w_2) + w_2\right]$$

$$-\frac{\left[n\int_0^\tau \mu(y)q(y)dy + F_m(T-\tau)(c_f - c_p) + c_p\right]}{\left[\tau + \int_0^{T-\tau} \overline{F}_m(t)dt + F_m(T-\tau)(w_1 - w_2) + w_2\right]^2} \times$$

$$\left[1 + \int_0^{T-\tau} \frac{\partial \overline{F}_m(t)}{\partial \tau}dt - \overline{F}_m(T-\tau) + \frac{\partial F_m(T-\tau)}{\partial \tau}(w_1 - w_2)\right]$$

A necessary condition that a pair (τ^*,T^*) minimizes $L(\tau,T)$ is that it satisfies

$$\frac{\partial L(\tau,T)}{\partial \tau} = 0 \quad \text{and} \quad \frac{\partial L(\tau,T)}{\partial T} = 0$$

The optimal (τ,T) maintenance policy is obtained by solving the above equations.

8.3 Imperfect PM: Case 1

Section 8.2 assumes that PM is always perfect. In practice, however, this assumption may not be realistic in some cases. This section is different from Section 8.2 in that PM at time $T, 2T, 3T, \ldots$ is imperfect and after PM a k-out-of-n system is good as new with probability p (perfect PM) and is bad as old with probability $q = 1 - p$ (minimal PM). Other assumptions and notations are identical to those in Section 8.2.

According to renewal theory of stochastic processes, the times between consecutive perfect maintenance, preventive or corrective, constitute a renewal cycle. From the classical renewal reward theory, the long-run expected system maintenance cost per unit time, or cost rate with parameter p, is

$$L(\tau,T \mid p) = \frac{C(\tau,T \mid p)}{D(\tau,T \mid p)}$$

where $C(\tau,T \mid p)$ is the expected system maintenance cost per renewal cycle and $D(\tau,T \mid p)$ is the expected duration of a renewal cycle.

Let $Z_1, Z_2, \ldots$ be *i.i.d.* random variables with distribution function $F_m(y)$, and $Z_i^* = \min(Z_i, kT \mid k = \text{number of PM until the first perfect one})$, $i = 1,2,\ldots$ Note that a renewal cycle is completed either by any CM together with PM or by a perfect PM at time kT, and the probability that a PM alone is perfect is p. Then a renewal cycle consists of maintenance time and the Z_i^* duration. It follows from above arguments that

$$D(\tau,T \mid p) = E\left[Z_i^*\right] + \text{Expected maintenance time} \qquad (8.7a)$$

Let T_p be the first perfect PM-alone time point. Note that events $\{T_p = T\}$, $\{T_p = 2T\}$, $\{T_p = 3T\}, \ldots$ are mutually disjoint events satisfying sample space $\Omega = \bigcup_{j=1}^{\infty} \{T_p = jT\}$. Note also that a renewal cycle is completed either by any CM combined with PM or by a perfect PM at time kT. Note also that the probability that a PM alone is perfect is p. Pham and Wang (2000) prove

$$E[Z_i^*] = E[Z_i^* \mid T_p = T] \cdot I_{[T]}(T_p) + E[Z_i^* \mid T_p = 2T] \cdot I_{[2T]}(T_p) + \cdots$$

$$= p\left[\tau + \int_0^{T-\tau} \overline{F}_m(t)dt\right] + qp\left[\tau + \int_0^{2T-\tau} \overline{F}_m(t)dt\right] + q^2 p\left[\tau + \int_0^{3T-\tau} \overline{F}_m(t)dt\right] + \cdots$$

$$= \tau + p(1+q+q^2+\cdots)\int_0^{T-\tau} + pq(1+q+q^2+\cdots)\int_{T-\tau}^{2T-\tau} + \cdots$$

$$= \tau + \int_0^{T-\tau} \overline{F}_m(t)dt + \sum_{j=2}^{\infty} q^{j-1} \int_{(j-1)T-\tau}^{jT-\tau} \overline{F}_m(t)dt \qquad (8.7b)$$

Let CM be the event that CM together with PM is performed in a renewal cycle. Pham and Wang (2000) show that the probability that CM combined with PM is

$$P(CM) = P(CM \mid T_p = T) \cdot I_{[T]}(T_p) + P(CM \mid T_p = 2T) \cdot I_{[2T]}(T_p) + \cdots$$

$$= 1 - p \sum_{j=1}^{\infty} q^{j-1} \overline{F}_m(jT - \tau) \qquad (8.7c)$$

and the probability that PM alone occurs is

$$P(PM) = \sum_{j=1}^{\infty} q^{j-1} \overline{F}_m(jT - \tau) \qquad (8.7d)$$

The above expression at Equation (8.7d) also has direct meaning. For example, term $q \cdot \overline{F}_m(2T - \tau)$ represents the probability that less than m components have failed in the interval $(\tau, 2T)$ and the first PM turns out to be not perfect (with probability q). Obviously,

$$\text{Expected Maintenance Time} = w_1 \cdot P(CM) + w_2 \cdot P(PM) \qquad (8.7e)$$

It follows from Equations (8.7a – e) that

$$D(\tau, T \mid p) = E\left[Z_i^*\right] + \text{Expected maintenance time}$$

$$= \tau + \int_0^{T-\tau} \overline{F}_m(t)dt + \sum_{j=2}^{\infty} q^{j-1} \int_{(j-1)T-\tau}^{jT-\tau} \overline{F}_m(t)dt$$

$$+ w_1 \left[1 - p \sum_{j=1}^{\infty} q^{j-1} \overline{F}_m(jT - \tau) \right] + w_2 \left[\sum_{j=1}^{\infty} q^{j-1} \overline{F}_m(jT - \tau) \right] \qquad (8.7)$$

Next we determine expected system maintenance cost per renewal cycle $C(\tau, T \mid p)$, which consists of three parts: minimal repair cost, PM cost and cost of CM together with PM. The total minimal repair cost in one cycle is the same as the one in Equation (8.4). Again note that a renewal cycle is completed either by any CM together with PM or by a perfect PM, and that the probability that a PM alone is perfect is p. Similarly to the derivation of the expected maintenance time, it is easy to show that the total cost of PM alone and CM combined with PM is given by

$$C_{pf} = c_p \sum_{j=1}^{\infty} q^{j-1} \overline{F}_m(jT - \tau) + c_f \left[1 - p \sum_{j=1}^{\infty} q^{j-1} \overline{F}_m(jT - \tau) \right]$$

Thus,

$$C(\tau, T \mid p) = C_{smr} + C_{pf}$$

$$= n \int_0^{\tau} \mu(y)q(y)dy + c_p \sum_{j=1}^{\infty} q^{j-1} \overline{F}_m(jT - \tau) + c_f \left[1 - p \sum_{j=1}^{\infty} q^{j-1} \overline{F}_m(jT - \tau) \right]$$

$$(8.8)$$

From Equations (8.7) and (8.8) the following proposition follows:

PROPOSITION 8.4 *If the PM is perfect with probability p and minimal with probability* $q = 1 - p$, *then the long-run expected system maintenance cost per unit time, or cost rate, for a k-out-of-n system is given by*

$$L(\tau, T \mid p) =$$

$$\frac{n \int_0^\tau \mu(y)q(y)dy + c_p \sum_{j=1}^\infty q^{j-1} \overline{F}_m(jT - \tau) + c_f \left[1 - p \sum_{j=1}^\infty q^{j-1} \overline{F}_m(jT - \tau) \right]}{\tau + \int_0^{T-\tau} \overline{F}_m(t)dt + \sum_{j=2}^\infty q^{j-1} \int_{(j-1)T-\tau}^{jT-\tau} \overline{F}_m(t)dt + w_2 [\sum_{j=1}^\infty q^{j-1} \overline{F}_m(jT - \tau)] + w_1 \left[1 - p \sum_{j=1}^\infty q^{j-1} \overline{F}_m(jT - \tau) \right]}$$

$$(8.9)$$

and the limiting average system availability is

$$A(\tau, T \mid p) =$$

$$\frac{\tau + \int_0^{T-\tau} \overline{F}_m(t)dt + \sum_{j=2}^\infty q^{j-1} \int_{(j-1)T-\tau}^{jT-\tau} \overline{F}_m(t)dt}{\tau + \int_0^{T-\tau} \overline{F}_m(t)dt + \sum_{j=2}^\infty q^{j-1} \int_{(j-1)T-\tau}^{jT-\tau} \overline{F}_m(t)dt + w_2 [\sum_{j=1}^\infty q^{j-1} \overline{F}_m(jT - \tau)] + w_1 \left[1 - p \sum_{j=1}^\infty q^{j-1} \overline{F}_m(jT - \tau) \right]}$$

$$(8.9a)$$

Obviously, if we set $p = 1$ in Equations (8.9) and (8.9a) then we obtain Equations (8.6) and (8.6a). The optimal (τ, T) maintenance policy with parameter p can be obtained in the same method as in Section 8.2.

8.4 Imperfect PM: Case 2

The model in this section is exactly like the model in Section 8.2 except that after PM with probability p the system is good as new, and with probability q_i exactly i components become failed (all other components become good as new) and are subject to perfect CMs immediately where $i = 1, 2, ..., n$ and $\sum_{i=1}^n q_i = 1 - p$. Obviously, the latter case may happen in practice (Nakagawa 1987) and consequently a longer maintenance time and a larger maintenance cost are incurred since an additional CM on the failed component(s) due to PM is needed. It should be noted that only the components which have failed due to PM will be repaired immediately. Notice also that more than m components may fail due to PM since PM may cause adjacent damage (Nakagawa 1987) and becomes a worst PM.

Now we discuss modeling of system maintenance cost rate and availability. According to renewal theory, the times between consecutive perfect maintenance, preventive or corrective, constitute a renewal cycle. The long-run expected system maintenance cost per unit time, or maintenance cost rate with parameters p and q_i for $i = 1,2,...,n$, is

$$L(\tau,T \mid p,q_i) = \frac{C(\tau,T \mid p,q_i)}{D(\tau,T \mid p,q_i)}$$

where $C(\tau,T \mid p,q_i)$ is the expected system maintenance cost per renewal cycle and $D(\tau,T \mid p,q_i)$ is the expected duration of a renewal cycle.

Let $Z_1, Z_2,...$ be *i.i.d.* random variables with distribution function $F_m(y)$, and $Z_i^* = \min(Z_i,T)$, $\forall i, i = 1,2,...$ Let w_{1i} represent their total CM time when exactly i components failed after PM at time T. Note that the a renewal cycle is completed either by any CM together with PM, by CM alone right after T, or by a perfect PM; and that the probability that a PM alone is perfect is p. Thus, a renewal cycle consists of maintenance time and the Z_i^* duration. It follows from the above arguments that

$$D(\tau,T \mid p) = E\left[Z_i^*\right] + \text{maintenance time}$$

$$= \tau + \int_0^{T-\tau} \overline{F}_m(t)dt + w_1 F_m(T - \tau) + \sum_{i=1}^{n} q_i \cdot w_{1i} \cdot \overline{F}_m(T - \tau) + w_2 \overline{F}_m(T - \tau)$$

(8.10)

Next we evaluate the expected system maintenance cost per renewal cycle $C(\tau,T \mid p,q_i)$, which consists of four parts: minimal repair cost, cost of PM alone, cost of CM alone right after T, and cost of CM together with PM. The total minimal repair cost in one cycle is the same as the one in Equation (8.4). Let c_{fi} represent their total subsequent CM cost when exactly i components have failed due to PM. It is noted that a renewal cycle is completed either by a CM combined with a PM, by a CM alone right after T, or by a perfect PM at time T, and that the probability that a PM is perfect is p. It follows that the total cost of PM and CM after τ is given by

$$C_{pf} = c_p \overline{F}_m(T - \tau) + c_f F_m(T - \tau) + \sum_{i=1}^{n} c_{fi} \cdot q_i \cdot F_m(T - \tau)$$

Thus,

$$C(\tau,T \mid p,q_i) = C_{smr} + C_{pf}$$

$$= n \int_0^{\tau} \mu(y)q(y)dy + c_p \overline{F}_m(T - \tau) + c_f F_m(T - \tau) + \sum_{i=1}^{n} c_{fi} \cdot q_i \cdot F_m(T - \tau) \quad (8.11)$$

We may assume that c_{fi} above has such a cost structure:

$$c_{fi} = c_{00} + i \cdot c_s$$

where c_{00} represents one-time shut-off cost and c_s represents the cost of parts and labor and incremental system-unavailable-for-work cost. However, if c_{fi} has other cost structures, the results in this section are still valid.

From Equations (8.10) and (8.11) the following proposition follows:

PROPOSITION 8.5 *If PM is perfect with probability p and causes exactly i components to fail where* $\sum_{i=1}^{n} q_i - 1 = p$ *and the failed components due to PM are subject to perfect CM immediately, then the long-run expected system maintenance cost rate for the system is given by*

$$L(\tau, T \mid p, q_i) =$$

$$\frac{n \int_0^\tau \mu(y)q(y)dy + c_p \overline{F}_m(T-\tau) + c_f F_m(T-\tau) + \sum_{i=1}^{n} c_{fi} \cdot q_i \cdot \overline{F}_m(T-\tau)}{\tau + \int_0^{T-\tau} \overline{F}_m(t)dt + w_1 F_m(T-\tau) + \sum_{i=1}^{n} q_i \cdot w_{1i} \cdot \overline{F}_m(T-\tau) + w_2 \overline{F}_m(T-\tau)}$$

(8.12)

and the limiting average system availability is

$$A(\tau, T \mid p, q_i) =$$

$$\frac{\tau + \int_0^{T-\tau} \overline{F}_m(t)dt}{\tau + \int_0^{T-\tau} \overline{F}_m(t)dt + w_1 F_m(T-\tau) + \sum_{i=1}^{n} q_i \cdot w_{1i} \cdot \overline{F}_m(T-\tau) + w_2 \overline{F}_m(T-\tau)}$$

(8.12a)

Obviously, if we set $p = 1$ in Equations (8.12) and (8.12a) then we obtain Equations (8.6) and (8.6a). The optimal (τ, T) maintenance policy with parameters p and q_i can be obtained in the same way as in Section 8.2.

8.5 Special Cases

The three models in Sections 8.2, 8.3 and 8.4 include some previous maintenance models as special cases. A summary is given below. Since Proposition 8.3 is a special case of Propositions 8.4 and 8.5, the discussions in this section will be focused on Proposition 8.3. The following special cases are in terms of Equation (8.6) except case 10 and case 11.

Case 1 ($n = k = m = 1$, $w_1 = w_2 = 0$, $\tau = 0$): this is the classical age-replacement policy which was called policy I in Barlow and Hunter (1960). If we set parameters $n = k = m = 1, w_1 = w_2 = 0$, and $\tau = 0$ in Equation (8.6), then we obtain the well-known result by Barlow and Hunter (1960):

$$L(0, T) = \frac{F(T)c_f + c_p \overline{F}(T)}{\int_0^T \overline{F}(t)dt}$$

Case 2 ($m = n - k + 1$, $w_1 = w_2 = 0$, $\tau = 0$, $\overline{G}(y) = e^{-\lambda y}$): Nakagawa (1985) investigates this case. If we set parameters $m = n - k + 1, w_1 = w_2 = 0, \tau = 0$ and $\overline{G}(y) = e^{-\lambda y}$ in Equation (8.6) then the cost rate becomes

$$L(0,T) = \frac{c_p + (c_f - c_p) \sum_{i=0}^{k-1} \binom{n}{i} e^{-i\lambda T} \left[1 - e^{-\lambda T}\right]^{n-i}}{\int_0^T \sum_{i=k}^{n} \binom{n}{i} e^{-i\lambda t} \left[1 - e^{-\lambda t}\right]^{n-i} dt}$$

which agrees with Equation (8) in Nakagawa (1985).

Case 3 ($n = k = m = 1$, $w_1 = w_2 = 0$, $\tau = T$, $g(c_1(t,i), c_2(t)) = c$): this is policy II discussed by Barlow and Hunter (1960), i.e., the classical periodic replacement with minimal repair at failures. If we set parameters $n = k = m = 1$, $w_1 = w_2 = 0$, $\tau = T$, and $g(c_1(t,i), c_2(t)) = c$ in Equation (8.6), the system maintenance cost rate becomes

$$L(T,T) = \frac{cQ(T) + c_p}{T}$$

which is the same as the well-know result by Barlow and Hunter (1960).

Case 4 ($n = k = m = 1$, $w_1 = w_2 = 0$, $\tau = T$, $g(c_1(t,i), c_2(t)) = c$, $c_p = c(T)$): This is the case treated by Tilquin and Cleroux (1975). If we set $n = k = m = 1$, $w_1 = w_2 = 0, \tau = T, g(c_1(t,i), c_2(t)) = c$ and $c_p = c_0 + a(T)$ in Equation (8.6), then the system maintenance cost rate becomes

$$L(T,T) = \frac{cQ(T) + c_0 + a(T)}{T}$$

which is the same as the cost rate in Tilquin and Cleroux (1975).

Case 5 ($n = k = m = 1$, $w_1 = w_2 = 0$, $\tau = T$, $g(c_1(t,i), c_2(t)) = c(y)$): this is the case investigated by Boland (1982).

Case 6 ($n = k = m = 1$, $w_1 = w_2 = 0$, $\tau = T$, $g(c_1(t,i), c_2(t)) = c_i$): Boland and Proschan (1982) study this case. In particular, they considered the cost structure $c_i = a + ic$ in which minimal repair cost is increasing with the number of minimal repairs.

Case 7 ($n = k = m = 1$, $w_1 = w_2 = 0$, $g(c_1(t,i), c_2(t)) = c$): this is the policy considered by Tahara and Nishida (1975). If we set $n = k = m = 1, w_1 = w_2 = 0$, and $g(c_1(t,i), c_2(t)) = c$ in Equation (8.6), then the expected systems maintenance cost rate is

$$L(\tau,T) = \frac{cQ(\tau) + G(T-\tau)(c_f - c_p) + c_p}{\tau + \int_0^{T-\tau} \overline{G}(t)dt}$$

which agree with Equation (23) in Tahara and Nishida (1975).

It is noted that Tahara and Nishida (1975) discuss the optimality of the (τ,T) policy for a one-unit system by means of dynamic programming techniques and showed that the (τ,T) maintenance policy is optimal.

Case 8 ($n = k = m = 1$, $w_1 = w_2 = 0$, $T = \infty$, $g(c_1(t,i),c_2(t)) = c$): Muth (1977) studies this case. If we set parameters $n = k = m = 1$, $w_1 = w_2 = 0$, $T = \infty$, and $g(c_1(t,i),c_2(t)) = c$ in Equation (8.6), we obtain the same result as in Muth (1977)

$$L(\tau,T) = \frac{cQ(\tau) + c_f}{\tau + \int_0^\infty \overline{G}(t)dt}$$

Case 9 ($n = k = m = 1$, $w_1 = w_2 = 0$, $T = \infty$, $g(c_1(t,i),c_2(t)) = c(t)$): Yun (1989) considers this case. If we set parameters $n = k = m = 1$, $w_1 = w_2 = 0$, $T = \infty$, and $g(c_1(t,i),c_2(t)) = c(t)$ in Equation (8.6), we obtain the same result as in Yun (1989):

$$L(\tau,\infty) = \frac{\int_0^\tau c(t)q(y)dy + c_f}{\tau + \int_0^\infty e^{-Q(\tau+x)+Q(\tau)}dt}$$

Case 10 ($n = k = m = 1$, $w_1 = w_2 = 0$, $\tau = 0$) for Equation (8.9): Nakagawa (1979) deals with this case. If we set $n = k = m = 1$, $w_1 = w_2 = 0$, and $\tau = 0$ in Equation (8.9), then the cost rate becomes

$$L(0,T \mid p) = \frac{c_p \sum_{j=1}^\infty q^{j-1}\overline{F}(jT) + c_f[1 - p\sum_{j=1}^\infty q^{j-1}\overline{F}(jT)]}{\sum_{j=1}^\infty q^{j-1}\int_{(j-1)T}^{jT} \overline{F}(t)dt}$$

which is the same as Equation (1) in Nakagawa (1979).

Case 11 ($n = k = m = 1$, $\tau = 0$) for Equation (8.12a): Chan and Downs (1978) study this case. If we set $n = k = m = 1$, $w_{11} = w_1$, $q_1 = q$ and $\tau = 0$ in Equation (8.12a), then the limiting system availability becomes

$$A(0,T \mid p,q_i) = \frac{\int_0^T \overline{F}(t)\,dt}{w_1 F(T) + qw_1 \overline{F}(T) + \int_0^T \overline{F}(t)\,dt + w_2 \overline{F}(T)}$$

which is the same as Equation (1) in Chan and Downs (1978).

Case 12 ($n = k$): this case corresponds to optimal (τ, T) maintenance policy of a series system. If we set $n = k$ and $m = 1$ in Equation (8.6), it follows that the long-run expected system maintenance cost rate for a series system with n component is

$$L(\tau,T) = \frac{n\int_0^\tau \mu(y)q(y)\,dy + \left[1 - \overline{F}^{\,n}(T)/\overline{F}^{\,-n}(\tau)\right](c_f - c_p) + c_p}{\tau + \left[\overline{F}(\tau)\right]^{-n} \int_0^{T-\tau} \left[\overline{F}(\tau+t)\right]^n dt + \left[1 - \overline{F}^{\,n}(T)/\overline{F}^{\,-n}(\tau)\right](w_1 - w_2) + w_2}$$

Case 13 ($k = 1, n > 1$): in this case the k-out-of-n system is reduced to a parallel system. If we let $k = 1$ and $m = n$, it follows that the long-run expected system maintenance cost rate for a parallel system with n components is

$$L(\tau,T) =$$

$$\frac{n\int_0^\tau \mu(y)q(y)\,dy + (c_f - c_p)\left[\overline{F}(\tau) - \overline{F}(T)\right]^n \overline{F}^{\,-n}(\tau) + c_p}{\tau + \int_0^{T-\tau} \left\{1 - \left[\overline{F}(\tau) - \overline{F}(\tau+t)\right]^n \overline{F}^{\,-n}(\tau)\right\}dt + (w_1 - w_2)\left[\overline{F}(\tau) - \overline{F}(T)\right]^n \overline{F}^{\,-n}(\tau) + w_2}$$

If we further set $\tau = 0$ and $w_1 = w_2 = 0$, then the above equation becomes

$$L(0,T) = \frac{(c_f - c_p)F^n(T) + c_p}{\int_0^T \left[1 - F^n(t)\right]dt}$$

which is the same as the result in Yasui *et al.* (1988).

8.6 Optimization Problems

In Sections 8.2, 8.3 and 8.4 we investigate expected system maintenance cost rate and availability. In some cases, the optimal maintenance policies may be required that while some availability requirements are satisfied the maintenance cost rate is minimized, or while maintenance cost rate is less than some predetermined value the system availability is maximized. For example, for the maintenance model in Section 8.2 the following optimization problem can be formulated in terms of decision variables T and τ :

Maximize $A(\tau,T) = \dfrac{\tau + \displaystyle\int_0^{T-\tau} \overline{F}_m(t)dt}{\tau + \displaystyle\int_0^{T-\tau} \overline{F}_m(t)dt + F_m(T-\tau)(w_1 - w_2) + w_2}$

Subject to $L(\tau,T) = \dfrac{n\displaystyle\int_0^{\tau} \mu(y)q(y)dy + F_m(T-\tau)(c_f - c_p) + c_p}{\tau + \displaystyle\int_0^{T-\tau} \overline{F}_m(t)dt + F_m(T-\tau)(w_1 - w_2) + w_2} \le L_0$ (8.13)

where constant L_0 is the predetermined maintenance cost rate requirement.

For maintenance model in Section 8.3 the following optimization problem can be formulated in terms of decision variables T and τ :

Minimize

$L(\tau,T \mid p) =$

$$\frac{n\displaystyle\int_0^{\tau} \mu(y)q(y)dy + c_p\sum_{j=1}^{\infty} q^{j-1}\overline{F}_m(jT - \tau) + c_f\left[1 - p\sum_{j=1}^{\infty} q^{j-1}\overline{F}_m(jT - \tau)\right]}{\tau + \displaystyle\int_0^{T-\tau} \overline{F}_m(t)dt + \sum_{j=2}^{\infty} q^{j-1}\int_{(j-1)T-\tau}^{jT-\tau} \overline{F}_m(t)dt + w_2\left[\sum_{j=1}^{\infty} q^{j-1}\overline{F}_m(jT - \tau)\right] + w_1\left[1 - p\sum_{j=1}^{\infty} q^{j-1}\overline{F}_m(jT - \tau)\right]}$$

Subject to $\begin{cases} A(\tau,T \mid p) \ge A_0 \\ \tau \ge 0 \\ T > \tau \end{cases}$

(8.14)

where constant A_0 is the predetermined availability requirement.

The optimal maintenance policy (T^*, τ^*) can be determined from it by using nonlinear programming software. Similarly, other optimization models can be formulated.

8.7 Numerical Example

Consider a 2-out-of-3 system. Assume that the time to failure of each unit follow a Weibull distribution with shape parameter β and scale parameter θ which has *pdf* given by

$$f(y) = \frac{\beta}{\theta}\left(\frac{y}{\theta}\right)^{\beta-1} \exp\left[-\left(\frac{y}{\theta}\right)^{\beta}\right] \qquad y > 0, \ \beta,\theta > 0$$

and has failure rate

$$q(y) = \frac{\beta}{\theta}\left(\frac{y}{\theta}\right)^{\beta-1}$$

Suppose that $\beta = 2$ and $\theta = 500$ (days) and $g(c_1(y,i),c_2(y)) = c_1(y) + c_2(y)$ in this example. Since $\beta = 2 > 1$ the lifetime of each unit has IFR and by Proposition 8.1 life of this system has IFR. Assume that $c_1(y) = 1 + \sqrt{y}$ and $c_2(y)$ follow the normal distribution with mean 1. Then

$$\begin{aligned}
\mu(y) &= E_{N(y)}\left[E_{C_2(y)}[g(c_1(y,N(y)+1),c_2(y))]\right] \\
&= E[c_1(y) + c_2(y)] \\
&= 2 + \sqrt{y}
\end{aligned}$$

Since the aircraft unit system is considered to be critical and very important, m is taken to be 1. The following parameters are assumed:

$w_1 = 5$ days	$w_2 = 2$ days	$c_f = 59$
$c_p = 40$	$p = 1$	$A_0 = 0.99$

Let

$$\begin{aligned}
\zeta &= \tau + 500 \times 6^{-\frac{1}{2}} \exp\left[3(0.002\tau)^2\right]\int_{0.002\tau\sqrt{6}}^{0.002T\sqrt{6}} \exp\left[-\tfrac{1}{2}u^2\right]du \\
&\quad - \exp\left[-3(0.002T)^2 + 3(0.002\tau)^2\right] + 5
\end{aligned}$$

Substituting the above data and parameters into optimization model (8.14) in Section 8.6 results in:

Minimize

$$L = \frac{6(\tau^2 + \tfrac{2}{5}\tau^{5/2})/500^2 - 19 \cdot \exp\left[-3(0.002T)^2 + 3(0.002\tau)^2\right] + 59}{\zeta}$$

Subject to

$$\begin{cases}
A = \dfrac{\tau + 500 \times 6^{-\frac{1}{2}} \exp\left[3(0.002\tau)^2\right]\int_{0.002\tau\sqrt{6}}^{0.002T\sqrt{6}} \exp\left[-\tfrac{1}{2}u^2\right]du}{\zeta} \geq 0.99 \\[2em]
\tau \geq 0 \\
T \geq \tau
\end{cases}$$

Various kinds of approximations for the integral in the above optimization model have been developed and a simple approximation with high accuracy is by Zelen and Severo (1964):

$$\int_{-\infty}^{t} \exp\left(\tfrac{1}{2}u^2\right)du \approx \sqrt{2\pi}\left[1 - \tfrac{1}{2}\left(1 + .196854t - .115194t^2 + .000344t^3 + 0.019527t^4\right)^{-4}\right]$$

The error, for $t \geq 0$, is less than 2.5×10^{-4} (see Johnson and Kotz 1970).

By nonlinear optimization software the optimal solution for the above optimization model is found to be

$$\tau^* = 320.24 \text{ days} \qquad T^* = 410.86 \text{ days}$$

which results in the minimum system maintenance cost rate given by

$$L(\tau^*, T^*) = 0.182$$

That is, the optimal maintenance policy from optimization model (8.13) is that before time $\tau^* = 320.24$ (days) only minimal repairs are performed; after $\tau^* = 320.24$ the failed unit will be subject to perfect repair and the two unfailed will undergo PM once any unit fails, or PM at $T^* = 410.86$ (days), whichever comes first (*i.e.*, CM combined with PM or PM only at *T*, whichever occurs first). Or, we can say that after $\tau^* = 320.24$: the failed unit will be subject to perfect repair together with PM on the remaining two once any unit fails; If no unit fails until time $T^* = 410.86$ (days), PM is carried out at $T^* = 410.86$.

Other numerical examples can be found in Pham and Wang (2000).

8.8 Concluding Discussions

This chapter deals with the opportunistic maintenance of *k*-out-of-*n* systems. In many applications, the optimal maintenance actions for one component often depend on the states of the other components and system reliability requirements. Three new (τ, T) opportunistic maintenance models with consideration of reliability requirements are investigated. In these models only minimal repairs are performed on failed components before time τ and CM of all failed components are combined with PM of all functioning but deteriorated components after τ. If the system survives to time *T* without a perfect maintenance it will be subject to PM at time *T*. Considering the maintenance time, system asymptotic cost rate and availability are derived and the results obtained generalize and unify some previous research in this area.

In all three maintenance models PM on non-failed but deteriorated components is also carried out at the moment when CM activities are called for after τ. Such maintenance policies may reduce the number of unexpected CM activities at fairly low costs, since CM together with PM can be performed without substantial additional expenses. Besides, the following points should be noted:

1) Some group replacement policies (see Section 3.3.1) do not consider system reliability structure and reliability requirements. For a *k*-out-of-*n* system, when one

of its components is down if we repair or replace it there will be a one-time shut-off during which system will not be available. In fact, even if we do not take any action on this component the system may still operate as long as the number of failed components does not exceed $n - k$. However, once the number of failed components surpasses $n - k$ the system fails. Thus, most maintenance actions may start earlier than that moment in practice.

2) Equations (8.6), (8.6a), (8.9), (8.9a), (8.12), and (8.12a) are still valid if CM and PM costs as well as maintenance time of a system are random variables. In this case, c_f, c_p, w_1 and w_2 represent expected costs of CM combined with PM and PM alone, expected maintenance time of CM combined with PM and PM alone, respectively in Equations (8.6), (8.6a), (8.9), (8.9a), (8.12) and (8.12a).

3) m could be a decision variable according to different situations. Its optimal value, together with the optimal values of τ and T, can be found by minimizing the system maintenance cost rate in Equations (8.6), (8.6a), (8.9), (8.9a), (8.12) and (8.12a) in terms of decision variables τ, T and m.

4) m could take a natural number greater than $(n - k + 1)$ depending on different reliability and cost requirements.

9

Reliability and Optimal Inspection-maintenance Models of Multi-degraded Systems

In practice, the failure rate of a system may depend not only on the time, but also upon the state of the system. The system may not fail fully, but can degrade. Its operating condition can be characterized by a finite number of states: states of degradation. The failure rate transition process from one degradation state to the next degradation state might take faster rate as system reaches the last stages of degradation. Therefore, the state-dependent transition rates for the degradation process should be considered. In some cases if the degradation level exceeds a particular limit the system may not operate successfully and fail. This chapter investigates multi-state degraded systems subject to multiple competing failure processes including two independent degradation processes and random shocks, following Li and Pham (2005a, 2005b). We first discuss system reliability models and then optimal inspection-maintenance. This chapter also presents a methodology to generate the system states when there exists multi-failure processes. The system reliability model can be used not only to determine the reliability of the degraded systems in the context of multi-state functions but also to obtain the states of the systems by calculating the system state probabilities. A generalized condition-based inspection-maintenance model, consisting of the time sequence for inspection and PM threshold levels, is presented. An average long-run maintenance cost rate function is derived based on expressions for degradation paths and cumulative shock damage. A quasi-renewal process introduced in Chapter 4 is employed to develop the inter-inspection sequence. The PM thresholds for degradation processes and inspection sequence are the decision variables of the proposed inspection-maintenance model. The optimum solution to minimize the average long-run maintenance cost rate is discussed. Numerical examples are given to illustrate these system reliability and inspection-maintenance models.

In recent years, increasing research on degraded systems and their reliability as well as inspection and maintenance has been conducted. Pham *et al.* (1996) present a model for predicting the reliability of a k-out-of-n:G system in which components are subject to multi-stage degradation as well as catastrophic failures. Due to the aging effect, the failure rate of the component will increase. Pham *et al.* (1996)

consider the state-dependent transition rates for the degradation process. Because of degradation, the catastrophic state-dependent failure rate may increase as time progress as well based on the Markov approach. Sim and Endrenyi (1993) propose a Markov model for a continuously operating device with deterioration and Poisson failures. In the Markov diagram, the distribution of the inter-arrival between successive degradation stages is assumed to be exponentially distributed with constant rates.

Lam and Yeh (1994b) study state-age-dependent replacement policies for a multi-state system subjected to both deterioration and random shocks. The deteriorating process of the system was modeled based on a semi-Markov process. They assumed that the inter-arrival time between two successive states follows a continuous distribution $F_i(t)$ with a finite mean.

Other reliability models considering the degradation and catastrophic failures are developed. Zuo et al. (1999) present a mixture model assuming that the whole population can be divided into two independent sub-populations where one sub-population is subjected to degradation and the other is subjected to catastrophic failure. Hosseini et al. (2000) develop a condition-based maintenance model for a system subject to deterioration-failures and to Poisson-failures using the Generalized Stochastic Petri Nets. Xue and Yang (1995) model the lifetime distribution of the multi-state deterioration systems based on the continuous-time Markov process and semi-Markov process.

Pham et al. (1997) create models for predicting the availability and mean lifetime of multistage degraded systems with partial repairs. The transition (degradation, partial failure, and repair rates) rates are assumed constant. Klutke and Yang (2002) study the availability of maintained systems subject to both the effects of the degradation and random shocks. They assumed that shocks occur according to a Poisson process and the shock magnitudes are independent and identically distributed random variables. Recently, Pham and Xie (2002) investigate a generalized surveillance model consisting of dual mutually dependent stochastic processes for surveillance systems. Their model can be used to understand better both the inspection process and the repair unit itself and to provide information that can be used to assist inspectors in scheduling and prioritizing their future inspections.

This chapter first discusses the reliability model and then optimal inspection-maintenance model of a degraded system subject to multiple competing failure processes including two degradation processes and random shocks. We assume that these three processes are independent and any of them would cause the system to fail based on the threshold value of each process. Applications of such systems can be found in the space shuttle computer complex due to critical mission phases such as boost, reentry and landing and in the electric generator power systems. More applications related to this can be found in Pham (1991). The system can also fail catastrophically whether it is either in a good state or in any of the degraded states due to random shocks. It should be noted that each competing process can be considered as a component in a series system in which system failure occurs when any component fails. In other word, the system fails whichever cause occurs first. This chapter also models the performance of the systems from a perfect condition to degradation stages.

The following notation and symbols will be used throughout this chapter:

Notation

$Y_i(t)$	The ith degradation process, $\forall i, i = 1, 2$
$D(t)$	Cumulative random shock damage by the time t. $D(t) = \sum_{i=1}^{N(t)} X_i$
	where X_is are $i.i.d.$ with $pdf\ f_X(x)$, $cdf\ F_X(x)$ and the kth
	convolution $F_X^{(k)}(x)$
S	Critical threshold value for the shock process. The system will fails due to random shocks when $D(t) > S$
G_i	Critical value for degradation process i for $i = 1, 2$ where the system will fail due to degradation when $Y_i(t) > G_i$
Ω_U	$\{M, ..., 1, 0, F\}$ a system state space
M	Perfect (good) state
0	Degraded failure state
$M-1, ..., 1$	Intermediate degradation states
F	Catastrophic failure state
Ω	$\{M, ..., 1, 0\}$ a system degradation state space without catastrophic failure
Ω_i	$\{M_i, ..., 1_i, 0_i\}$ a state space corresponding to degradation process i
0_i	Degraded failure state due to the ith degradation process
M_i	Good state of degradation process i, $\forall i, i = 1, 2$
R	$\Omega_1 \times \Omega_2$ Cartesian product of Ω_1 and Ω_2
R_i	The ith equivalence class, $i = 0, 1 ..., M$
$R(t)$	Reliability function
$r.v.$	Random variable
C_c	Cost per CM action
C_p	Cost per PM action
C_m	Loss per unit idle time
C_i	Cost per inspection
L_1	PM critical threshold value for degradation process 1
L_2	PM critical threshold value for degradation process 2
$C(t)$	Cumulative maintenance cost up to time t
$E[C_1]$	Average total maintenance cost during a cycle
$E[W_1]$	Mean cycle length
$E[N_i]$	Mean number of inspections during a cycle
$E[\xi]$	Mean idle time during a cycle
$\{I_i\}_{i \in N}$	Inspection sequence
$\{U_i\}_{i \in N}$	Inter-inspection sequence
$\{W_i\}_{i \in N}$	Renewal times
T	Time to failure

P_{i+1} Probability that there are a total of $(i+1)$ inspections in a renewal cycle

P_p Probability that a renewal cycle ends by a PM action

P_c Probability that a renewal cycle ends by a CM action

$EC(L_1, L_2, I_1)$ Expected long-run cost rate function

9.1 Reliability Modeling

In this section, we discuss models for evaluating the reliability of multi-state degraded systems subject to multiple competing failure processes. Two of them are the continuous and increasing degradation processes (processes 1 and 2) and the third one is random shocks. The performance of the systems can vary from a perfect condition as good as new to degradation stages as time passes since the multi-state reliability model in this section, from the multi-state perspective, can be capable of handling a wide range of performance (Li and Pham 2005b). The remaining of this section is organized as follows. Section 9.1.1 describes the multi-state system description, modeling assumptions and methodologies. It also presents a method to determine the system state and to view degradation process in terms of multi-state. In Section 9.1.2, we present a model for evaluating the reliability of multi-state degraded systems with random shocks. Section 9.1.3 delivers numerical examples to illustrate the obtained results.

9.1.1 System Description and Modeling Methodologies

Assume a system is subject to a variety of three independent competing failure processes in which two of them are degradation processes: degradation process 1 measured by the function $Y_1(t)$ and degradation process 2 measured by $Y_2(t)$), and the third is a random shock process $D(t)$; whichever occurred first would cause the system to fail.

Initially, the system is considered to be in a good state (i.e., M_1 and M_2). As time passes, it can either go to the first degraded state (i.e., $(M-1)_1$ or $(M-1)_2$) upon degradation or can go to a catastrophic failed state (state F) subject to random shocks. When a system reaches the first degraded state, it can either stay in that state until the mission time, or it can go to the second degradation state (i.e., $(M-2)_1$ or $(M-2)_2$) upon degradation, or can go to a failed state (F state) upon random shocks.

The same process will be continued for all stages of degradation except the last degradation state (i.e., either stage 0_1 or stage 0_2). If the system reaches the last degradation state, it cannot perform its functions satisfactorily (it considers to reach an unacceptable limit) and be treated as a failure (state 0). Figure 9.1 shows the system state transition diagram of the multiple competing transition processes. In Figure 9.1, the top portion represents degradation process 1; the bottom represents degradation process 2; F represents a catastrophic failure state due to random shocks.

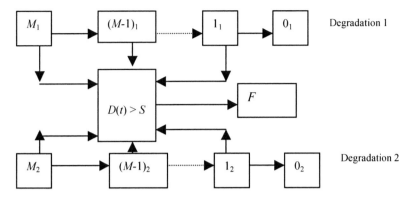

Figure 9.1. Flow diagram of the system subjected to multiple failure processes

In this section, we have the following assumptions:

i) The system consists of $(M + 2)$ states where state 0 and state F are both complete failure states, state i is a degradation state, and $1 < i < M$.

ii) No repair or maintenance is performed on the system.

iii) $Y_i(t)$, $i = 1,2$ is a non-negative non-decreasing function at time t, since degradation is an irreversible accumulation of damage.

iv) $Y_i(t)$, $i = 1,2$ and $D(t)$ are statistically independent. The independence assumption implies that the state of one process will have no effect on the state of the others.

v) At time $t = 0$, the system is at state M.

vi) The system can fail either due to each degradation process when $Y_i(t) > G_i$, $i = 1,2$ or due to random shocks (it goes to a catastrophic failure state F) when $D(t) = (\sum_{i=1}^{N(t)} X_i) > S$

vii) The critical threshold value G_i depends upon the function of states of the degraded systems.

In this section, degradation paths are modeled by some continuous probabilistic functions. Note that the operating condition of the system is characterized by a finite number of states: system state space Ω_U. Therefore, we need to discretize continuous processes. In Step 1 below, we discuss a procedure how to discretize the two degradation processes in order to obtain Ω_1 and Ω_2 which correspond to degradation process 1 and 2, respectively. After obtaining the degradation process space Ω_1 and Ω_2, Step 2 presents a methodology how to establish a relationship between the system state space Ω_U, and the degradation and random shock state spaces $\{\Omega_1, \Omega_2, F\}$.

Step 1: Characterizing Degradation Processes into Discrete State Sets
The two degradation processes cases are considered here. A general situation is to allow each degradation process to be discretized into a number of different states. The state space denoted by $\Omega_1 = \{M_1,...,1_1,0_1\}$ corresponds to degradation process 1 with $(M_1 + 1)$ states. Similarly, the state space denoted by $\Omega_2 = \{M_2,...,1_2,0_2\}$ associates with degradation process 2 having $(M_2 + 1)$ states. M_1 and M_2 may or may not be the same and, $M_i < \infty$, $i = 1,2$.

We view the degradation process from the perspective of a finite number of states. For example, when the value $Y_1(t)$ of degradation process 1 falls into a pre-defined interval, then its corresponding state will be determined. Let us define as follows: $[0,W_M],...,(W_2,W_1]$ are the intervals on the degradation 1 curve (see Figure 9.2a) corresponding to state M_1, 0_1, where $W_M < W_{M-1} < \cdots < W_1$ and $[0, A_M],...,(A_2, A_1]$ are intervals associated with degradation process 2 curve (see Figure 9.2b) corresponding to state M_2, 0_2, where $A_M < A_{M-1} < ... < A_1$.

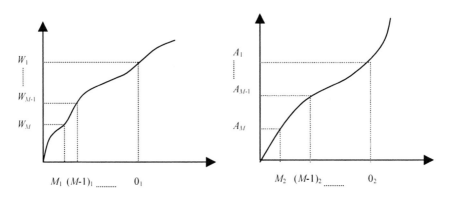

a) States for degradation 1   b) States for degradation 2

Figure 9.2. Degradation process function in terms of multi-state

Mathematically, the relationship between the degradation process states $\Omega_1 = \{M_1,...,1_1,0_1\}$, $\Omega_2 = \{M_2,...,1_2,0_2\}$ and their corresponding degradation intervals are given as follows:

Degradation Process 1 | Degradation Process 2

$$0 < Y_1(t) \leq W_M \quad \Rightarrow \text{state } M_1 \qquad\qquad 0 < Y_2(t) \leq A_M \Rightarrow \text{ state } M_2$$
$$W_M < Y_1(t) \leq W_{M-1} \Rightarrow \text{state } (M-1)_1 \qquad A_M < Y_2(t) \leq A_{M-1} \Rightarrow \text{state } (M-1)_2$$
$$\vdots \qquad\qquad\qquad\qquad\qquad\qquad\qquad \vdots$$
$$W_2 < Y_1(t) \leq W_1 \quad \Rightarrow \text{state } 1_1 \qquad\qquad A_2 < Y_2(t) \leq A_1 \quad \Rightarrow \text{state } 1_2$$
$$G_1 = W_1 < Y_1(t) \Rightarrow \text{state } 0_1 \qquad\qquad G_2 = A_1 < Y_2(t) \Rightarrow \text{state } 0_2$$

Step 2: Constructing System State Space based on Degradation States

The system state space is defined as $\Omega_U = \{M,...,1,0,F\}$, consisting of $M+2$ states. This step will create a method to develop a function to generate a relationship between the system state space Ω_U and degradation state spaces $\{\Omega_1,\Omega_2,F\}$. For example, at a given time t, suppose that degradation process 1 is at state $i_1 \in \Omega_1$, and degradation process 2 is at state $j_2 \in \Omega_2$, what is the system state?

Assume that at the current time the system is not in a catastrophic failure state. So state F can be ignored for the time being. So, we can simply look at ways to define a function that represents relationship between Ω_U and $\{\Omega_1,\Omega_2\}$ instead of Ω_U and $\{\Omega_1,\Omega_2,F\}$. The operation can be described by a mapping function $f : R = \Omega_1 \times \Omega_2 \rightarrow \Omega = \{M,..,1,0\}$ where $R = \Omega_1 \times \Omega_2 = \{(i_1,j_2)\,|\,i_1 \in \Omega_1, j_2 \in \Omega_2\}$ is a *Cartesian* product as the input space domain and shown in Figure 9.3.

Figure 9.3. A mapping function

The matrix H_c given below is an output space consisting of $M+1$ elements corresponding each input space domain through the function f:

$$
H_c =
\begin{array}{c}
\\
0_2 \\
1_2 \\
\vdots \\
M_2
\end{array}
\begin{array}{c}
\begin{array}{cccc}
0_1 & 1_1 & \cdots & M_1
\end{array} \\
\begin{bmatrix}
\times & 0 & \cdots & 0 \\
0 & \ddots & & \vdots \\
\vdots & & \ddots & \vdots \\
0 & \cdots & \cdots & M
\end{bmatrix}
\end{array}
$$

The top row of H_c represents the state from degradation process 1. The very left column represents the state from degradation process 2. The elements of H_c represent $f(i_1,j_2) = k$ where $i_1 \in \Omega_1$, $j_2 \in \Omega_2$ and $k \in \Omega$. Note that in matrix H_c, all the elements in the first row and first column are zeros except the one denoted by $\times$ (will explain this later) since the system will go to a degraded failure state (state 0) when either of degradations reaches state $0_i, i = 1,2$. Besides, some elements in matrix H_c are also zeros since we define when degradation 1 is in some low state $l_1 ((0_1 < l_1 < M_1)$ and degradation 2 is also in some low state l_2 $(0_2 < l_2 < M_2)$. Hence, we consider it as degradation failure. It is also observed that $f(M_1,M_2) = M$, since initially the system is in a brand new state (state M).

As mentioned above, the first element in H_c is marked by $\times$ which means that it does not exist. The reason is as follows. Note that we define the time-to-failure as

$$T = \inf\{t : Y_1(t) > G_1, Y_2(t) > G_2 \text{ or } D(t) > S\} \tag{9.1}$$

It should be noted that all the three processes are competing with each other for the life of a system. However, there is only one of the three processes, whichever occur first, that when exceeding its corresponding critical threshold value, will cause the system to fail. Hence, the following events will not happen:

$$P\{Y_1(t) > G_1, Y_2(t) > G_2, D(t) > S\} = 0$$

$$P\{Y_1(t) > G_1, Y_2(t) > G_2, D(t) > S\} = 0$$

$$P\{Y_1(t) > G_1, Y_2(t) < G_2, D(t) > S\} = 0$$

$$P\{Y_1(t) < G_1, Y_2(t) > G_2, D(t) > S\} = 0.$$

Since $f(0_1, 0_2) = P\{Y_1(t) > G_1, Y_2(t) > G_2, D(t) \le S\}$, the combination of $f(0_1, 0_2)$ does not exist. The function $f : R = \Omega_1 \times \Omega_2 \to \Omega = \{M, ..., 1, 0\}$ is defined to satisfy following conditions:

i) $f(0_1, b) = f(a, 0_2) = 0$ where $b \in \Omega_2, a \in \Omega_1$, $f(M_1, M_2) = M$

ii) f is a monotonic non-decreasing in each variable. For instance,

$f(a_1, b_2) \ge f(l_1, b_2)$ if $a_1 \ge l_1$ and $f(a_1, b_2) \ge f(a_1, l_2)$ if $b_2 \ge l_2$

Figure 9.4 demonstrates the system state generating box. There are two inputs i_1 and j_2 and an output k. The inside mapping mechanism is performed by the function f. At time t, suppose that degradation 1 is at state i_1 and degradation 2 is at state j_2; i_1 and j_2 are as inputs; via matrix H_c, system state k is then generated as output.

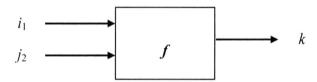

Figure 9.4. A representation of system state generating box

It is observed that in matrix H_c different state combination inputs can generate the same results of the system state. To explain this, we need the following definition on the equivalence class:

DEFINITION The ith equivalence class R_i is defined as follows:

$$R_i = \{(k_1, j_2) \text{ where } k_1 \in \Omega_1, j_2 \in \Omega_2 \mid f(k_1, j_2) = i\}, \text{ i=0,1,...,M} \tag{9.2}$$

R_i represents all possible state combinations which generate the system state i. $R_0, ..., R_M$ are disjointed sets which partition R into $M+1$ equivalence classes, so that

$$R = \bigcup_{i=0}^{M} R_i .$$ (9.3)

Next we give an example to illustrate the concepts. Assume the state spaces for degradation process 1 and degradation process 2 are: $\Omega_1 = \{4_1, 3_1, 2_1, 0_1\}$ and $\Omega_2 = \{3_2, 2_2, 1_2, 0_2\}$ respectively. The system state space is: $\Omega_U = \{3, 2, 1, 0, F\}$. The matrix H_c is defined as follows:

$$H_c = \begin{array}{c} \\ 0_2 \\ 1_2 \\ 2_2 \\ 3_2 \end{array} \begin{array}{c} 0_1 \; 1_1 \; 2_1 \; 3_1 \; 4_1 \\ \begin{bmatrix} \times & 0 & 0 & 0 & 0 \\ 0 & 0 & 1 & 2 & 2 \\ 0 & 1 & 2 & 2 & 2 \\ 0 & 2 & 2 & 2 & 3 \end{bmatrix} \end{array}$$

R is numerated as follows:

$R = \{(0_1, 1_2), (0_1, 2_2), (0_1, 3_2), (1_1, 0_2), (2_1, 0_2), (3_1, 0_2), (4_1, 0_2), (1_1, 1_2), (1_1, 2_2)$
$\quad (2_1, 1_2), (3_1, 1_2), (4_1, 1_2), (2_1, 2_2), (3_1, 2_2), (4_1, 2_2), (1_1, 3_2), (2_1, 3_2), (3_1, 3_2), (4_1, 3_2)\}$

According to the H_c, the equivalence classes can be obtained as follows:

$R_0 = \{(0_1, 1_2), (0_1, 2_2), (0_1, 3_2), (1_1, 0_2), (2_1, 0_2), (3_1, 0_2), (4_1, 0_2), (1_1, 1_2)\}$
$R_1 = \{(1_1, 2_2), (2_1, 1_2)\}$
$R_2 = \{(3_1, 1_2), (4_1, 1_2), (2_1, 2_2), (3_1, 2_2), (4_1, 2_2), (1_1, 3_2), (2_1, 3_2), (3_1, 3_2)\}$
$R_3 = \{(4_1, 3_2)\}$

and $R = \bigcup_{i=0}^{3} R_i .$

9.1.2 Reliability Modeling

Now we are ready to derive the *pdf* and the system mean time to failure based on the state probabilities given in Section 9.1.1. First we derive the probability in each state. Initially, the system is in a brand-new state, *i.e.*, in state $M = f(R_M)$. The probability for state M is given by

$$P_t(M) = P_t(f(R_M))$$ (9.4)

As defined in Section 9.1.1, R_i represents all possible state combinations generating the system state i. The probability in state i is the union of all the elements in R_i:

$$P_t(i) = P\{f(R_i)\}$$ (9.5)

The probability for a catastrophic failure state F is given by

$$P_t(F) = P\{Y_1(t) \le G_1, Y_2(t) \le G_2, D(t) > S\} \tag{9.6}$$

The reliability $R(t)$ can be calculated as follows:

$$R(t) = P\{\text{system state} \ge 1\}$$

$$= \sum_{i=1}^{M} P_t(i) \tag{9.7}$$

where $P_t(t)$ is the probability in state i.
The mean time to failure is expressed as

$$E[T] = \int_0^\infty P\{T > t\}dt$$

Li and Pham (2005a) prove that

$$E[T] = \sum_{j=0}^{\infty} \frac{F_X^{(j)}(S)}{j!} \int_0^\infty P\{Y_1(t) \le G_1\} P\{Y_2(t) \le G_2\} (\lambda_2 t)^j e^{-\lambda_2 t} dt \tag{9.8}$$

Equation (9.8) shows that the mean time to failure $E(T)$ would depend on the expression of $P\{Y_1(t) \le G_1\} P\{Y_2(t) \le G_2\}$. The *pdf* of time to failure, $f_T(t)$ is, therefore, as follows:

$$f_T(t) = -\frac{d}{dt}\big[P\{T > t\}\big]$$

$$= -\frac{d}{dt}\bigg[P\{Y_1(t \le G_1\} P\{Y_2(t) \le G_2\} \sum_{j=0}^{\infty} \frac{(\lambda_2 t)^j e^{-\lambda_2 t}}{j!} F_X^{(j)}(S) \bigg] \tag{9.9}$$

9.1.3 Numerical Examples

Consider a system subjected to two degradation processes and random shocks. Assume that degradation process 1 is described as the function $Y_1(t) = A + Bg(t)$ where the random variables A and B are independent and both follow the normal distributions with mean 90 and variance 2.5, and mean 78 and variance 6, respectively, *i.e.*, $A \sim N(90,2.5)$, $B \sim N(78,6)$. The degradation function is assumed as $g(t) = t^3$. Suppose that critical threshold values: $G_1 = 2500$ and $W_3 = 1500$, $W_2 = 2000$, $W_1 = 2500$.

Assume that degradation process 2 is described by $Y_2(t) = W \cdot e^{BBt}/(AA + e^{BBt})$ where the random variables AA and BB are independent and follow the uniform distribution with interval [0,100] and exponential distribution with parameter 0.1 respectively: $AA \sim U[0,100]$, $BB \sim Exp(0.01)$. Assume critical values $G_2 = 5000$, $A_2 = 2600$, $A_1 = 500\,0$, and $W = 7000$.

Suppose that the random shock is represented by $D(t) = \sum_{i=0}^{N(t)} X_i$ with critical value $S = 200$, where $X_i \sim Exp(0.1)$ and X_is are *i.i.d.*

Assume that the states associated with degradation process 1 and degradation 2 are, respectively, $\Omega_1 = \{3_1, 2_1, 1_1, 0_1\}$ and $\Omega_2 = \{2_2, 1_2, 0_2\}$. We define the system state space as $\Omega_U = \{3, 2, 1, 0, F\}$ and the matrix H_c is given as follows:

$$
H_c = \begin{array}{c} \\ 0_2 \\ 1_2 \\ 2_2 \end{array}
\begin{array}{cccc} 0_1 & 1_1 & 2_1 & 3_1 \\ \end{array}
\left[\begin{array}{cccc}
\times & 0 & 0 & 0 \\
0 & 0 & 2 & 3 \\
0 & 1 & 2 & 3
\end{array} \right]
\tag{9.10}
$$

Then we obtain

$$R = \{(0_1, 1_2), (0_1, 2_2), (1_1, 0_2), (2_1, 0_2), (3_1, 0_2), (1_1, 1_2), (2_1, 1_2), (3_1, 1_2),$$
$$(1_1, 2_2), (2_1, 2_2), (3_1, 2_2)\}$$

The equivalence classes can be listed as follows:

$$R_0 = \{(0_1, 1_2), (0_1, 2_2), (1_1, 0_2), (2_1, 0_2), (3_1, 0_2), (1_1, 1_2)\}$$
$$R_1 = \{ (1_1, 2_2) \}$$
$$R_2 = \{(2_1, 1_2), (2_1, 2_2)\}$$
$$R_3 = \{(3_1, 1_2), (3_1, 2_2)\}$$

$$R = \sum_{i=0}^{3} R_i$$

According to the above H_c, the probability of the system in state 3 is the sum of the probability $f(3_1, 2_2)$ and of probability $f(3_1, 1_2)$ and is calculated as follow:

$$P_t(3) = P_t(f(R_3))$$

$$= \Phi\left(\frac{1500 - (90 + 78t)}{\sqrt{2.5 + 6t^6}} \right) \left\{ 1 - \frac{1}{100}(0.4)^{\frac{0.01}{t}} \left(\frac{t}{t - 0.01} \right) \left(0.01^{1-\frac{0.01}{t}} \right) \right\} \bullet$$

$$e^{-\lambda_2 t} \sum_{j=0}^{\infty} \left(\frac{\lambda_2 t}{j!} \right) F_X^{(j)}(200)$$

$$\tag{9.11}$$

Figure 9.5 shows the probability for the system in state 3 as a function of time t where the solid line represents the compound Poisson process $D(t) = \sum_{i=0}^{N(t)} X_i$ with rate $\lambda = .04$ and the dotted line represents the compound Poisson process with rate $\lambda = .8$. In Figure 9.5, we observe in this example that, as t reaches to 50 the system probability in state 3 quickly approaches to zero when the rate is given as $\lambda = .8$. and as a stable condition with $\lambda = .04$.

Since $R_2 = \{(2_1, 1_2), (2_1, 2_2)\}$, the probability in state 2 is given by

$$P_t(2) = P_t\{f(2_1,1_2)\} + P_t\{f(2_1,2_2)\}$$

$$= (UV)e^{-\lambda_2 t}\sum_{j=0}^{\infty}\left(\frac{\lambda_2 t}{j!}\right)F_X^{(j)}(200) \tag{9.12}$$

where $U = \Phi\left(\dfrac{2000 - (90 + 78t)}{\sqrt{2.5 + 6t^6}}\right) - \Phi\left(\dfrac{1500 - (90 + 78t)}{\sqrt{2.5 + 6t^6}}\right)$,

$$V = 1 - \frac{1}{100}\left(\frac{t}{t - 0.01}\right)(0.4)^{\frac{0.01}{t}}\left(\frac{t}{t - 0.01}\right)(0.01)^{1 - \frac{0.01}{t}}$$

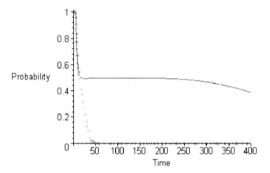

Figure 9.5. Probability plot for state 3 *vs.* time

Figure 9.6 shows the probability in state 2 as a function of time t where the solid line represents the compound Poisson process $D(t) = \sum_{i=0}^{N(t)} X_i$ with rate $\lambda = .04$, and the dotted line represents the compound Poisson process with rate $\lambda = .8$.

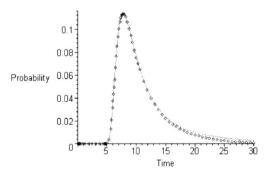

Figure 9.6. Probability plot for state 2 *vs.* time

From Figure 9.6 we observe that before the time t progresses to 5 the probability in state 2 stays close to zero for both the rate $\lambda = .8$. and $\lambda = .04$. It should be noted that the two curves are almost the same for rate $\lambda = .8$ and $\lambda = .04$. Similar observations are true for the probability in state 1 except that the curve rising starting point is 15 not 5. For details, see Li and Pham (2005b).

We can also easily obtain the probability in state 0 as follows:

$$P_t(0) = P\{f(0_1,1_2) + f(0_1,2_2) + f(1_1,0_2) + f(2_1,0_2) + f(3_1,0_2) + f(1_1,1_2)\}$$

$$= (X_1 Y_1 + X_2 Y_2 + X_3 Y_3) e^{-\lambda_2 t} \sum_{j=0}^{\infty} \left(\frac{\lambda_2 t}{j!} \right) F_X^{(j)}(200) \qquad (9.13)$$

where

$$X_1 = 1 - \Phi\left(\frac{2500 - (90 + 78t)}{\sqrt{2.5 + 6t^6}} \right),$$

$$X_2 = \Phi\left(\frac{2500 - (90 + 78t)}{\sqrt{25 + 6t^2}} \right)$$

$$Y_1 = 1 - \frac{1}{100}(0.4)^{\frac{0.01}{t}} \left(\frac{t}{t - 0.01} \right) \left(0.01^{1 - \frac{0.01}{t}} \right),$$

$$Y_2 = \frac{1}{100}(0.4)^{\frac{0.01}{t}} \left(\frac{t}{t - 0.01} \right) \left(0.01^{1 - \frac{0.01}{t}} \right),$$

$$X_3 = \Phi\left(\frac{2500 - (90 + 78t)}{\sqrt{2.5 + 6t^6}} \right) - \Phi\left(\frac{2000 - (90 + 78t)}{\sqrt{2.5 + 6t^6}} \right),$$

$$Y_3 = 1 - \frac{1}{100} \left[\left(\frac{22}{13} \right)^{\frac{0.01}{t}} + (0.4)^{\frac{0.01}{t}} \right] \left(\frac{t}{t - 0.01} \right) \left(0.01^{1 - \frac{0.01}{t}} \right)$$

Figure 9.7 shows the probability in state 0 $vs.$ time t where the solid line represents the compound Poisson process $D(t) = \sum_{i=0}^{N(t)} X_i$ with rate $\lambda = .04$, and the dotted line represents the compound Poisson process with rate $\lambda = .8$. From Figure 9.7, we observe that the probability in state 0 is almost close to zero as t reaches 100 or higher for the rate $\lambda = .8$.

The probability in state F is calculated as

$$P_t(F) = P\{Y_1(t) \leq G_1, Y_2(t) \leq G_2, D(t) > S\}$$

$$= KL\left\{ 1 - e^{-\lambda_2 t} \sum_{j=0}^{\infty} \left(\frac{\lambda_2 t}{j!} \right) F_X^{(j)}(200) \right\} \qquad (9.14)$$

where $K = \Phi\left(\frac{2500 - (90 + 78t)}{\sqrt{2.5 + 6t^6}} \right)$, $L = 1 - \frac{1}{100}(0.4)^{\frac{0.01}{t}} \left(\frac{t}{t - 0.01} \right) \left(0.01^{1 - \frac{0.01}{t}} \right)$

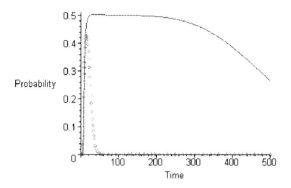

Figure 9.7. Probability plot for state 0

Figure 9.8 shows the probability in state F as a function of time t where the solid line represents the compound Poisson process $D(t) = \sum_{i=0}^{N(t)} X_i$ with rate $\lambda = .04$, and the dotted line represents the compound Poisson process with rate $\lambda = .8$. The two curves exhibit quite different shapes.

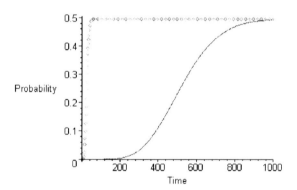

Figure 9.8. Probability plot for state F

Finally, the system reliability $R(t)$ is given by

$$R(t) = P\{\text{system state} \geq 1\}$$

$$= \sum_{i=1}^{3} P_t(i)$$

$$= X_3 Y_3 e^{-\lambda_2 t} \sum_{j=0}^{\infty} \left(\frac{\lambda_2 t}{j!} \right) F_X^{(j)}(200) \qquad (9.15)$$

where

$$X_3 = \Phi\left(\frac{2000-(90+78t)}{\sqrt{2.5+6t^6}}\right)\left\{1-\frac{1}{100}\left[(0.4)^{\frac{0.01}{t}}+\left(\frac{22}{13}\right)^{\frac{0.01}{t}}\right]\left(\frac{t}{t-0.01}\right)\left(0.01^{1-\frac{0.01}{t}}\right)\right\},$$

$$Y_3 = \left\{\Phi\left(\frac{2500-(90+78t)}{\sqrt{2.5+6t^6}}\right)-\Phi\left(\frac{2000-(90+78t)}{\sqrt{2.5+6t^6}}\right)\right\}$$

$$\times\left\{1-\frac{1}{100}\left(\frac{t}{t-0.01}\right)\left(\frac{22}{13}\right)^{\frac{0.01}{t}}0.01^{1-\frac{0.01}{t}}\right\}$$

Figure 9.9 shows the system reliability *vs.* time t where the solid line represents the compound Poisson process with rate $\lambda = .04$, and the dotted line represents the compound Poisson process with rate $\lambda = .8$. As for the rate $\lambda = .8$. we can see that the system likely will fail after time t equals to 50.

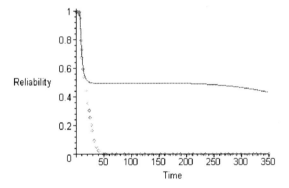

Figure 9.9. Reliability function

This section discusses a generalized model for evaluating the reliability of multi-state degraded systems, without maintenance and repair, considering two degradation processes and random shocks. The model can be used not only to determine the reliability of the degraded systems in the context of muli-state functions but also to obtain the states of the systems by calculating the system state probabilities. Next section will incorporate maintenance and repair strategies into the developed model and investigate the trade-off between the maintenance cost and reliability.

9.2 Optimal Inspection-maintenance

In this section, we will present a generalized condition-based maintenance model subject to multiple competing failure processes including two degradation processes and random shocks. An average long-run maintenance cost rate function is derived based on the expressions for the degradation paths and cumulative shock damage, which are measurable. A quasi-renewal process is employed to develop the inter-inspection sequence. Upon inspection, a decision will be made as to whether one needs to perform maintenance, either preventive or corrective, or to do nothing. The PM thresholds for degradation processes and inspection sequence are decision variables. This section also discusses an algorithm based on the Nelder-Mead downhill simplex method (Rardin 1998) to obtain the optimum solution minimizing the average long-run maintenance cost rate. Numerical examples are given to illustrate the results using the optimization algorithm.

In some production systems failures are not possible to detect but can only be determined by an inspection (Bris *et al*. 2003). Various inspection policies and models for systems with a degradation process have been proposed. Grall *et al*. (2002a) study a system subject to a random deterioration process. They develop a model based on a stationary process to determine both the PM threshold and inspection dates that minimized the average long-run cost rate.

Grall *et al*. (2002b) recently study the inspection-maintenance strategy for a single unit deteriorating system based on a Gamma process in which it has the stationary and independent increment property. They create the inspection-maintenance strategy consisting of PM threshold and inspection schedule that minimized the maintenance cost function based on regenerative and semi-regenerative properties.

Chelbi and Ait-Kadi (1999) address the optimal inspection strategies for deteriorating equipment subject to PM and CM. Klutke and Yang (2002) study the availability of maintained systems subject to both the effects of the degradation and random shocks. They considered the degradation process as a deterministic function of time *t* and the shocks occur according to a Poisson process in which the shock magnitudes are independent and identically distributed random variables.

This section considers systems with inspection-based maintenance subject to three failure processes that are competing for the life of such systems. Same as in Section 9.1, two of them are degradation processes (degradation process i measured by $Y_i(t)$ for $i = 1,2$) and the third is a random shock process measured by the function $D(t)$. The three processes are independent and any of them would cause the system to fail. The failure of the system is defined as when $Y_1(t) > G_1, Y_2(t) > G_2$ or $D(t) > S$ whichever occurs first. Note that the state of the systems can only be revealed through an inspection,

This section discusses optimal inspection-maintenance policies, consisting of the time sequence for inspection and PM threshold levels for both degradation processes, to determine trade-off between the failure frequency and system total cost.

The optimal inspection-maintenance policies in this section differ from others. First, a system with three competing processes instead of one is considered. Second, it is assumed that the degradation and shock damage are measurable. Otherwise there are some parameters associated with the processes that can be traced (such as vibration analysis for tool wear and failure). The maintenance decision is made based upon the threshold levels of the degradations and cumulative shock damage, not on the distribution parameters or transition probability as in other studies (Grall *et al.* 2002b; Chelbi and Ait-Kadi 1999). Third, the modeling method is other than Markov, semi-Markov and the stationary processes.

This section uses the two degradation path functions as follows:

i) Function $Y(t) = A + Bg(t)$: random-coefficient degradation path, where $A > 0$ and $B > 0$ are independent random variables and $g(t)$ is an increasing function of time t. The random variable A represents a measure of the initial degradation value and B represents the measure of the coefficient of the degradation.

ii) Function $Y(t) = We^{Bt}/(A + e^{Bt})$: the randomized logistic degradation path function, where A and B are independent non-negative random variables, and W is a constant. The random variable A represents the initial threshold level of degradation and B describes the rate at which degradation accumulates.

Note that logistic function $y(t) = e^{bt}/(a + e^{bt})$ for $a > 0$ and $b > 0$, as an *S*-shaped curve, for example, describes well the degradation process and it matches the path of the cumulative degradation of many systems in practice. The *S*-shaped curve reflects an initial run-in period of low usage, following by a period of steady rate of the usage, and finally ending with an increasing rate of use due to the aging of the system. Unlike most other work, the degradation path in this section is a stochastic process associated with the two random variables, not a deterministic. This section assumes the shock process is modeled according to a compound Poisson process $D(t) = \sum_{i=0}^{N(t)} X_i$ where X_is are *i.i.d.* and $N(t)$ follows a Poisson distribution with parameter λ_1: $N(t) \sim Poisson(\lambda_1)$.

Maintenance has evolved from simply the model which reacts to machine breakdowns, to the time-based model, and to today's condition-based model. This section considers condition-based maintenance where there are two possible maintenance actions: PM or CM. The need for PM or CM is determined upon each inspection and the inspection cycles is reduced according to a quasi-renewal process as the system ages. The system is inspected at times $I_1, ..., I_n$. Upon inspection, one of the following two choices has to be made:

i) Do nothing but determine the time for the next inspection.
ii) The system has failed and a maintenance (PM or CM) action is begun instantaneously.

Since the state of the system can only be determined through inspection, the determination of the inspection times $\{I_1, ..., I_i, ...\}$ and PM thresholds (L_1, L_2) will certainly make a great influence on the maintenance cost rate as well as the total system cost. This section uses a condition-based maintenance model to determine the optimal inspection schedule and PM thresholds (L_1, L_2) for complex repairable systems. Both the decision variables – inspection times and PM thresholds are important for trading off the cost between the maintenance (both PM and CM), inspection and the losses due to system idle.

The rest of Section 9.2 is organized as follows. The inspection-maintenance policy is described in Section 9.2.1. A new mathematical cost rate model is derived in Section 9.2.2. An optimization algorithm based on Nelder-Mead downhill simplex method is presented in Section 9.2.3. Numerical examples are given in Section 9.2.4.

9.2.1 A General Inspection-maintenance Policy

This section assumes:

i)   The system failure is only detected by each inspection. Inspections are assumed to be instantaneous, perfect and non-destructive. Since the system is not continuously monitored, if the system fails it will remain failed until the next inspection, which causes a loss of C_m per unit time. In that case, a maintenance action is begun instantaneously at the inspection's time.

ii) After a maintenance action, either PM or CM, the system state will become as good as new.

iii) A CM action will cost more than a PM action. Similarly, a PM action will cost much more than an inspection itself. This implies that $C_c > C_p > C_i$.

iv) The three processes: $Y_1(t), Y_2(t)$ and $D(t)$ are independent.

v)   No continuous monitoring is performed on the system.

vi) CM or PM time is negligible.

Although continuous monitoring process to some systems is feasible, the cost to monitor the process and the labor extensive would not make it realistic in practices. Therefore the criteria we consider in this section is to improve the system performance by performing periodic inspections with a maintenance action if necessary as the same token by minimizing the total system maintenance cost. Since deterioration while running leads to system failure, it proves to be better to assume that, as we take into account in this section, the degradation paths are continuous and increasing functions.

The length of the inspection will be reduced as the system ages. In other words, the intervals between successive inspections become shorter as the system ages. A quasi-renewal process is applied in this section to develop the inter-inspection sequence. Inspection time is constructed as $I_n = \sum_{j=1}^{n} \alpha^{j-1} I_1$, where $0 < \alpha \le 1$ and I_1 is the first inspection time. We define $U_n = I_n - I_{n-1} = \alpha^{n-1} I_1$ as the inter-inspection interval and $\{U_i\}_{i \in N}$ are a decreasing geometric sequence. According to the state detected at the inspection $I_n, n = 1,...,$ one of the following actions is taken:

i)   If both degradation values are below their PM thresholds and the shock damage value is less than its threshold, in other words, if $\{Y_1(I_n) \le L_1, Y_2(I_n) \le L_2\} \cap \{D(I_n) \le S\}$, then the system is still in a good condition. In this case, we do nothing but determine the next

inspection at $I_{n+1} = I_n + U_n$ where U_n is the inter-inspection time betwe-
en n^{th} and $(n+1)^{th}$ inspection interval.

ii)   If there is a degradation process that falls into the PM zone
$(L_i < Y_i(I_n) \le G_i, i = 1,2)$ and the other two processes are less than their
corresponding critical thresholds, then the system is called for a PM action
and it is instantaneously performed accordingly.

iii)   If any of the processes is exceeding its corresponding critical threshold
value $(Y_i(t) > G_i, i = 1,2, \text{ or } D(t) > S)$, then the system is called for a CM
action and it is instantaneously performed. In this case, the system has
failed and a CM is performed on the system.

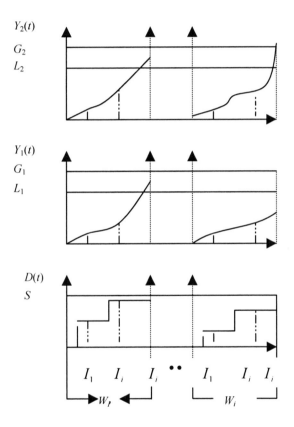

Figure 9.10. Evolution of the system condition

We assume that after a maintenance action, *i.e.*, PM or CM, the system will
again be as good as new. A new sequence of inspection begins which is defined in
the same way and the system maintenance follows the same decision rules outlined
above. Figure 9.10 shows the evolution of the system where $Y_1(t)$ and $Y_2(t)$

represent the degradation process 1 and process 2, respectively, and $D(t)$ represents a cumulative shock damage. $\{W_i\}_{i\in N}$ is a renewal sequence.

Figure 9.11 shows the maintenance zones on the $Y_1(t), Y_2(t)$ plane. G_i and L_i are the CM and PM critical thresholds for $Y_i(t)$, respectively, $i = 1,2$.

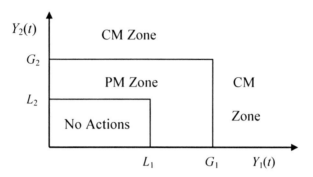

Figure 9.11. Maintenance zone projected on $Y_1(t), Y_2(t)$

9.2.2 Average Long-run Maintenance Cost Analysis

In this section we derive the average long-run maintenance cost per unit time, and next section will discuss minimizing the average long-run maintenance cost rate by determining the PM critical threshold values (L_1, L_2) and inspection sequence.

9.2.2.1 Expected Maintenance Cost in a Cycle
Suppose that the time horizon is infinity. From the basic renewal theory, we have

$$\lim_{t\to\infty} \frac{C(t)}{t} = \frac{E[C_1]}{E[W_1]}.$$

The expected total maintenance cost per cycle, $E[C_1]$, is given as

$$E[C_1] = C_i E[N_I] + C_p P_p + C_c P_c + C_m E[\zeta] \tag{9.16}$$

where C_i is the cost associated with each inspection, C_p is the cost associated with a PM action, and C_c is the CM action cost. Since failure is not self-announcing and it can be occurred at any given instant time T within the inspection time interval $[I_i, I_{i+1}]$, the system will remain idle during the interval $[T, I_{i+1}]$. The cost coefficient C_m is defined as the penalty cost per unit time associated with such an event.

1) Let $P\{N_I = i+1\}$ be the probability that there are a total of $(i+1)$ inspections in the cycle. The expected number of inspections during a cycle, $E[N_I]$, is given by

$$E[N_I] = \sum_{i=0}^{\infty}(i+1)P_{i+1} \tag{9.17}$$

where $P_{i+1} = P\{N_I = i+1\}$.

It can be obtained that

$$P_{i+1} = P\{N_I = i+1\}$$
$$= \bigcup_{j=1}^{17} P\{E_j^{(i+1)}\}$$

where $E_j^{(i+1)}$ ($j = 1,...,17$) denotes the renewal cycle ending at the jth possibility at time I_{i+1}:

$E_1^{(i+1)} = \{Y_1(I_i) \le L_1, Y_2(I_i) \le L_2, D(I_i) \le S\} \cap \{L_1 < Y_1(I_{i+1}) \le G_1, L_2 < Y_2(I_{i+1}) \le G_2,$
$\qquad D(I_{i+1}) \le S\}$

$E_2^{(i+1)} = \{Y_1(I_i) \le L_1, Y_2(I_i) \le L_2, D(I_i) \le S\} \cap \{L_1 < Y_1(I_{i+1}) \le G_1, Y_2(I_{i+1}) \le L_2,$
$\qquad D(I_{i+1}) \le S\}$

$E_3^{(i+1)} = \{Y_1(I_i) \le L_1, Y_2(I_i) \le L_2, D(I_i) \le S\} \cap \{Y_1(I_{i+1}) \le L_1, L_2 < Y_2(I_{i+1}) \le G_2,$
$\qquad D(I_{i+1}) \le S\}$

$E_4^{(i+1)} = \{Y_1(I_i) \le L_1, Y_2(I_i) \le L_2, D(I_i) \le S\} \cap \{L_1 < Y_1(I_{i+1}) \le G_1, Y_2(I_{i+1}) > G_2,$
$\qquad D(I_{i+1}) \le S\}$

$E_5^{(i+1)} = \{Y_1(I_i) \le L_1, Y_2(I_i) \le L_2, D(I_i) \le S\} \cap \{L_1 < Y_1(I_{i+1}) \le G_1,$
$\qquad L_2 < Y_2(I_{i+1}) \le G_2, D(I_{i+1}) > S\}$

$E_6^{(i+1)} = \{Y_1(I_i) \le L_1, Y_2(I_i) \le L_2, D(I_i) \le S\} \cap \{L_1 < Y_1(I_{i+1}) \le G_1, Y_2(I_{i+1}) > G_2,$
$\qquad D(I_{i+1}) > S\}$

$E_7^{(i+1)} = \{Y_1(I_i) \le L_1, Y_2(I_i) \le L_2, D(I_i) \le S\} \cap \{Y_1(I_{i+1}) > G_1, L_2 > Y_2(I_{i+1}),$
$\qquad D(I_{i+1}) \le S\}$

$E_8^{(i+1)} = \{Y_1(I_i) \le L_1, Y_2(I_i) \le L_2, D(I_i) \le S\} \cap \{Y_1(I_{i+1}) > G_1, L_2 < Y_2(I_{i+1}) \le G_2,$
$\qquad D(I_{i+1}) \le S\}$

$E_9^{(i+1)} = \{Y_1(I_i) \le L_1, Y_2(I_i) \le L_2, D(I_i) \le S\} \cap \{Y_1(I_{i+1}) > G_1, L_2 > Y_2(I_{i+1}),$
$\qquad D(I_{i+1}) > S\}$

$E_{10}^{(i+1)} = \{Y_1(I_i) \le L_1, Y_2(I_i) \le L_2, D(I_i) \le S\} \cap \{Y_1(I_{i+1}) > G_1, L_2 < Y_2(I_{i+1}) \le G_2,$
$\qquad D(I_{i+1}) > S\}$

$E_{11}^{(i+1)} = \{Y_1(I_i) \le L_1, Y_2(I_i) \le L_2, D(I_i) \le S\} \cap \{Y_1(I_{i+1}) > G_1, Y_2(I_{i+1}) > G_2,$
$\qquad D(I_{i+1}) \le S\}$

$$E_{12}^{(i+1)} = \{Y_1(I_i) \le L_1, Y_2(I_i) \le L_2, D(I_i) \le S\} \cap \{Y_1(I_{i+1}) > G_1, Y_2(I_{i+1}) > G_2,$$
$$D(I_{i+1}) > S\}$$

$$E_{13}^{(i+1)} = \{Y_1(I_i) \le L_1, Y_2(I_i) \le L_2, D(I_i) \le S\} \cap \{L_1 < Y_1(I_{i+1}) \le G_1, L_2 > Y_2(I_{i+1}),$$
$$D(I_{i+1}) > S\}$$

$$E_{14}^{(i+1)} = \{Y_1(I_i) \le L_1, Y_2(I_i) \le L_2, D(I_i) \le S\} \cap \{L_1 > Y_1(I_{i+1}), L_2 > Y_2(I_{i+1}),$$
$$D(I_{i+1}) \le S\}$$

$$E_{15}^{(i+1)} = \{Y_1(I_i) \le L_1, Y_2(I_i) \le L_2, D(I_i) \le S\} \cap \{L_1 > Y_1(I_{i+1}), L_2 > Y_2(I_{i+1}),$$
$$D(I_{i+1}) > S\}$$

$$E_{16}^{(i+1)} = \{Y_1(I_i) \le L_1, Y_2(I_i) \le L_2, D(I_i) \le S\} \cap \{L_1 > Y_1(I_{i+1}), L_2 < Y_2(I_{i+1}) \le G_2,$$
$$D(I_{i+1}) > S\}$$

$$E_{17}^{(i+1)} = \{Y_1(I_i) \le L_1, Y_2(I_i) \le L_2, D(I_i) \le S\} \cap \{L_1 > Y_1(I_{i+1}), Y_2(I_{i+1}) > G_2,$$
$$D(I_{i+1}) \le S\}$$

$$E_{18}^{(i+1)} = \{Y_1(I_i) \le L_1, Y_2(I_i) \le L_2, D(I_i) \le S\} \cap \{Y_1(I_{i+1}) \le L_1, Y_2(I_{i+1}) \le L_2,$$
$$D(I_{i+1}) \le S\}$$

Note that $E_j^{(i+1)}$ s are mutually disjoined events for $j = 1, ..., 17$.

There are a total of 18 system state combinations that can be revealed at any given interval $(I_i, I_{i+1}]$ where as there is only one state event $E_{18}^{(i+1)}$ representing the fact that the system is in a good condition and that no maintenance action will be required. Any other remaining state events will trigger either a PM or a CM action at time I_{i+1}.

After some simplifications, we have

$$P_{i+1} = P\{Y_1(I_i) \le L_1, Y_2(I_i) \le L_2, D(I_i) \le S\} - P\{Y_1(I_{i+1}) \le L_1, Y_2(I_{i+1}) \le L_2,$$
$$D(I_{i+1}) \le S\}$$

$$(9.18)$$

Therefore,

$$E[N_1] = \sum_{i=0}^{\infty} (i+1)\{P\{Y_1(I_i) \le L_1, Y_2(I_i) \le L_2, D(I_i) \le S\}$$
$$- P\{Y_1(I_{i+1}) \le L_1, Y_2(I_{i+1}) \le L_2, D(I_{i+1}) \le S\}\}$$

2) There will be either a PM or CM action ending a renewal cycle. It is obviously that the two events (PM and CM) are mutually exclusive at renewal time points, *i.e.*, $P_p + P_c = 1$. We now evaluate P_p as follows:

$$P_p = P\{\text{the cycle ends due to an PM action}\}$$

$$= \sum_{i=0}^{\infty} \sum_{j=1}^{3} P\{E_j^{(i+1)}\}$$

After some simplifications, we obtain

$$P_p = \sum_{i=0}^{\infty} \{ P\{Y_1(I_i) \leq L_1, L_1 < Y_1(I_{i+1}) \leq G_1\} P\{Y_2(I_i) \leq L_2, Y_2(I_{i+1}) \leq G_2\} P\{D(I_{i+1}\}$$

$$+ P\{Y_1(I_{i+1}) \leq L_1\} P\{Y_2(I_i) \leq L_2, L < Y_2(I_{i+1}) \leq G_2\} P\{D(I_{i+1})\}\}$$

$$(9.19)$$

and $P_c = 1 - P_p$.

We can obtain the joint probability density function $f_{Y(I_i),Y(I_{i+1})}(y_1, y_2)$ of $Y(I_i)$ and $Y(I_{i+1})$ by computing probabilities $P\{Y_1(I_i) \leq L_1, Y_1(I_{i+1}) \leq G_1\}$ and $P\{Y_2(I_i) \leq L_2, Y_2(I_{i+1}) \leq G_2\}$.

We consider two different degradation functions for $Y(t)$ as follows:

CASE 1: Assume $Y(t) = A + Bg(t)$ where $A > 0$ and $B > 0$ are two independent random variables and its corresponding *pdf* are $f_A(a)$ and $f_B(b)$, respectively, and $g(t)$ is an increasing function. Let

$$\begin{cases} y_1 = a + bg(I_i) \\ y_2 = a + bg(I_{i+1}) \end{cases}$$

After solving the above two equations in terms of y_1 and y_2, we have

$$a = \frac{y_1 g(I_{i+1}) - y_2 g(I_i)}{g(I_{i+1}) - g(I_i)} = h_1(y_1, y_2),$$

$$b = \frac{y_2 - y_1}{g(I_{i+1}) - g(I_i)} = h_2(y_1, y_2)$$

The *Jacobian J* is given by

$$J = \begin{vmatrix} \dfrac{\partial h_1}{\partial y_1} & \dfrac{\partial h_1}{\partial y_2} \\ \dfrac{\partial h_2}{\partial y_1} & \dfrac{\partial h_2}{\partial y_2} \end{vmatrix} = \begin{vmatrix} \dfrac{1}{g(I_i) - g(I_{i+1})} \end{vmatrix}$$

Therefore, the joint continuous *pdf* of a random vector $(Y(I_i), Y(I_{i+1}))$ can be calculated as follows:

$$f_{Y(I_i),Y(I_{i+1})}(y_1, y_2) = |J| f_A(h_1(y_1, y_2)) f_B(h_2(y_1, y_2))$$

$$(9.20)$$

CASE 2: Suppose $Y(t) = We^{At}/(B + e^{At})$ where $A > 0$ and $B > 0$ are two independent *r.v.*s and its corresponding *pdf* are $f_A(a)$ and $f_B(b)$, respectively. Let

$$\begin{cases} y_1 = \dfrac{We^{aI_i}}{b + e^{aI_i}} \\[2mm] y_2 = \dfrac{We^{aI_{i+1}}}{b + e^{aI_{i+1}}} \end{cases}$$

The solutions for a and b can be easily solved from the above equations in terms of y_1 and y_2 as follows:

$$\begin{cases} a = \dfrac{\ln\left(\dfrac{y_2(y_1 - W)}{y_1(y_2 - W)}\right)}{I_{i+1} - I_i} = h_1(y_1, y_2), \text{ where } y_2 \neq W \\[6mm] b = -\dfrac{(y_2 - W)}{y_2} \exp\left(\dfrac{\ln\left(\dfrac{y_2(y_1 - W)}{y_1(y_2 - W)}\right)I_{i+1}}{I_{i+1} - I_i}\right) = h_2(y_1, y_2), \text{ where } y_2 \neq W \end{cases}$$

Similarly, the joint continuous *pdf* of random vector $(Y(I_i), Y(I_{i+1}))$ can be obtained:

$$f_{Y(I_i),Y(I_{i+1})}(y_1, y_2) = |J| f_A(h_1(y_1, y_2)) f_B(h_2(y_1, y_2)) \tag{9.21}$$

where J is given by

$$J = \frac{y_1(y_2 - W)\left(\dfrac{y_2}{y_1(y_2 - W)} - \dfrac{y_2(y_1 - W)}{y_1^2(y_2 - W)}\right)(-d(y_1, y_2) - d_1(y_1, y_2) + d_2(y_1, y_2))}{y_2(y_1 - W)(I_{i+1} - I_i)}$$
$$+ d_3(y_1, y_2)$$

where

$$d(y_1, y_2) = \frac{\left(\dfrac{y_1 - W}{y_1(y_2 - W)} - \dfrac{y_2(y_1 - W)}{y_1(y_2 - W)^2}\right) y_1(y_2 - W)^2 I_{i+1} \, e^{\frac{\ln\left(\frac{y_2(y_1-W)}{y_1(y_2-W)}\right)I_{i+1}}{I_{i+1} - I_i}}}{y_2^2(y_1 - W)(I_{i+1} - I_i)},$$

where $y_1 \neq W, y_2 \neq W$

$$d_1(y_1,y_2) = \frac{1}{y_2}\exp\left[\frac{\ln\left(\frac{y_2(y_1-W)}{y_1(y_2-W)}\right)I_{i+1}}{I_{i+1}-I_i}\right], \quad y_2 \neq W,$$

$$d_2(y_1,y_2) = \frac{(y_2-W)\exp\left[\frac{\ln\left(\frac{y_2(y_1-W)}{y_1(y_2-W)}\right)I_{i+1}}{I_{i+1}-I_i}\right]}{y_2^2}, \quad y_2 \neq W,$$

$$d_3(y_1,y_2) = \frac{d_{31}(y_1,y_2)d_{32}(y_1,y_2)}{y_2^3(y_1-W)^2(I_{i+1}-I_i)^2}, \quad y_1 \neq W,$$

$$d_{31}(y_1,y_2) = \left(\frac{y_1-W}{y_1(y_2-W)} - \frac{y_2(y_1-W)}{y_1(y_2-W)^2}\right)y_1^2(y_2-W)^3, \quad y_2 \neq W,$$

$$d_{32}(y_1,y_2) = \left(\frac{y_2}{y_1(y_2-W)} - \frac{y_2(y_1-W)}{y_1^2(y_2-W)}\right)I_{i+1}\cdot\exp\left[\frac{\ln\left(\frac{y_2(y_1-W)}{y_1(y_2-W)}\right)I_{i+1}}{I_{i+1}-I_i}\right]$$

where $y_2 \neq W$,

3) Let T denote the time to system failure, *i.e.*, mathematically, $T = \inf\{t : Y_1(t) > G_1, Y_2(t) > G_2$ or $D(t) > S\}$. If $I_i < T \leq I_{i+1}$, the unit will be idle during the interval $[T, I_{i+1}]$. Let $E[\zeta]$ denote the average idle time between the failure occurrence epoch and its inspection during the cycle. Li and Pham (2005a) obtain that

$$E[\xi] = \sum_{i=0}^{\infty} E[(I_{i+1}-T)I_{I_i<T\leq I_{i+1}}]$$

$$= \sum_{j=0}^{\infty} R_j \int_{I_i}^{I_{i+1}} (I_{i+1}-t)dF_T(t)$$

(9.22)

where

$$R_j = \{P\{Y_1(I_i)\leq L_1, L_1 < Y_1(I_{i+1})\leq G_1\}P\{Y_2(I_i)\leq L_2, L_1 < Y_1(I_{i+1})\}$$
$$+ P\{Y_2(I_i)\leq L_2\}P\{Y_1(I_i)\leq L_1, Y_1(I_{i+1})> G_1\} + P\{Y_1(I_{i+1})\leq L_1\}$$
$$P\{Y_2(I_i)\leq L_2\}\}P\{D(I_i)\leq S\}$$

$$F(t) = P\{Y_1(t) > G_1, Y_2(t) \le G_2, D(I_i) \le S\} + P\{Y_1(t) \le G_1, Y_2(t) > G_2, D(I_i) \le S\}$$
$$+ P\{Y_1(t) \le G_1, Y_2(t) \le G_2, D(I_i) > S\}$$

and $I_{I_i < T \le I_{i+1}}$ is an indicator function.

9.2.2.2 Expected Cycle Length

The expected cycle length $E[W_1]$ is given as follows:

$$E[W_1] = E[E[W_1 \mid N_I]]$$

$$= \sum_{i=0}^{\infty} E[W_1 \mid N_I = i] P\{N_i = i\}$$

$$= \sum_{i=0}^{\infty} I_{i+1} P_{i+1}$$

(9.23)

where P_{i+1} is given in Equation (9.18).

Therefore, the average long-run maintenance cost rate function $EC(L_1, L_2, I_1)$:

$$EC(L_1, L_2, I_1) = \frac{E[C_1]}{E[W_1]}$$

is a function of the inspection times $\{I_1, ... I_i, ...\}$ and the PM critical threshold values (L_1, L_2) through functions P_p, P_c, $E[N_I]$, $E[\zeta]$ and $E[W_1]$. It can be obtained by solving the two functions given in Equations (9.16) and (9.23).

9.2.3 Algorithms for Optimal Inspection-maintenance Policy

This section will discuss a step-by-step algorithm based on the Nelder-Mead downhill simplex method (Rardin 1998) to obtain the optimum decision variables (I_1, L_1, L_2) such that the long-run average system maintenance cost rate $EC(L_1, L_2, I_1)$ is minimized. Note that the inspection sequence $\{I_1, ... I_i, ...\}$, where $I_n = \sum_{j=1}^{n} \alpha^{j-1} I_1$, depends on I_1 for given α. Mathematically, the optimization problem for the maintenance cost rate can be formulated as follows:

Optimization Problem:

Find I_1, L_1 and L_2 $(0 < L_1 \le G_1, 0 < L_2 \le G_2)$ to minimize

$$EC(L_1, L_2, I_1) = \left\{ C_i \sum_{i=0}^{\infty} (i+1) \left[P\{Y_1(\sum_{j=1}^{i} \alpha^{j-1} I_1) \le L_1, Y_2(\sum_{j=1}^{i} \alpha^{j-1} I_1) \le L_2, D(\sum_{j=1}^{i} \alpha^{j-1} I_1) \le S\} \right. \right.$$

$$\left. \left. - P\{Y_1(\sum_{j=1}^{i+1} \alpha^{j-1} I_1) \le L_1, Y_2(\sum_{j=1}^{i+1} \alpha^{j-1} I_1) \le L_2, D(\sum_{j=1}^{i+1} \alpha^{j-1} I_1) \le S \right] \right.$$

$$+ C_p \sum_{i=0}^{\infty} \left[P\{Y_1(\sum_{j=1}^{i}\alpha^{j-1}I_1) \le L_1, Y_2(\sum_{j=1}^{i}\alpha^{j-1}I_1) \le G_2\} \right.$$

$$\left. \times P\{Y_2(\sum_{j=1}^{i}\alpha^{j-1}I_1) \le L_2, Y_2(\sum_{j=1}^{i+1}\alpha^{j-1}I_1) \le G_2\} P\{D(\sum_{j=1}^{i+1}\alpha^{j-1}I_1) \le S\} \right]$$

$$+ C_c \left[1 - \sum_{i=0}^{\infty} \left(P\{Y_1(\sum_{j=1}^{i}\alpha^{j-1}I_1) \le L_1, Y_2(\sum_{j=1}^{i}\alpha^{j-1}I_1) \le G_2\} \right. \right.$$

$$\left. \left. \times P\{Y_2(\sum_{j=1}^{i}\alpha^{j-1}I_1) \le L_2, Y_2(\sum_{j=1}^{i+1}\alpha^{j-1}I_1) \le G_2\} P\{D(\sum_{j=1}^{i+1}\alpha^{j-1}I_1) \le S\} \right) \right]$$

$$+ C_m \sum_{i=0}^{\infty} \left[(R_{1i} + R_{2i} + R_{3i}) P\{D(\sum_{j=1}^{i+1}\alpha^{j-1}I_1) \le S\} \right) \int_{\sum_{j=1}^{i}\alpha^{j-1}I_1}^{\sum_{j=1}^{i+1}\alpha^{j-1}I_1} \left(\sum_{j=1}^{i+1}\alpha^{j-1}I_1 - t \right) dF_T(t) \right\}$$

$$\left/ \left\{ \sum_{i=0}^{\infty} \left(\sum_{j=1}^{i+1}\alpha^{j-1}I_1 \right) \left(P\{Y_1(\sum_{j=1}^{i}\alpha^{j-1}I_1) \le L_1, Y_2(\sum_{j=1}^{i}\alpha^{j-1}I_1) \le L_2, D(\sum_{j=1}^{i}\alpha^{j-1}I_1) \le S\} \right. \right. \right.$$

$$\left. \left. \left. - P\{Y_1(\sum_{j=1}^{i+1}\alpha^{j-1}I_1) \le L_1, Y_2(\sum_{j=1}^{i+1}\alpha^{j-1}I_1) \le L_2, D(\sum_{j=1}^{i+1}\alpha^{j-1}I_1) \le S\} \right) \right\}$$

where

$$R_{1i} = P\{Y_1(\sum_{j=1}^{i}\alpha^{j-1}I_1) \le L_1, L_1 < Y_1(\sum_{j=1}^{i+1}\alpha^{j-1}I_1) \le G_1\}$$

$$\times P\{Y_2(\sum_{j=1}^{i}\alpha^{j-1}I_1) \le L_2, L_1 < Y_1(\sum_{j=1}^{i+1}\alpha^{j-1}I_1)\}$$

$$R_{2i} = P\{Y_1(\sum_{j=1}^{i}\alpha^{j-1}I_1) \le L_1, G_1 < Y_1(\sum_{j=1}^{i+1}\alpha^{j-1}I_1)\} P\{Y_2(\sum_{j=1}^{i}\alpha^{j-1}I_1) \le L_2\}$$

$$R_{3i} = P\{Y_1(\sum_{j=1}^{i}\alpha^{j-1}I_1) \le L_1\} P\{Y_2(\sum_{j=1}^{i}\alpha^{j-1}I_1) \le L_2\}$$

The objective function above is a complex nonlinear function. Nelder-Mead downhill simplex method (Rardin 1998) is one of the most popular direct search methods for obtaining the optimum solution of unconstrained nonlinear function, which does not require the calculation of derivatives.

We use the Nelder-Mead method based on the comparison of the function values at the $(n+1)$ vertices for n-dimensional decision variables. Each iteration will generate a new vertex for the simplex. If this new point is better than at least one of the existing vertices, it replaces the worst vertex. The simplex vertices are changed through reflection, expansion and contraction operations in order to find an improving solution. The step-by-step algorithm based on Nelder-Mead downhill simplex method is given as follows:

Step 1: *Choose (n+1) distinct vertices as an initial set* $\{Z^{(1)},...,Z^{(n+1)}\}$. *Then calculate*

function value $f(Z)$ *for* $i = 1,2,...,(n+1)$ *where* $f(Z) = EC(L_1, L_2, I_1)$. *Putting* *the values $f(Z)$ in an increasing order where $f(Z^{(1)}) = \min\{EC(L_1, L_2, I_1)\}$ and $f(Z^{(n+1)}) = \max\{EC(L_1, L_2, I_1)\}$. Set $k = 0$.*

Step 2: *Compute the best-n centroid* $X^{(k)} = \dfrac{1}{n}\sum_{i=1}^{n} Z^{(i)}$.

Step 3: *Use the centroid $X^{(k)}$ in Step 2 to compute away-from-worst move direction*

$$\Delta X^{(k+1)} = X^{(k)} - Z^{(n+1)}.$$

Step 4: *Set $\lambda = 1$ and compute $f(X^{(k)} + \lambda\Delta X^{(k+1)})$.*

If $f(X^{(k)} + \lambda\Delta X^{(k+1)}) \le f(Z^{(1)})$ then go to Step 5.

Otherwise, if $f(X^{(k)} + \lambda\Delta X^{(k+1)}) \ge f(Z^{(n)})$ then go to Step 6.

Else, fix $\lambda = 1$ and go to Step 8.

Step 5: *Set $\lambda = 2$ and compute $f(X^{(k)} + 2\Delta X^{(k+1)})$.*

If $f(X^{(k)} + 2\Delta X^{(k+1)}) \le f(X^{(k)} + \Delta X^{(k+1)})$ then set $\lambda = 2$.

Otherwise set $\lambda = 1$.

Then go to Step 8.

Step 6: *If $f(X^{(k)} + \lambda\Delta X^{(k+1)}) \le f(Z^{(n+1)})$ then set $\lambda = 1/2$. Compute*

$$f(X^{(k)} + \frac{1}{2}\Delta X^{(k+1)}).$$

If $f(X^{(k)} + \frac{1}{2}\Delta X^{(k+1)}) \le f(Z^{(n+1)})$ then set $\lambda = 1/2$ and go to Step 8.

Otherwise, set $\lambda = -1/2$ and if $f(X^{(k)} - \frac{1}{2}\Delta X^{(k+1)}) \le f(Z^{(n+1)})$ then set

$\lambda = -1/2$ *and go to Step 8. Otherwise, go to Step 7.*

Step 7: *Shrinking the current solution set toward best $Z^{(1)}$ by*

$$Z^{(i)} = \frac{1}{2}(Z^{(1)} + Z^{(i)}), i = 2,...,n+1. \text{ Compute the new } f(Z^{(2)}),...,f(Z^{(n+1)}),$$

let $k = k+1$, and return to Step 2.

Step 8: *Replace the worst $Z^{(n+1)}$ by $X^{(k)} + \lambda\Delta X^{(k+1)}$. If $\sqrt{\dfrac{1}{n+1}\sum_{i=1}^{n+1}[f(Z^{(i)}) - \bar{f}]^2} < 0.5$,*

where $\bar{f}$ is an average value, then STOP. Otherwise, let $k = k+1$ and return to Step 2.

Note that the criterion in Step 8 is not unique but will depend on how you would like the algorithm to stop when the vertices function values are close. In this section, reference value is when the difference between the maximum and the minimum values of f is less than 0.5.

9.2.4 Numerical Example

Assume degradation process 1 is described by function $Y_1(t) = We^{B_1 t}/(A_1 + e^{B_1 t})$ where the random variables A_1 and B_1 are independent and follow the uniform distribution with parameter interval [0,40] and exponential distribution with parameter 1, respectively, $i.e.$, $A_1 \sim U[0,40]$ and $B_1 \sim Exp(1)$ Degradation process 2 is modeled as $Y_2(t) = A_2 + B_2 g(t)$ where $A_2 \sim U[0,2]$, $B_2 \sim Exp(0.2)$ and $g(t) = \sqrt{t}e^{0.01t}$. Suppose that the random shock is represented by the function $D(t) = \sum_{i=0}^{N_2(t)} X_i$, where $X_i \sim Exp(.04)$ and $N(t) \sim Poisson(.1)$. Critical threshold values are $G_1 = 300$, $G_2 = 70$ and $S = 100$.

Suppose the cost parameters are given as follows: $C_c = 560$ units/CM, $C_p = 400$ units/PM, $C_i = 100$ units/inspection, $C_m = 500$ units/unit time, and $\alpha = 0.97$.

The inspection sequence $\{I_1, ..., I_n, ...\}$ is per $I_n = \sum_{j=1}^{n} \alpha^{j-1} I_1$. The objective is to determine the values of I_1, L_1, and L_2 so that the average long-run maintenance cost rate per unit time is minimized.

Following are the computing details for this example using the optimization algorithm in Section 9.2.3:

Step 1: There are three decision variables: L_1, L_2, I_1, and so we need $(n+1) = 4$ distinct vertices as an initial set of values which are

$Z^{(1)} = (270,56,76)$, $Z^{(2)} = (280,60,72)$, $Z^{(3)} = (290,52,66)$ and $Z^{(4)} = (300,50,57)$. Set $k=0$.

Then calculate the function value $f(Z)$ corresponding to each vertices and put them in an increasing order of the objective value $EC(L_1, L_2, I_1)$ from smallest to highest.

Step 2: Compute the centroid: $X^{(0)} = \frac{1}{3}(Z^{(1)} + Z^{(2)} + Z^{(3)}) = (280,56,71.3)$.

Step 3: Search for away-from-worst direction: $\Delta X = X^{(0)} - Z^{(4)} = (-20,6,14.3)$.

Step 4: Set $\lambda = 1$; it will generate a new minimal $EC(260,60,85.6) = 291.9$ which leads to try an expansion with $\lambda = 2$ that is $(240,60,99.9)$.

Step 5: Set $\lambda = 2$. Similarly, compute $f(Z)$ that leads to 247.9. Go to Step 8.

This result turns out to be a better solution, hence $(300,50,57)$ is replaced by $(240,60,99.9)$. The iteration continues and stops at $k = 4$ (see Table 9.1) since

$$\sqrt{\frac{1}{4}\sum_{i=1}^{4}\left[EC(Z^{(i)}) - \overline{EC(L_1, L_2, I_1)}\right]^2} = 0.449 < 0.5$$

where $\overline{EC(L_1, L_2, I_1)}$ is the average value.

Table 9.1. Nelder-Mead algorithm results

k	$Z^{(1)} = (L_1, L_2, I_1)$	$Z^{(2)}$	$Z^{(3)}$	$Z^{(4)}$	Search results
0	$(270,56,76)$ $\frac{E[C_1]}{E[W_1]} = 300.7$	$(280,60,72)$ $\frac{E[C_1]}{E[W_1]} = 332.2$	$(290,52,66)$ $\frac{E[C_1]}{E[W_1]} = 360.4$	$(300,50,57)$ $\frac{E[C_1]}{E[W_1]} = 388.2$	$\lambda = 2$ $\frac{E[C_1]}{E[W_1]} = 247.9$
1	$(240,60,99.9)$ $\frac{E[C_1]}{E[W_1]} = 247.9$	$(270,56,76)$ $\frac{E[C_1]}{E[W_1]} = 300.7$	$(280,60,72)$ $\frac{E[C_1]}{E[W_1]} = 332.2$	$(290,52,66)$ $\frac{E[C_1]}{E[W_1]} = 360.4$	$\lambda = 1$ $\frac{E[C_1]}{E[W_1]} = 248.0$
2	$(236,60,99.2)$ $\frac{E[C_1]}{E[W_1]} = 247.9$	$(240,60,99.9)$ $\frac{E[C_1]}{E[W_1]} = 248.0$	$(270,56,76)$ $\frac{E[C_1]}{E[W_1]} = 300.7$	$(280,60,72)$ $\frac{E[C_1]}{E[W_1]} = 332.2$	$\lambda = 2$ $\frac{E[C_1]}{E[W_1]} = 246.7$
3	$(187,56,131)$ $\frac{E[C_1]}{E[W_1]} = 246.7$	$(236,60,99.2)$ $\frac{E[C_1]}{E[W_1]} = 247.9$	$(240,60,99.9)$ $\frac{E[C_1]}{E[W_1]} = 248.0$	$(270,56,76)$ $\frac{E[C_1]}{E[W_1]} = 300.7$	$\lambda = 1$ $\frac{E[C_1]}{E[W_1]} = 245.9$
4	$\mathbf{(172,60,144)}$ $\frac{E[C_1]}{E[W_1]} = 245.9$	$(187,56,131)$ $\frac{E[C_1]}{E[W_1]} = 246.7$	$(236,60,99.2)$ $\frac{E[C_1]}{E[W_1]} = 247.9$	$(240,60,99.9)$ $\frac{E[C_1]}{E[W_1]} = 248.0$	Stop

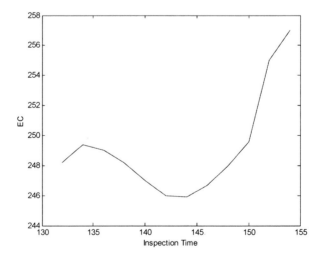

Figure 9.12. $EC(L_1, L_2, I_1)$ vs. I_1

Table 9.1 shows the optimal solution for (L_1, L_2, I_1): $L_1^* = 172$, $L_2^* = 60$, $I_1^* = 144$ and the corresponding minimum average long-run maintenance cost rate is $EC(L_1^*, L_2^*, I_1^*) = 245.9$. Figure 9.12 depicts the average long-run maintenance cost rate curve $EC(L_1, L_2, I_1)$ as a function of the inspection time interval I_1 given that $L_1 = 172$ and $L_2 = 60$.

Table 9.2 exhibits a sensitivity analysis in terms of the probability that the cycle will end due to a PM action, P_p, for various values of (L_1, L_2) for given $\alpha = 0.97$ and $I_1 = 144$. Table 9.2 shows that probability P_p slightly increases as both L_1 and L_2 decrease. This suggests that one would perform more PMs than CMs when L_1 and L_2 are both getting smaller.

Table 9.2. The effect of (L_1, L_2) on P_p for a given inspection sequence

L_1	L_2	P_p
200	60	.5910
190	58	.5928
180	56	.5936
170	54	.5948
160	52	.5950
150	50	.5968

Similarly, Table 9.3 shows the probability that the cycle will end due to a PM action, P_p, for various values of I_1 given $L_1 = 172, L_2 = 60$ and $\alpha = 0.97$. From Table 9.3, we observe that the probability P_p decreases as I_1 increases. In other words, the maintenance cycle likely will be ended due to a CM rather than a PM if one should delay the inspection. This result can help the maintenance managers or inspectors to allocate resources and time.

Table 9.3. The effect of inspection sequence on P_p for fixed PM values

I_1	P_p
110	.642
120	.610
130	.578
140	.510
150	.480
160	.430

This section has presented a generalized inspection-maintenance model with multiple competing processes based on different degradation paths and cumulative shock damage. A step-by-step algorithm based on Nelder-Mead downhill simplex method is presented to obtain the optimum decision variables that minimize the long-run average maintenance system cost rate. The decision variables are the inspection sequences as well as the PM critical threshold values.

The models in this section can be used to help the maintenance managers and inspectors in particular and marketing managers in general to allocate the resources as well as promotion strategies for the new products.

10

Warranty Cost Models with Dependence and Imperfect Repair

Today most equipment offers some level of warranty, and warranty cost could be a significant portion of overall product cost (Wang *et al.* 2004). Warranty is an obligation attached to products that requires the warranty issuers (manufacturers or sellers) to provide compensation for consumers (buyers) according to the warranty terms when the warranted products fail to perform their pre-specified functions under normal usage within the warranty coverage period, per Blischke and Murthy (1996). It can be considered as a contractual agreement between buyer and seller which becomes effective upon the sale of the warranted products.

A warranty contract should contain at least three characteristics: the coverage period (fixed or random), the method of compensations, and the conditions under which such compensations would be offered. The last characteristic is closely related to warranty execution since it clarifies consumers' rights and protects warranty issuers from excessive false claims. From the cost perspective, the first two characteristics are more important to manufacturers because they determine the depth of the protection against premature failures and the direct cost related to those failures.

From the manufacturer's point of view, one of the main roles of warranty is protection (Blischke and Murthy 1996). Warranty terms may, and often do, specify the use and conditions of use for which the product is intended and provide for limited coverage or no coverage at all in the event of misuse of the product. A second important purpose of warranty for the manufacturer is promotion. Since buyers often infer a more reliable product when a longer warranty is offered, warranty has been used as an effective advertising tool. This is often particularly important when marketing new and innovative products, which may be viewed with a degree of uncertainty by many potential consumers. In addition, warranty has become an instrument, similar to product performance and price, used in competition with competitors.

From the consumers' point of view, warranty provides protection against low-quality products. It helps consumers minimize the financial risk of using warranted products under the normal usage conditions specified in warranty contracts. Also warranty is informative since it serves as a signal of product quality and reliability.

However, the link between warranty and product quality may not always be strong since manufacturers sometimes use it so aggressively as a marketing tool such that even relative inferior products may carry a generous' warranty.

One of the basic characteristics of warranties is whether they are renewable or not. For a regular renewable policy with warranty period w, whenever a product fails within w, the buyer is compensated according to the terms of the warranty contract and the warranty policy is renewed for another period w. As a result, a warranty cycle T, starting from the date of sale, ending at the warranty expiration date, is a random variable whose value depends on w, the total number of failures under the warranty, and the actual failure inter-arrival times. Renewable warranties are often offered for inexpensive, non-repairable consumer electronic products such as a microwave, a coffee maker, and so forth, either implicitly or explicitly. One should note that theoretically the warranty cycle for a renewable policy could be arbitrarily large. This might be one of the reasons why such policies are not as popular as non-renewable ones for warranty issuers. The majority of warranties in the market are non-renewable for which the warranty cycle, which is the same as the warranty period, is not random, but pre-determined (fixed) since the warranty obligation will be terminated as soon as w units of time passes after the sale. These types of policies are also known as fixed period warranties.

In this chapter, we discuss warranty cost models of repairable complex systems from manufacturers' point of view by considering a comprehensive set of warranty cost factors, such as warranty policies, system reliability structure, product failure mechanism, warranty service cost, impact of warranty service, value of time, warranty service time, and warranty claim related factors, under three types of existing warranty policies: free repair warranty (FRPW), free replacement warranty (FRW), and pro-rata warranty (PRW), and two new warranty policies: renewable full service warranty (RFSW) and repair-limit risk-free warranty (RLRFW), based on Bai and Pham (2004, 2005, 2006a).

A traditional way to model warranty cost for warranted multi-component systems is the black-box approach that does not utilize the information of system structure or architecture. By studying the RFSW policies with explicit consideration of four types of systems: series, parallel, series-parallel (s-p) and parallel-series (p-s), this chapter demonstrates the importance of system reliability architecture information.

Previous research has studied the discounted warranty cost (DWC), a measure incorporating the value of time, for black-box systems under various warranty policies. This chapter presents models to extend the previous studies by deriving the moments of DWC for minimally repaired series systems with an FRPW or a PRW while considering both continuous and discrete time discounting. The optimal warranty reserve level and the optimal warranty duration are obtained for lot sales.

Most studies on warranties of repairable products assume perfect repair. This chapter models warranty cost for RLRFW policies given minimal repair or imperfect repair. The exact expressions of the expectation and variance of warranty cost are obtained based on truncated quasi-renewal processes and truncated non-homogeneous Poisson processes.

Due to the random nature of warranty cost factors such as product failure times, warranty cost is also a random variable whose statistical behavior can be determined by establishing mathematical links between warranty factors and warranty cost. There are many factors that may affect warranty cost, and this chapter considers the following factors:

i)   Characteristics of warranty policies
ii)   Warranty service cost per failure
iii)   Product failure mechanism
iv)   Impact of warranty service on product reliability
v)   Warranty service time
vi)   Warranty claim related factors

The following acronyms and notations will be used in the chapter:

ACRONYMS

cdf	Cumulative distribution function
CMW	Combination warranty
DWC	Discounted warranty cost
EWC	Expected warranty cost
FRPW	Free repair warranty
FRW	Free replacement warranty
FSW	Full service warranty
NHPP	Non-homogeneous Poisson process
pdf	Probability density function
PM	Preventive maintenance
pmf	Probability mass function
p-s	Parallel-series
PRW	Pro-rata warranty
PV	Present value
RCLW	Repair-cost-limit warranty
RFSW	Renewable full service warranty
RLRFW	Repair-limit risk-free warranty
RNLW	Repair-number-limit warranty
RRLRFW	Renewable repair-limit risk-free warranty
RTLW	Repair-time-limit warranty
r.v.	Random variables
s-p	Series-parallel
WPD	Warranty period determination
WRD	Warranty reserve determination

NOTATION

w	Length of a warranty period ($w > 0$)
m	Upper limit of the number of repairs under warranty
$N_a(w), N_b(w)$	Number of free repairs and replacements within w respectively
c_a, c_b	unit repair and replacement cost respectively, both constant

T_p, t_p	Pivot points. Capital letter indicates a *r.v.*
X_i	Inter-occurrence times of a truncated quasi-renewal process
F, f	*cdf* and *pdf* of the first failure time of a new product
F_i, f_i	*cdf* and *pdf* of the ith failure time of a (truncated) quasi-renewal process
α	Parameter for a (truncated) quasi-renewal process
$G^{(n)}$	*cdf* of the nth occurrence time of a (truncated) quasi-renewal process
$C(w)$	Warranty cost per product sold with warranty duration of w
$M_q(\cdot), M_{q,2}(\cdot)$	First and second moments of a truncated quasi-renewal process
$M_d(\cdot), M_{d,2}(\cdot)$	First and second moments of a delayed renewal process
$N(w)$	Number of system failures within w
$N_i(w)$	Number of failures of component i within w
$\lambda_i(\cdot), \Lambda_i(\cdot)$	Failure intensity function and accumulative failure intensity Function of component i
$H(\cdot)$	Discounting function
δ	Discount rate
$C_{d_1}(w)$	Discounted warranty costs per product sold under continuous discounting
$C_{d_2}(w)$	Discounted warranty costs per product sold under discrete discounting
q	Number of components in a system
c_i	Repair cost per failure of component i
S_{ij}	j^{th} failure time of component i
L	Size of a single lot sale
TC_{wr}	Required warranty reserve level for a single lot sale
c_0	Warranty budget level, a predetermined constant
Θ	Set of warranty parameters, $\Theta = \{\theta_1, \theta2, \cdots, \theta_n\}$
$\Psi(\Theta)$	A warranty policy with parameters in Θ
B	Consumer base of a product with warranty
D	Demand of the product sold with warranty
T	Length of a warranty cycle
π	Total profit of the producer
$U(\cdot)$	Utility function of the producer
ΔP	Profit per product not accounting for the warranty cost
p_c	Probability of accepting a product with warranty
N_a	Number of renewals of an RRLRFW
N_b	Number of minimal repairs under an RRLRFW

N'_b	Number of minimal repairs in the last warranty period of an RRLRFW
c_f	Fixed warranty cost per warranty service, a constant
$\lambda(\cdot), \Lambda(\cdot)$	Failure rate function and cumulative failure rate function in RRLRFW respectively
S_i	Occurrence times of an NHPP or a truncated quasi-renewal process
R_0	Warranty budget level
ε	Acceptance level of a producer towards the risk that the total warranty cost is over the warranty budget level
G_i, g_i	cdf and pdf of the ith occurrence time of an NHPP respectively
C	Warranty cost per product sold
TC	Total warranty cost of the producer

10.1 RFSW Policies for Multi-component Systems

In this section, we discuss a new warranty model, namely renewable full-service warranty (RFSW), for repairable multi-component systems, following Bai and Pham (2006a). Under the RFSW policy, if a warranted product fails, the failed component(s) or subsystem(s) that cause the system failure will be replaced; besides, a preventive maintenance (PM) action will be carried out to reduce the chance of future failure. Both consumers and manufacturers will benefit from the policy because consumers will receive better warranty service compared to the traditional free repair policy. As to manufacturers, it may boost sales as well as reduce the overall warranty cost as a result of better warranty service.

10.1.1 Background

Many researchers have incorporated PM actions in designing and analyzing product warranties. Among them, Chun (1992) considers periodic PM actions during the warranty period. Jack and Dagunar (1994) generalize Chun's idea by allowing unequal PM intervals. Yeh (2001) further extends the work by considering the degree of maintenance as a decision variable. The RFSW policy in this chapter is different from others in two aspects: (a) the PM action will only take place when a system failure happens within the warranty coverage while others consider the case that maintenance actions are not necessarily failure-dependent; (b) the warranty policy considered in this chapter is renewable.

   Warranty service cost per product failure, which includes diagnosis cost, repair or replacement cost, labor cost, and possible PM cost, is often assumed to be constant. In particular, most warranty cost models in the literature use an aggregated cost parameter, which may be estimated from historical data, to approximate the true warranty service cost per failure. However, for a reparable multi-component product, the constant cost assumption may not hold since in general the system repair cost is random due to the randomness in the combination of failed components upon a system failure. To incorporate the randomness, this section decomposes the warranty service cost into two parts: replacement/repair cost and system PM cost. System PM cost is assumed to be constant, which may be interpreted as the aggregated average cost per PM action. However, the replacement/repair cost per system failure is considered as a random variable, whose value depends on component level replacement cost and system failure mechanism.

A simple way to model system warranty cost of multi-component systems is the so-called black-box approach which ignores system reliability structure. As a result, warranty cost models for single-component products can be applied directly. The disadvantage of black-box approach lies in the fact that it does not utilize the information on system structure. Therefore, the resulting warranty cost models should be only used as an approximation.

Warranty analysis of complex systems is relatively new and there are few systematic and explicit analyses on warranty policies for complex systems. Ritchken (1986) models warranty of a two-component parallel system under a two-dimensional warranty. Hussain and Murthy (1998) also discuss warranty cost estimation for parallel systems under the setting that uncertain quality of new products may be a concern for the design of warranty programs. Balachandran *et al.* (1981) use Markovian approach to model warranty cost for a three-component system. Chukova and Dimtrov (1996) provide several warranty cost models for simple series systems and parallel systems under a free replacement warranty based on renewal theory, but only the expected warranty cost is addressed there.

This section discusses the RFSW policy for complex systems: series, parallel, s-p, and p-s. Section 10.1.2 addresses model considerations and assumptions. Section 10.1.3 presents a warranty cost model for series systems. In Section 10.1.4 we analyze warranty cost per system sold for parallel systems. Sections 10.1.5 and 10.1.6 generalize the ideas for simple series and parallel systems and present warranty cost analysis for complex systems with p-s and s-p structure. A numerical example is given in Section 10.1.7.

10.1.2 Model Details

This section provides model descriptions, assumptions, and some preliminary results.

10.1.2.1 RFSW Policy
The warranty policy under study is a RFSW with a pre-specified warranty period denoted by w. For systems under such a warranty, upon a system failure, manufacturers are responsible for replacing the failed component(s) or subsystem(s) that cause the failure. After the repair, a PM action will be performed to ensure that the system is in good working condition. Due to the renewable nature of the warranty, the restored system will automatically carry the same warranty as for the original one. For example, in Figure 10.1, t_1 is the first system failure time. Since $t_1 < w$, the system will receive the warranty service free of charge to consumers, but it will cost the manufacturer C_1, a random variable in nature, which is composed of two parts: the replacement cost for the failed

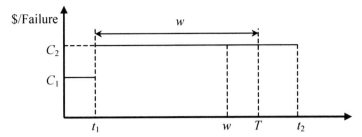

Figure 10.1. Warranty service cost per failure and system failure times

component(s) or subsystem(s), and the system maintenance cost. Starting from t_1, the restored system will have the same warranty with duration w again.

Let's define the warranty cycle T as follows: T is a time interval starting from the date of sale, ending at the warranty expiration date. It is obvious that for a non-renewable warranty, a warranty cycle coincides with a warranty period w. However, for a renewable policy, T is a random variable whose value depends on w, the total number of system failures under the warranty and the actual failure inter-arrival times. Denote $N_s(w)$ as the total number of system failures under the RFSW, and let t_1, $t_2, \dots, t_{N_s(w)}$ be the corresponding inter-arrival failure times, then T can be expressed as

$$T = t_1 + t_2 + \dots + t_{N_s(w)}$$

For the example exhibited in Figure 10.1, $T = t_1 + w$ since the inter-arrival time of the second system failure is $t_2 - t_1$, which is longer than w, therefore, the warranty expires exactly at the time point $t_1 + w$.

10.1.2.2 Assumptions
In this section, we assume perfect PM such that after each warranty service, the restored system is as good as new. The corresponding maintenance cost, denoted by CM, is assumed to be constant, which may be interpreted as the aggregated average cost per PM action. Assume that the maintenance cost is a random variable. Note that this assumption may make the computation of higher moments of warranty cost analytically intractable unless other assumptions such as statistical independence, which is not realistic, are adopted. It is also assumed that all warranty claims are valid, all system failures under warranty are claimed, and any warranty service is instant.

As mentioned before, systems under consideration could be series, parallel, s-p, and p-s. For the p-s system, it is also assumed that no other working components in a failed subsystem (in series) can fail before a system failure. As to the s-p system, it is supposed that only the failed subsystem (in parallel) that causes a system failure is replaced, and thus they are as good as new upon replacement. For all the systems under study, we assume that their components are statistically independent.

10.1.2.3 Distribution of N_s

For a system under the RFSW policy, to derive the statistical properties of warranty cost per cycle or per product sold, it is necessary to obtain the distribution of N_s, the number of system failures within T. The following lemma gives the probability mass function (*pmf*) of N_s.

LEMMA 10.1 *Under the perfect PM assumption, for a system under the RFSW policy with parameter w, the pmf of N_s is*

$$P[N_s = n_s] = [F_s(w)]^{n_s} R_s(w) \qquad \forall n_s, \, n_s = 0,1,2,...$$

where $F_s(\cdot)$ is the cumulative distribution function (cdf) of the system failure times under the warranty, which is assumed to be known, and $R_s(\cdot)$ is the system reliability function.

Proof. Let t_1, t_2,..., be the subsequent system failure times within T, *i.i.d.*, and follow the distribution F_s. It's easy to see that for $i \in \{1,2,...\}$,

$$N_s = \min\{i : t_i > w\} - 1$$

Therefore $\forall n_s, n_s = 0,1,2,...$,

$$P[N_s \geq n_s] = P[\min\{i : t_i > w\} \geq n_s + 1]$$
$$= P[\bigcup_{i \leq n_s} t_i \leq w)]$$
$$= \prod_{i=1}^{n_s} P[t_i \leq w]$$
$$= [F_s(w)]^{n_s}$$

Hence

$$P[N_s = n_s] = P[N_s \geq n_s] - P[N_s \geq n_s + 1]$$
$$= [F_s(w)]^{n_s} - [F_s(w)]^{n_s+1}$$
$$= [F_s(w)]^{n_s} \cdot R_s(w)$$

Note that although the result in Lemma 10.1 coincides with the well-known result for renewable free replacement warranty, it is necessary to provide a formal mathematical proof. Intuitively, for any systems under the RFSW, N_s being geometrically distributed is simply because the warranty will not terminate until the original system or a restored system survive a period of w for the first time and it is assumed that after each warranty service the system is as good as new.

10.1.3 RFSW for Series Systems

This section will discusses the distribution, the first and second centered moments of the warranty cost per cycle for series systems under the RFSW policy.

Figure 10.2. q-Component series system

Define $\Omega \equiv \{1,2,...,q\}$, and let $F_i(\cdot)$ and $R_i(\cdot)$ be the *cdf* and the reliability function of the failure times of component $i, i \in \Omega$, respectively, then for the series system shown in Figure 10.2, the system reliability function is given by

$$F_s(w) = 1 - R_s(w) = 1 - \prod_{i=1}^{q}[1 - F_i(w)] = 1 - \prod_{i=1}^{q} R_i(w) \tag{10.1}$$

Denote N_i be the number of failures within T for component i in the series system. Let TC be the system warranty cost per cycle, then TC can be formulated as

$$TC = \sum_{i=1}^{q}(c_i + c_m)N_i \tag{10.2}$$

Equation (10.2) shows that the distribution of TC can be determined as long as one knows the joint distribution of $N_1, N_2, ..., N_q$. To derive the joint distribution, we first define two quantities and present some useful properties.

LEMMA 10.2. *Define* $p_i(w) \equiv P[T_i \leq Y, T_i \leq w]$ *and* $\alpha_i(w) \equiv p_i(w)/F_s(w)$, *where* T_i *is a failure time of component i,* $Y = \min(T_j, \forall j, j \in \Omega, j \neq i)$, *then*

$$p_i(w) = \int_0^w \frac{h_i(t)}{h_s(t)} f_s(t)dt \tag{10.3}$$

$$\sum_{i=1}^{q} p_i(w) = F_s(w) \tag{10.4}$$

$$\alpha_i(w) = \frac{1}{F_s(w)} \int_0^w \frac{h_i(t)}{h_s(t)} f_s(t)dt \tag{10.5}$$

$$\sum_{i=1}^{q} \alpha_i(w) = 1 \tag{10.6}$$

where subscript s represents the system, subscripts i or j represents a component in the system, while $h(\cdot)$ *and* $f(\cdot)$ *are the hazard rate function and the probability density function (pdf) respectively.*

Proof. From the definition of $p_i(w)$, we have

$$p_i(w) = \int_0^\infty P[T_i \leq Y, T_i \leq w \mid T_i = t]dF_i(t)$$

$$= \int_0^w P[Y \geq t]\,dF_i(t) \qquad\qquad \text{(since } Y \text{ and } T_i \text{ are independent)}$$

$$= \int_0^w P\big[\min(T_j, \forall j, j \in \Omega, j \neq i) \geq i) \geq t\big]\,dF_i(t)$$

$$= \int_0^w \prod_{j\in\Omega, j\neq i} P[T_j \geq t]\,dF_i(t)$$

$$= \int_0^w R_s(t)\frac{f_i(t)}{R_i(t)}\,dt \qquad\qquad \text{(since } R_s(t) = \prod_{i=1}^q R_i(t) \quad \text{and } h(t) = \frac{f(t)}{R(t)} \text{)}$$

$$= \int_0^w \frac{h_i(t)}{h_s(t)} f_s(t)\,dt.$$

To prove Equation (10.4), use $\sum_{j=1}^q h_i(t) = h_s(t)$, and then the result then follows. The proof of Equations (10.5) and (10.6) is straightforward.   ◆

Remarks. $p_i(w)$ can be interpreted as the probability that component i in a series system causes a system failure before the end of a warranty period w. Similarly, α_i can be interpreted as the conditional probability that a failure of component i is the cause of a series system failure given that the system fails within w. Interestingly, $p_i(w)$ is also the partial expectation up to time w of $\zeta_i(t)$, denoted as $E_{T_s}[\zeta_i(t), w]$, with regard to the system failure time T_s, where $\zeta_i(t) \equiv h_i(t)/h_s(t)$. Depending on $\zeta_i(t)$ and $F_s(t)$, $p_i(w)$ or $\alpha_i(w)$ may have to be obtained numerically. It should be noted that if the hazard rate functions of components in series are proportional (proportional-hazard-in-series), *i.e.*, $h_i(t) = \lambda_i g(t)$, $\forall i, i \in \Omega,$, where $g(\cdot)$ is a positive function, then we have $\alpha_i(w) = \lambda_i / \sum_{j=1}^q \lambda_j$, a constant, not depending on w. As a result, probability $p_i(w) = F_s(w)\lambda_i / \sum_{j=1}^q \lambda_j$.

LEMMA 10.3 *The conditional joint distribution of* $N_1, N_2, ..., N_q$, *given* $N_s = n_s$, *is*

$$P(N_1 = n_1, N_2 = n_2, ..., N_q = n_q \mid N_s = n_s) = \binom{n_s}{n_1, n_2, ..., n_q} \prod_{i=1}^q [\alpha_i(w)]^{n_i}$$

The joint distribution of $N_1, N_2, ..., N_q$ *is given by*

$$P(N_1 = n_1, N_2 = n_2, ..., N_q = n_q) = R_s(w) \binom{n_s}{n_1, n_2, ..., n_q} \prod_{i=1}^q [p_i(w)]^{n_i}$$

where $\sum_{i=1}^q n_i = n_s$ *and* $n_i \in \{0, 1, ..., n_s\}, \forall i, i \in \Omega$.

Proof. Given $N_s = n_s$, we know that there are exactly n_s system failures (*i.i.d.*) before the end of T, which implies that the failure times of all such failures are within a period of length w. Hence for each of these system failures, the probability that it is caused by component i is simply $\alpha_i(w)$ according to its definition. As a result, the conditional joint distribution of $N_1, N_2, ..., N_q$, given $N_s = n_s$, is multinomial with parameters $n_s, \alpha_1(w), \alpha_2(w), ...,$ and $\alpha_{q-1}(w)$. Unconditioning on N_s and using $N_s \sim$ geometric$[F_s(w)]$, we then have

$$P(N_1 = n_1, N_2 = n_2, ..., N_q = n_q) = (F_s(w))^{n_s} R_s(w) \binom{\sum_{i=1}^{q} n_i}{n_1, n_2, ..., n_q} \prod_{i=1}^{q} \left(\frac{p_i(w)}{F_s(w)} \right)^{n_i}$$

$$= \binom{\sum_{i=1}^{q} n_i}{n_1, n_2, ..., n_q} \prod_{i=1}^{q} [p_i(w)]^{n_i} R_s(w)$$

We are now ready to derive the distribution of N_i.

PROPOSITION 10.1 N_i *follows a geometric distribution with parameter* $R_s(w)/[R_s(w) + p_i(w)]$, $\forall i, i \in \Omega$. *The corresponding pmf is*

$$P[N_i = n_i] = \left[\frac{p_i(w)}{R_s(w) + p_i(w)} \right]^{n_i} \frac{R_s(w)}{R_s(w) + p_i(w)}, \quad n_i = 0, 1, 2, ... \qquad (10.7)$$

The covariance, $COV(N_i, N_j), i, j \in \Omega, i \neq j$, *is given by*

$$COV(N_i, N_j) = \frac{p_i(w) p_j(w)}{R_s^2(w)} \qquad (10.8)$$

Proof. First we prove Equation (10.7). From Lemma 10.3 and the properties of multinomial distribution, we have that $\forall i, i \in \Omega, N_i \mid N \sim$ Binomial$(N, \alpha_i(w))$. So the moment generating function (*mgf*) of N_i is

$$E[e^{tN_i}] = E[E[e^{tN_i} \mid N]]$$

Bai and Pham (2006a) prove that

$$E[e^{tN_i}] = \frac{\dfrac{R_s(w)}{R_s(w) + p_i(w)}}{1 - \dfrac{p_i(w)}{R_s(w) + p_i(w)} e^t}$$

By realizing that the last expression is nothing but the *mgf* of a geometric distribution with parameter $R_s(w)/[R_s(w)+p_i(w)]$, we complete the proof for Equation (10.7).

Now we prove (10.8). Since $COV(N_i,N_j)=E(N_iN_j)-E(N_i)E(N_j)$, and by the properties of the geometric distribution, $E(N_i)E(N_j)=p_i(w)p_j(w)/R_s^2(w)$ it is sufficient to show that $E(N_iN_j)=2p_i(w)p_j(w)/R_s^2(w)$. For $q\geq 3$, Define N_k as the number of system failures within T due to the components other than i or j, then by the properties of multinomial distribution and from Lemma 10.3, for N_i and N_j, denote $P[N_i=n_i,N_j=n_j]$ by $P[n_i,n_j]$, we obtain

$$p[n_i,n_j]=\sum_{n_k=0}^{\infty}P(N_i=n_i,N_j=n_j,N_k=n_k\,|\,N_s=n_i+n_j+n_k)$$

$$\cdot (F_s(w))^{n_i+n_j+n_k}R_s(w)$$

$$=\sum_{n_k=0}^{\infty}\binom{n_i+n_j+n_k}{n_i,n_j,n_k}[\alpha_i(w)]^{n_i}[\alpha_j(w)]^{n_j}[\alpha_k(w)]^{n_k}[F_s(w)]^{n_i+n_j+n_k}R_s(w)$$

$$=\binom{n_i+n_j}{n_i}\left[\frac{p_i(w)}{R_s(w)+p_i(w)+p_j(w)}\right]^{n_i}\left[\frac{p_j(w)}{R_s(w)+p_i(w)+p_j(w)}\right]^{n_j}$$

$$\cdot\frac{R_s(w)}{R_s(w)+p_i(w)+p_j(w)},$$

where the last step is due to

$$\sum_{n_k=0}^{\infty}\binom{n_i+n_j+n_k}{n_i,n_j,n_k}x^{n_k}=\binom{n_i+n_j}{n_i}(1-x)^{-(n_i+n_j+1)},\qquad \forall x,x\in(0,1),$$

and $1-p_k(w)=R_s(w)+p_i(w)+p_j(w)$.
Bai and Pham (2004) obtain that

$$E(N_iN_j)=\sum_{n_i=0}^{\infty}\sum_{n_j=0}^{\infty}\binom{n_i+n_j}{n_i}\left[\frac{p_i(w)}{R_s(w)+p_i(w)+p_j(w)}\right]^{n_i}$$

$$\cdot\left[\frac{p_j(w)}{R_s(w)+p_i(w)+p_j(w)}\right]^{n_j}\cdot\frac{R_s(w)}{R_s(w)+p_i(w)+p_j(w)}$$

$$=\frac{2p_i(w)p_j(w)}{(R_s(w))^2}$$

The proof of Equation (10.8) for $q=2$ is similar, and is omitted here.

Applying Proposition 10.1 to Equation (10.2), we can then conclude that for a series system under the RFSW policy, the distribution of TC is simply a mixture of dependent random variables each of which follows a geometric distribution. The pmf of TC may be written as

$$P[TC = x] = \begin{cases} R_s(w) \sum_{\{n_1,n_2,...n_q\} \& \sum_{i=1}^q (c_i+c_m)n_i = x} \left(\dfrac{\sum_{i=1}^q n_i}{n_1,n_2,...n_q} \right) \prod_{i=1}^q (p_i(w))^{n_i}, \\ \qquad \text{if } x \in \left\{ \sum_{i=1}^q (c_i + c_m)n_i \right\} \text{ and } n_i \in \{0,1,...\}, \forall i, i \in \Omega \\ 0 \qquad \text{otherwise} \end{cases}$$

COROLLARY 10.1 *The expected warranty cost per cycle for the q-component series system under the RFSW policy is given by*

$$E[TC] = \frac{1}{R_s(w)} \sum_{i=1}^q (c_i + c_m) p_i(w) \tag{10.9}$$

The corresponding variance of TC is

$$Var[TC] = \frac{1}{[R_s(w)]^2} \left\{ \sum_{i=1}^q (c_i + c_m)^2 p_i(w)[p_i(w) + R_s(w)] \right.$$
$$\left. + 2 \sum_{i<j,i,j \in \Omega} (c_i + c_m)(c_j + c_m) p_i(w) p_j(w) \right\} \tag{10.10}$$

where c_i is the replacement cost of component i, and c_m is the system PM cost.

Proof. From Equation (10.2), it follows that

$$E[TC] = \sum_{i=1}^q (c_i + c_m) E[N_i] \quad \text{and}$$

$$Var[TC] = \sum_{i=1}^q (c_i + c_m)^2 Var(N_i) + 2 \sum_{i<j,i,j \in \Omega} (c_i + c_m)(c_j + c_m) \text{cov}(N_i, N_j)$$

By Proposition 10.1 and the properties of the geometric distribution, the results follow. $\blacklozenge$

10.1.4 RFSW for Parallel Systems

For a parallel system, it won't fail unless all the components in the system fail. As a result, under the RFSW policy, the warranty service cost per system failure for the system shown in Figure 10.3 is simply $C_m + \sum_{i=1}^q c_i$. Again let N_s be the number of system failures within T, then the corresponding system warranty cost TC per system sold is

$$TC = N_s \left(c_m + \sum_{i=1}^q c_i \right) \tag{10.11}$$

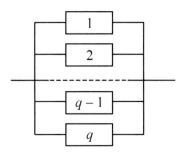

Figure 10.3. q-Component parallel system

Not surprisingly, N_s again follows a geometric distribution, but $F_s(w)$ is the failure time *cdf* of the parallel system evaluated at w, which is given by

$$F_s(w) = \prod_{i=1}^{q} F_i(w) \tag{10.12}$$

COROLLARY 10.2 *Under the RFSW policy, the pmf of the system warranty cost per cycle is*

$$P[TC = x] = \begin{cases} F_s(w))^{n_s}(1 - F_s(w)), & \text{if } x \in \left\{ \left(c_m + \sum_{i=1}^{q} c_i \right) n_s \right\}, n_s \in \{0,1,...\} \\ 0, & \text{otherwise} \end{cases}$$

The expected system warranty cost is

$$E[TC] = \frac{F_s(w)}{R_s(w)}(c_m + \sum_{i=1}^{q} c_i) \tag{10.13}$$

The corresponding warranty cost variance is

$$Var[TC] = \frac{F_s(w)}{R^2{}_s(w)}(c_m + \sum_{i=1}^{q} c_i)^2 \tag{10.14}$$

10.1.5 RFSW for Series-parallel Systems

This section discusses the RFSW policy for s-p systems. For the s-p system composed of q subsystems in series drawn in Figure 10.4, denote the number of components in subsystem i that are in parallel as r_i. Let C_i be the warranty service cost for subsystem i, and let c_{ij} be the replacement cost of component j in subsystem i; then we have $C_i = c_m + \sum_{j=1}^{r_i} c_{ij}, \forall i, i \in \Omega.$.

For the s-p system, denote N_i as the number of failures of subsystem i within T. Similar to Equation (10.2), the total system warranty cost per cycle can be formulated as

$$TC = \sum_{i=1}^{q} N_i (c_m + \sum_{j=1}^{r_i} c_{ij})$$

(10.15)

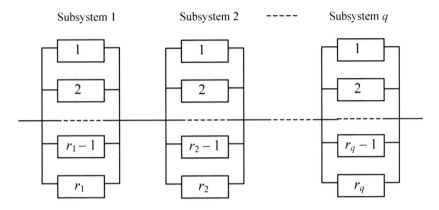

Figure 10.4. s-p System with q subsystems

The *cdf* of failure times of the s-p system under the warranty is given by

$$F_s(w) = 1 - \prod_{i=1}^{q} R_i(w)$$

$$= 1 - \prod_{i=1}^{q} [1 - \prod_{j=1}^{r_i} F_{ij}(w)]$$

(10.16)

where $F_{ij}(\cdot)$ is the *cdf* of the failure times of component j in subsystem i.

Under the RFSW policy, we define $p_i(w)$ and $\alpha_i(w)$ the same way as that for simple series systems except that in this case, i refers to a subsystem instead of a single component. It is obvious that all the properties of $p_i(w)$ and $\alpha_i(w)$ in Lemma 10.2 still hold.

COROLLARY 10.3 *For the s-p system under the RFSW policy, the pmf of the warranty cost per cycle TC is*

$$P[TC = x] = \begin{cases} R_s(w) \sum_{\{n_1, n_2, \dots n_q\} \& \sum_{i=1}^{q}(c_m + \sum_{j=1}^{r_i} c_{ij})n_i = x} \left(\dfrac{\sum_{i=1}^{q} n_i}{n_1, n_2, \dots n_q} \right) \prod_{i=1}^{q} [p_i(w)]^{n_i}, \\ \qquad \text{if } x \in \left\{ \sum_{i=1}^{q}(c_m + \sum_{j=1}^{r_i} c_{ij})n_i \right\} \text{ and } n_i \in \{0,1,\dots\}, \forall i, i \in \Omega \\ \\ 0 \qquad \text{otherwise} \end{cases}$$

(10.17)

The first two centered moments of the warranty cost per cycle TC are as follows:

$$E[TC] = \frac{1}{R_s(w)} \sum_{i=1}^{q}(c_m + \sum_{j=1}^{r_i} c_{ij})p_i(w)$$ (10.18)

$$Var[TC] = \sum_{i=1}^{q}(c_m + \sum_{j=1}^{r_i} c_{ij})^2 \frac{p_i(w)[p_i(w) + R_s(w)]}{R_s^2(w)}$$

$$+ 2 \sum_{i < i', \, i, i' \in \Omega} (c_m + \sum_{j=1}^{r_i} c_{ij})(c_m + \sum_{j=1}^{r_{i'}} c_{i'j}) \frac{p_i(w)p_{i'}(w)}{R_s^2(w)}$$

(10.19)

Proof. The proof is similar to that for Corollary 10.1.   ◆

10.1.6 RFSW for Parallel-series Systems

Consider the system shown in Figure 10.5 with q subsystems in parallel, each of which consists of one or more components in series.

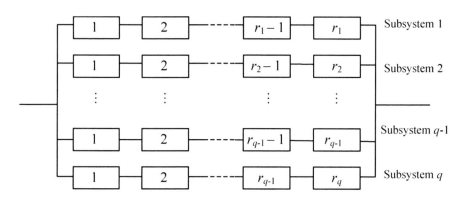

Figure 10.5. p-s System with q subsystems

In this section, we study the RFSW policies for p-s systems. The first two-centered warranty cost moments will be derived. Let r_i be the number of components in the ith subsystem and let N_s be the number of system failures within T. Under the perfect PM assumption, again we have that $N_s \sim$ geometric $(F_s(w))$. It's not difficult to verify that for the p-s system,

$$F_s(w) = \prod_{i=1}^{q}[1 - \prod_{j=1}^{r_i}[1 - F_{ij}(w)]]$$ (10.20)

Let N_{ij} be the number of failures of the jth component in subsystem i within T. Since each subsystem is in series, all subsystems are connected in parallel and it is assumed that no working components in a failed subsystem can fail before a system failure, we have that $\sum_{j=1}^{r_i} N_{ij} = N_s, \forall i, i = 1,2,...,q$. Denote c_{ij} as the replacement cost for component j in subsystem i, then the total system warranty cost per cycle, TC, can be written as

$$TC = \sum_{i=1}^{q} \sum_{j=1}^{r_i} \left(c_{ij} + \frac{c_m}{q} \right) N_{ij} \qquad (10.21)$$

To derive the expectation and the variance of TC for p-s systems, we need to obtain the distribution of $N_{ij}, \forall i, i \in \Omega, \forall j, j \in \Omega_i, \Omega_i \equiv \{1,2,...,r_i\}$, as well as the covariance between N_{ij} and N_{ik} for $j \neq k$ (covariance within a subsystem), and the covariance between N_{ij} and $N_{i'k}$ for $i \neq i'$ (covariance between subsystems). Similar to $p_i(w)$ and $\alpha_i(w)$ defined in Section 10.1.3, next we define $p_{ij}(w)$ and $\alpha_{ij}(w)$ and state the related properties in the following lemma.

LEMMA 10.4 *For the p-s system under the RFSW policy, let T_{ij} be the failure times of component j in subsystem i, define $p_{ij}(w) \equiv P[T_{ij} \leq Y_i, T_{ij} \leq w, T_s \leq w]$ and $\alpha_{ij}(w) \equiv p_{ij}(w)/F_s(w)$, where $Y_i = \min(T_{ik}, \forall k, k \in \Omega_i, k \neq j)$, then*

$$p_{ij}(w) = F_s^{\bar{i}}(w) \int_0^w \frac{h_{ij}(t)}{h_i(t)} f_i(t) dt \qquad (10.22)$$

$$\sum_{j=1}^{r_i} p_{ij}(w) = F_s(w) \qquad (10.23)$$

$$\alpha_{ij}(w) = \frac{1}{F_i(w)} \int_0^w \frac{h_{ij}(t)}{h_i(t)} f_i(t) dt \qquad (10.24)$$

$$\sum_{j=1}^{r_i} \alpha_{ij}(w) = 1 \qquad (10.25)$$

where $\bar{i}$ represents the new p-s system comprising $(q-1)$ subsystems of the original system except the subsystem i.

Proof. From the definition of $p_{ij}(w)$, we obtain

$p_{ij}(w) = P[Y_i \geq T_{ij}, T_{ij} \leq w, T_s^{\bar{i}} \leq w]$ (since all components are independent)

$\quad = P[Y_i \geq T_{ij}, T_{ij} \leq w] F_s^{\bar{i}}(w)$

$\quad = F_s^{\bar{i}}(w) \int_0^w P[\min(T_{ik}, \forall k, k \in \Omega_i, k \neq j) \geq t] dF_{ij}(t)$

$$= F_s^{\bar{i}}(w) \int_0^w \prod_{k \in \Omega_i, k \neq j} R_{ik}(t) dF_{ij}(t)$$

$$= F_s^{\bar{i}}(w) \int_0^w R_i(t) \frac{f_{ij}(t)}{R_{ij}(t)} dt \quad \text{(since } R_i(t) = \prod_{j=1}^{r_i} R_{ij}(t) \text{ and } h(t) = \frac{f(t)}{R(t)} \text{)}$$

$$= F_s^{\bar{i}}(w) \int_0^w \frac{h_{ij}(t)}{h_i(t)} f_i(t) dt$$

To prove Equation (10.23), use $F_i(t)F_s^{\bar{i}}(t) = F_s(t)$ and $\sum_{j=1}^{r_i} h_{ij} = h_i$; the result then follows. The proof for Equations (10.24) and (10.25) is straightforward.   ◆

Again, as a special case, if the proportional-hazards-in-series assumption is adopted, implying that the hazard rate function of component j in subsystem i has the form $\lambda_{ij} g_i(t), \forall j, j \in \Omega_i$, where $g_i(t)$ is a positive function, then it is easy to verify that $\alpha_{ij} = \lambda_{ij} / \sum_{k=1}^{r_i} \lambda_{ik}$ and $p_{ij} = F_s(w)\lambda_{ij} / \sum_{k=1}^{r_i} \lambda_{ik}$

Next, we derive the distribution of N_{ij} and some related properties.

Corollary 10.4 *For the p-s system under the RFSW, the pmf of* N_{ij}, $\forall i, i \in \Omega$, $\forall j, j \in \Omega_i$, *is given by*

$$P[N_{ij} = n_{ij}] = \left[\frac{p_{ij}(w)}{R_s(w) + p_{ij}(w)} \right]^{n_{ij}} \frac{R_s(w)}{R_s(w) + p_{ij}(w)}, n_{ij} = 0,1,2,... \quad (10.26)$$

The covariance between N_{ij}, N_{ik} *(within subsystem covariance) for* $j \neq k, j, k \in \Omega_i$ *is*

$$\text{cov}(N_{ij}, N_{ik}) = \frac{p_{ij}(w)p_{ik}(w)}{R_s^2(w)} \quad (10.27)$$

and the covariance between N_{ij}, $N_{i'k}$ *(between subsystem covariance) for* $i \neq i', i, i' \in \Omega$ *is*

$$\text{cov}(N_{ij}, N_{i'k}) = \sum_{n_{ij}=0}^{\infty} \sum_{n_{i'k}=0}^{\infty} n_{ij} n_{i'k} p(n_{ij}, n_{i'k}) - \frac{p_{ij}(w)p_{i'k}(w)}{R_s(w)^2} \quad (10.28)$$

where for $n_{i'k} \geq n_{ij}$

$$p(n_{ij}, n_{i'k}) = R_s(w) \sum_{n_{ij} \geq n_{i'k} - n_{ij}}^{\infty} \binom{n_{ij} + n_{i'j}}{n_{i'k}} [\alpha_{i'k}(w)]^{n_{i'k}} [1 - \alpha_{i'k}(w)]^{n_{ij} + n_{i'j} - n_{i'k}}$$

$$\cdot \binom{n_{ij} + n_{i'j}}{n_{ik}} [\alpha_{ij}(w)]^{n_{ij}} [1 - \alpha_{ij}(w)]^{n_{i'j}} [F_s(w)]^{n_{ij} + n_{i'j}} \quad (10.29)$$

Proof. The proof for Equations (10.26) and (10.27) is similar to that for Proposition 10.1 as long as one realizes that the joint distribution of $N_{i1}, N_{i2}, ..., N_{ir_i}, \forall i, i \in \Omega$ is given by

$$P[N_{i1} = n_{i1}, N_{i2} = n_{i2}, ..., N_{ir_i} = n_{ir_i}] = R_s(w) \binom{n_{i1} + n_{i2} + ... + n_{ir_i}}{n_{i1}, n_{i2}, ..., n_{ir_i}} \prod_{k=1}^{r_i} [p_{ik}(w)]^{n_{ik}}$$

Next we prove (10.29). Without loss of generality, let $n_{i'k} \geq n_{ij}$, and denote

$$p(n_{ij}, n_{\bar{ij}}, n_{i'k}) = P[N_{ij} = n_{ij}, \sum_{j' \neq j, j', j \in \Omega_i} N_{ij'} = n_{\bar{ij}}, N_{i'k} = n_{i'k}],$$

then

$$p(n_{ij}, n_{\bar{ij}}, n_{i'k}) = p[n_{i'k} \mid n_{ij}, n_{\bar{ij}}] p(n_{ij}, n_{\bar{ij}})$$

$$= \binom{n_{ij} + n_{\bar{ij}}}{n_{i'k}} [\alpha_{i'k}(w)]^{n_{i'k}} [1 - \alpha_{i'k}(w)]^{n_{ij} + n_{\bar{ij}} - n_{i'k}}$$

$$* \binom{n_{ij} + n_{\bar{ij}}}{n_{ij}} (\alpha_{ij}(w))^{n_{ij}} (1 - \alpha_{ij}(w))^{n_{\bar{ij}}} (F_s(w))^{n_{ij} + n_{\bar{ij}}} R_s(w)$$

Consequently,

$$p(n_{ij}, n_{i'k}) = \sum_{n_{\bar{ij}} \geq n_{i'k} - n_{ij}}^{\infty} \binom{n_{ij} + n_{\bar{ij}}}{n_{i'k}} (\alpha_{i'k}(w))^{n_{i'k}} (1 - \alpha_{i'k}(w))^{n_{ij} + n_{\bar{ij}} - n_{i'k}}$$

$$\cdot \binom{n_{ij} + n_{\bar{ij}}}{n_{ij}} [\alpha_{ij}(w)]^{n_{ij}} [1 - \alpha_{ij}(w)]^{n_{\bar{ij}}} [F_s(w)]^{n_{ij} + n_{\bar{ij}}} R_s(w).$$

The proof of Equation (10.28) is straightforward by using Equations (10.26) and (10.29).   ◆

COROLLARY 10.5 *For the p-s system under the RFSW policy, the expectation of the system warranty cost per cycle, TC, is given by*

$$E[TC] = \sum_{i=1}^{q} \sum_{j=1}^{r_i} (c_{ij} + \frac{c_m}{q}) \frac{p_{ij}(w)}{R_s(w)} \tag{10.30}$$

and the corresponding cost variance is

$$Var[TC] = \sum_{i=1}^{q} \sum_{j=1}^{r_i} (c_{ij} + \frac{c_m}{q})^2 \frac{p_{ij}(w)[R_s(w) + p_{ij}(w)]}{(R_s(w))^2}$$

$$+ 2 \sum_{i=1}^{q} \sum_{j<k, j,k \in \Omega_i} (c_{ij} + \frac{c_m}{q})(c_{ik} + \frac{c_m}{q}) \frac{p_{ij}(w) p_{ik}(w)}{R_s^2(w)}$$

$$+ 2 \sum_{i<i', i,i' \in \Omega} \sum_{j=1}^{r_i} \sum_{k=1}^{r_{i'}} (c_{ij} + \frac{c_m}{q})(c_{i'k} + \frac{c_m}{q})(\sum_{n_{ij}=0}^{\infty} \sum_{n_{i'k}=0}^{\infty} n_{ij} n_{i'k} p(n_{ij}, n_{i'k}))$$

$$-\frac{p_{ij}(w)p_{i'k}(w)}{R_s^2(w)})$$
(10.31)

Proof. The derivation of E[TC] is straightforward. To derive Var(TC), starting from Equation (10.21), we obtain that

$$Var(TC) = \sum_{i=1}^{q}\sum_{j=1}^{r_i}(c_{ij}+\frac{c_m}{q})^2 Var(N_{ij})$$

$$+2\sum_{i=1}^{q}\sum_{j<k}(c_{ij}+\frac{c_m}{q})(c_{ik}+\frac{c_m}{q})\mathrm{cov}(N_{ij},N_{ik})$$

$$+2\sum_{i<i'}\sum_{j=1}^{r_i}\sum_{k=1}^{r_{i'}}(c_{ij}+\frac{c_m}{q})(c_{i'k}+\frac{c_m}{q})\mathrm{cov}(N_{ij},N_{i'k})$$

Using Corollary 10.4, the results follow.   ◆

Next let's see a numerical example, a detailed sensitivity study, and a few possible further studies for this RFPW.

10.1.7 A Numerical Example and Sensitivity Study

Consider a three-component p-s system shown in Figure 10.6 under the RFSW policy. Suppose $w = 3$ and $c_m = \$220$. All other parameters are given in Table 10.1. The parameters of components' failure times were chosen such that $R_s(t) > 0.90$ for $t \le 20$. Assume that the failure times of components 1 and 3 follow Weibull distributions, and the failure time of component 2 are exponentially distributed.

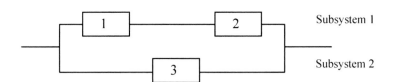

Figure 10.6. p-s System with three components

Table 10.1. Parameters for the three-component s-p system

Component number	1	2	3
c_i	\$200	\$250	\$550
$R(t)$	$\exp\left(-\frac{t^{0.59}}{28.33}\right)$	$\exp\left(-\frac{1}{121.68}\right)$	$\exp\left(-\frac{t^{0.37}}{9.16}\right)$

From Equations (10.30) and (10.31), the expected system warranty cost E[TC] per product sold is \$13.26, which only accounts for 1.33% of the total system production cost (the sum of all components' cost). The corresponding standard deviation Std(TC) is \$114.99, which is much higher than E[TC]. However, this is what one should expect since the distribution of TC is a mixture of geometrically distributed random variables whose standard deviation is always larger than its expectation. It is worth noting that most warranty models in warranty literature rely solely on expected warranty cost for the purpose of warranty cost modeling and analysis. This example shows the necessity of obtaining higher moments of warranty cost to evaluate the risk embedded in certain warranty policies for manufacturers.

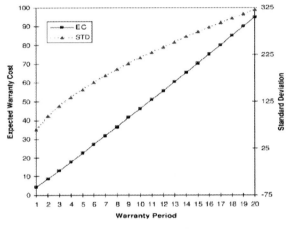

Figure 10.7. E[TC] and std(TC)

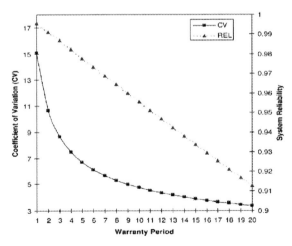

Figure 10.8. CV and system reliability

If one decomposes Var(TC) into three parts: the total variation due to each N_{ij}, the covariance within subsystems, and the covariance between subsystems, it seems that the first and the third parts are the dominant sources of Var(TC). In this example they accounts for 55.83% and 44.11% respectively.

To show how the moments of TC change while the warranty duration varies, we consider w in the range of one to twenty. From Figure 10.7 one can see that both the expectation and the standard deviation of TC increase monotonically over w. Coefficient of Variation (CV) is a standardized measure for the variability of random variables. Figure 10.8 shows that CV declines monotonically over w. For $w = 20$, CV = 3.39. We also plot the system reliability curve in Figure 10.8.

As indicated previously, expectation is probably the most commonly used measure for evaluating warranty programs. Nevertheless, expectation itself can reveal little information about the warranty cost risk for manufacturers. In contrast, prediction intervals or quantiles of TC may be a better measure to evaluate warranties. In theory, computing prediction intervals is equivalent as computing quantiles. Since usually warranty managers are more interested in controlling the upper limit of TC, the upper quantiles of TC for different $w, w \in [1,20]$ are computed at the 95% and 99% confidence level, denoted by $1 - \alpha$. It is possible to obtain the exact prediction intervals directly from the *pmf* of TC. But working with *pmf* of *TC* involves enumeration of all possible combinations of components' cost, which could be complicated and time consuming especially for large systems. Alternatively, Monte Carlo simulation is used to obtain estimates of those quantiles. The simulation procedure is given below:

i)   *Generate N_s from geometric $(F_s(w))$.*

ii) *If $N(w) = n > 0$, then for each i where $i = 1,2,...,n$, generate uniform random variables from $p_{ij}(w)$ to determine the exact combination of failed components, else $TC = 0$.*

iii) *Compute TC based on the combination of failed components, the corresponding component replacement cost and system maintenance cost.*

iv) *Repeat steps $i) - iii)$ for each w for 40,000 runs.*

The corresponding quantiles, or equivalently, the upper prediction confidence bounds are then computed from the generated data points. The results are reported in Table 10.2. Based on the results, for the manufacturers, the warranty cost per product sold would be no more than $1,020 for $w \leq 20$, roughly the same as the production cost, with the confidence level at 99%.

Section 10.1 has discussed a new warranty policy, the renewable full-service warranty, which offers extra incentive for consumers by including free PM into the traditional free repair service, for multi-component systems with series, parallel, s-p and p-s structure. Due to the more than ever fierce competition in markets, manufacturers are constantly in search of innovations in marketing strategies to promote their products. We believe that the RFSW policy can be used for marketing purpose since it provides extra compensation to consumers.

Table 10.2. Quantiles of TC for the p-s system

w	1	2	3	4	5	6	7	8	9	10
Q_{95}	0	0	0	0	0	0	0	0	0	0
Q_{99}	0	0	970	970	970	970	970	970	970	1020
w	11	12	13	14	15	16	17	18	19	20
Q_{95}	0	970	970	970	970	970	970	970	970	970
Q_{99}	1020	1020	1020	1020	1020	1020	1020	1020	1020	1020

There may be several potential extensions to the study of RFSW policies on complex systems. First is to relax the perfect PM assumption. Although this assumption is used widely in practice as well as in warranty and maintenance literature, we believe that more research should be done for RFSW policies considering minimal maintenance or imperfect maintenance. Second is to consider random system maintenance cost to fully take into account of the probabilistic behavior at component level. As indicated before, such a change could bring in much more complexity and the higher moments of warranty cost in general are analytically intractable. For s-p systems, it is assumed that only the components in a failed subsystem will be replaced under the warranty, thus the analysis of warranty cost under an RFSW becomes similar to that for simple series systems. In practice, consumers would prefer to have all failed components replaced upon a warranty service.

10.2 DWC for Minimally Repaired Series Systems

One of the primary questions to be answered in warranty analysis is how much a warranty program will cost. Due to the random nature of warranty cost, most warranty cost models would prefer to use the expected warranty cost (EWC) as the answer. Extensive research on modeling and estimating EWC has been done under various warranty policies, for example, in Blischke and Murthy (1994, 1996) and Singpurwalla and Wilson (1993).

In contrast to the EWC, the expected value of discounted warranty cost (DWC), which incorporates the value of time, may provide a better cost measure for warranties. This is because in general warranty cost can be treated as a random cash flow in the future. Warranty issuers do not have to spend all the money at the stage of warranty planning. Instead, they can allocate it over the life cycle of warranted products. Another reason that one should consider the value of time is for the purpose of determining warranty reserve. Warranty reserve is a fund set up specifically to meet future warranty claims. It is well-known that the present value (PV) of warranty liability or rebates to be paid in the future is less than the face value (Amato and Anderson 1976). Hence, for warranty issuers, it is desirable to determine the warranty reserve according to the PV of total warranty liability. DWC and warranty reserve related issues have been studied, from both manufacturers' and consumers' perspectives for single-component products, either repairable or non-repairable (Mamer 1969, 1987; Patankar and Mitra 1995; Thomas 1989).

In practice, most products are composed of several components. If warranties are offered for each component separately, then warranty models for single-component items can be applied directly. However, sometimes warranty terms are defined upon the entire systems. For such warranties, it is necessary to consider the system architecture as well as the component level warranty service cost (Blischke and Murthy 1996, p.543). Warranty analysis for multi-component systems based on system structure has been addressed in a few papers. Ritchken (1986) studies two-component parallel system under a two-dimensional warranty. Hussain and Murthy (1998) also discuss warranty cost estimation for parallel systems under the setting that uncertain quality of new products may be a concern for the design of warranty programs. A Markovian approach to the analysis of warranty cost for a three-component system can be found in Balachandran (1981). Chukova and Dimtrov (1996) derive the expected warranty cost for two-component series systems and parallel systems under a free replacement warranty. Bai and Pham (2004) study DWC for series systems. Section 10.1 presents the first two centered moments of renewable full-service warranties for complex systems with series, parallel, series-parallel and parallel-series structure.

There are many ways of modeling the impact on system failure times from repair actions. The simplest way is by assuming the as-good-as-new repair for which a single failure time distribution is sufficient to describe subsequent failure times of a system, no matter new or repaired. The advantage of such an approach is that it simplifies warranty cost models as renewal theory can be applied readily. However, it might lead to underestimated expected warranty cost due to the fact that repaired systems are usually not as reliable as a new one. A slight modification from as-good-as-new repair is by assuming a failure time distribution for all repaired systems, different from the failure time distribution of a new one, for which a delayed renewal process can be employed to model the product failure process (Blischke and Murthy 1996). For complex product, repair is often assumed to be minimal (as-good-as-old), *i.e.*, the failure rate of a repaired product is the same as that just before the most recent repair. Nguyen and Murthy (1984a) presents a general warranty cost model for single-component repairable items. Perfect repair, minimal repair and imperfect repair are covered, but the value of time is not addressed. Kulkarni *et al.* (2002) develop several warranty reserve models for single component products for non-stationary sales processes. Ja *et al.* (2001) present a warranty cost model on minimally repaired single-component systems with time dependent cost. One of the assumptions there is that repair costs at different times are statistically independent. Nevertheless, in the context of DWC, usually those time-dependent costs are not independent. For this reason, the assumption of independent repair costs is dropped in this section.

Another important issue in warranty analysis is the variability or the risk embedded in warranty cost for warranty issuers. It is often not sufficient for warranty managers to simply obtain the estimates of EWC or expected DWC. Additional information about higher warranty cost moments is essential for warranty risk management and decision making. Discussions on warranty cost variation for single-component or black-box systems can be seen in Ja *et al.* (2002), Blacer and Sahin (1986), and Patankar and Worm (1981). In this section, we will discuss the first two centered moments of DWC for minimally repaired series systems, following Bai and Pham (2004). This section assumes:

    i)   Components in a system are statistically independent.
    ii)   All repairs are instant within warranty.
    iii) All components in a system are repairable and any failed components are minimally repaired upon a system failure under warranty.
    iv) All warranty claims are executed and all claims are valid.

The rest of this section is organized as follows. Section 10.2.1 shows some preliminary results to be used in later sections. In Section 10.2.2, general expressions of expected DWC and its variance for a repairable series system under an FRW policy are discussed. Two special cases are also presented here. Section 10.2.3 discusses DWC of a PRW policy. Section 10.2.4 addresses two important warranty management problems.

10.2.1 Preliminary Results

Let $\{N(t), t > 0\}$ be a non-homogeneous Poisson process (NHPP) with intensity function $\lambda(t)$, the corresponding cumulative intensity function is $\Lambda(t)$, which is also referred as the mean value function, is given by $\Lambda(t) = \int_{s=0}^{t} \lambda(s)ds$. Let $S_i, i = 1,2,\cdots$, be the ith arrival time of the NHPP. Define a random variable $Y(t)$ as follows:

$$Y(t) = \begin{cases} \sum_{i=1}^{N(t)} H(S_i), & \text{if } N(t) \in \{1,2,\cdots\} \\ 0 & \text{if } N(t) = 0 \end{cases}$$

where $H(\cdot)$ is a non-negative bounded continuous function, then we have the following results:

PROPOSITION 10.2

$$E[Y(t)] = \Lambda(t)E[H(U)] = \int_0^t H(u)\lambda(u)du \tag{10.32}$$

$$V[Y(t)] = \Lambda(t)E[H^2(U)] = \int_0^t H^2(u)\lambda(u)du \tag{10.33}$$

where the pdf of U is given by $f_U(u) = \dfrac{\lambda(u)}{\Lambda(t)}, 0 \le u \le t.$

Proof. Define $\vec{S} \equiv (S_1, S_2, \cdots, S_n)$. Let $\vec{U} \equiv (U_{(1)}, U_{(2)}, \cdots, U_{(n)})$, where $U_{(i)}$'s are the order statistics of $U_1, U_2, \cdots, U_n$, which are *i.i.d.* samples of the random variable U. Conditioning on $N(t) = n$, we have that $\vec{S}$ is equal in distribution of $\vec{U}$ (Kulkarni 1995, p.228). Hence

$$E[Y(t)] = E\{E[\sum_{i=1}^{N(t)} H(S_i) \mid N(t) = n]\}$$

$$= E\{E[\sum_{i=1}^{n} H(U_{(i)})]\} \qquad \text{(since distribution}(\vec{U}) = \text{distribution}(\vec{S}))$$

$$= E\{E[\sum_{i=1}^{n} H(U_i)]\} \qquad \text{(for } \sum_{i=1}^{n} H(U_i) = \sum_{i=1}^{n} H(U_{(i)}))$$

$$= E[nE[H(U)]\}$$

$$= \Lambda(t)E[H(U)]. \qquad \text{(since } E[N(t)] = \Lambda(t))$$

$$= \int_0^t H(u)\lambda(u)du \qquad \text{(since } f_U(u) = \frac{\lambda(u)}{\Lambda(t)}, 0 \le u \le t)$$

Bai and Pham (2004) show that:

$$V[Y(t)] = E\{V[\sum_{i=1}^{N(t)} H(S_i) \mid N(t) = n]\} + V\{E[\sum_{i=1}^{N(t)} H(S_i) \mid N(t) = n]\}$$

$$= \int_0^t H^2(u)\lambda(u)du$$

Proposition 10.2 can also be proven by conditioning on the first failure time. Note that Ja *et al.* (2002) obtain the same results as in Proposition 10.2 except that they consider the special case of $H(t) = e^{-\alpha t}$ through the first failure time approach.

10.2.2 DWC Under an FRW Policy

In this section, we consider a series system with q statistically independent repairable components under an FRW policy with a fixed period w. We derive the mathematical expressions for expected DWC and the cost variance.

It is obvious that a series system will fail if and only if any of its components fails. Since simultaneous failures of components are impossible, each time a warranted system fails before w, there must be one and only one failed component in the system. Under the FRW policy, the failed component will be identified and repaired free of charge to consumers. For simplicity, we assume that every repair is instantaneous.

Let $\lambda_i(t)$ be the failure intensity function of component i. Due to instant repair assumption and the structure of series systems, the failure process for each component follows an NHPP on regular time scale, uniquely determined by the intensity function $\lambda_i(t)$ (Nguyen and Murthy 1984b). All these q NHPPs are statistically independent since all components in the system are independent.

Let the total DWC per system sold be $C(w)$ and define $Y_i(w), \forall i, i \in \{1, 2, \cdots, q\}$ as follows:

$$Y_i(w) = \begin{cases} \sum_{j=1}^{N_i(w)} c_i H(S_{ij}) & \text{if } N_i(w) \in \{1, 2, \cdots\} \\ 0 & \text{if } N_i(w) = 0 \end{cases}$$

(10.34)

where $c_i H(S_{ij})$ can be interpreted as the discounted repair cost upon a system failure due to the jth failure of component i at the time S_{ij}. Obviously $c_i H(S_{ij})$ depends on the repair cost related to the failed component, which is assumed to be deterministic, it also depends on the failure times of the component, which are random variables in nature and statistically dependent on each other. Since for a series system any component failure will cause a system failure and no simultaneous failures are allowed, we can write $C(w)$ in terms of $Y_i(w)$

$$C(w) = \sum_{i=1}^{q} Y_i(w)$$

(10.35)

From Equation (10.35), it is clear that the system DWC is simply a weighted superposition of q independent random variables $Y_i(w), i = 1, 2, \cdots, q$, and each of them has exactly the same structure as of the $Y(w)$ defined in Section 10.2.1. Therefore, it is easy to verify that

$$E[C(w)] = \sum_{i=1}^{q} c_i \Lambda_i(w) E[H(U_i)] = \sum_{i=1}^{q} c_i \int_0^w H(u) \lambda_i(u) du$$

(10.36)

$$V[C(w)] = \sum_{i=1}^{q} c_i^2 \Lambda_i(w) E[H^2(U_i)] = \sum_{i=1}^{q} c_i^2 \int_0^w (H(u))^2 \lambda_i(u) du$$

(10.37)

where $f_{U_i}(u) = \dfrac{\lambda_i(u)}{\Lambda_i(w)}$, $0 \le u \le w$, $\forall i, i = 1, 2, \dots, q$.

Now, let's determine the functional form of the discounting function $H(\cdot)$. There are two most popular ways of time discounting: the continuous discounting method and the discrete discounting method (Blischke and Murthy 1996, pp.282, 794). For the continuous time discounting, the discounting function is given by $H(S_{ij}) = e^{-\delta S_{ij}}$. Under the discrete time discounting, $H(S_{ij}) = (1 + \delta)^{-S_{ij}}$.

Let $C_{d_1}(w)$ be the DWC under the continuous discounting, we then have

$$E[C_{d_1}(w)] = \sum_{i=1}^{q} c_i \int_0^w e^{-\delta u} \lambda_i(u) du$$

(10.38)

$$V[C_{d_1}(w)] = \sum_{i=1}^{q} c_i^2 \int_0^w e^{-2\delta u} \lambda_i(u) du$$

(10.39)

Similarly, under the discrete time discounting, for the DWC $C_{d_2}(w)$,

$$E[C_{d_2}(w)] = \sum_{i=1}^{q} c_i \int_0^w (1+\delta)^{-u} \lambda_i(u)du \qquad (10.40)$$

$$V[C_{d_2}(w)] = \sum_{i=1}^{q} c_i^2 \int_0^w (1+\delta)^{-2u} \lambda_i(u)du \qquad (10.41)$$

Next we discuss three special cases based on the above results considering exponential and Weibull failure time distributions.

Case I: Exponential distribution

Suppose all components have constant failure rate, that is, $\lambda_i(t) = \lambda_i$, $\forall i, i = 1,2,\cdots,q$. This implies that the components' failure processes are independent Poisson processes with $\Lambda_i(w) = \lambda_i w$. So, the *pdf* of U_i is simply $f_{U_i}(u) = \dfrac{1}{w}, 0 \le u \le w$. Under the continuous discounting, from Equations (10.38) and (10.39), we obtain

$$E[C_{d_1}(w)] = \frac{(1-e^{-\delta w})}{\delta} \sum_{i=1}^{q} c_i \lambda_i$$

$$V[C_{d_1}(w)] = \frac{(1-e^{-2\delta w})}{2\delta} \sum_{i=1}^{q} c_i^2 \lambda_i$$

It is worth noting that, for $q = 1$, the expected DWC is exactly the same as that in Chun and Twang (1999), but our approach is simpler and more general. While they do not discuss the variance of DWC, we derive the explicit expression here, which is critical in evaluating warranty cost risk. Ja *et al.* (2002) obtain the non-centered first and second moments of the DWC of four types of warranty policies for one-component products. It turns out that for $q = 1$ the result agrees with theirs. For the case of discrete discounting, using Equations (10.40) and (10.41), upon simplification, we have

$$E[C_{d_2}(w)] = \frac{1-(1+\delta)^{-w}}{\ln(1+\delta)} \sum_{i=1}^{q} c_i \lambda_i \qquad (10.42)$$

$$V[C_{d_2}(w)] = \frac{1-(1+\delta)^{-2w}}{2\ln(1+\delta)} \sum_{i=1}^{q} c_i^2 \lambda_i \qquad (10.43)$$

Case II: Rayleigh distributions

Suppose that for component i, the failure time before any repairs follows a Rayleigh distribution, $\forall i, i = 1,2,\cdots,q$. Therefore, $\lambda_i(t) = \lambda_i t$, $\Lambda_i(w) = \lambda_i w^2 / 2$ and $f_{U_i}(u) = 2u/w^2$, $0 < u \le w$. So, under the continuous discounting, we obtain

$$E[C_{d_1}(w)] = \frac{1-e^{-w\delta} - \delta w e^{-w\delta}}{\delta^2} \sum_{i=1}^{q} c_i \lambda_i \qquad (10.44)$$

$$V[C_{d_1}(w)] = \frac{1 - e^{-2w\delta} - 2\delta w e^{-2w\delta}}{4\delta^2} \sum_{i=1}^{q} c_i^2 \lambda_i \tag{10.45}$$

Similarly, for the discrete discounting, the results are

$$E[C_{d_2}(w)] = \frac{1 - w(1+\delta)^{-w}\ln(1+\delta) - (1+\delta)^{-w}}{2\ln(1+\delta)} \sum_{i=1}^{q} c_i \lambda_i \tag{10.46}$$

$$V[C_{d_2}(w)] = \frac{1 - 2w(1+\delta)^{-2w}\ln(1+\delta) - (1+\delta)^{-2w}}{8\ln(1+\delta)} \sum_{i=1}^{q} c_i^2 \lambda_i \tag{10.47}$$

Case III: Weibull distributions

Now consider the case that all new components' failure times follow Weibull distribution. Or equivalently, $\lambda_i(t) = \frac{\alpha_i}{\beta_i} t^{\alpha_i - 1}, \alpha_i \geq 1, \beta_i > 0, \forall i, i = 1, 2, \cdots, q.$ Hence $\Lambda_i(w) = w^{\alpha_i} / \beta_i$ and $f_{U_i}(u) = \alpha_i u^{\alpha_i - 1} / w^{\alpha_i}, 0 \leq u \leq w.$ If the continuous discounting is considered, then

$$E[C_{d_1}(w)] = \sum_{i=1}^{q} c_i \int_0^w \frac{\alpha_i}{\beta_i} u^{\alpha_i - 1} e^{-\delta u} du \tag{10.48}$$

$$V[C_{d_1}(w)] = \sum_{i=1}^{q} c_i^2 \int_0^w \frac{\alpha_i}{\beta_i} u^{\alpha_i - 1} e^{-2\delta u} du \tag{10.49}$$

In general, it is necessary to evaluate $E[C_{d_1}(w)]$ and $V[C_{d_1}(w)]$ numerically. However, if α_i s are positive integers, close form expressions do exist. For example, let $\alpha_i = 3, \forall i, i = 1, 2, \cdots, q,$ simplifying (10.48) and (10.49), the results are

$$E[C_{d_1}(w)] = \frac{3(2e^{\delta w} - \delta^2 w^2 - 2\delta w - 2)}{\delta^3 e^{\delta w}} \sum_{i=1}^{q} \frac{c_i}{\beta_i}$$

$$V[C_{d_1}(w)] = \frac{3(e^{2\delta w} - 2\delta^2 w^2 - 2\delta w - 1)}{4\delta^3 e^{2\delta w}} \sum_{i=1}^{q} \frac{c_i^2}{\beta_i}$$

If instead the discrete discounting is adopted, for $\alpha_i = 3, \forall i, i = 1, 2, \cdots, q,$ the expected DWC and the cost variance are:

$$E[C_{d_2}(w)] = \frac{2((1+\delta)^w - w^2 \ln(1+\delta) - w\ln(1+\delta) - 1)}{(1+\delta)^w \ln(1+\delta)} \sum_{i=1}^{q} \frac{c_i}{\beta_i}$$

$$V[C_{d_2}(w)] = \frac{(1+\delta)^{2w} - 4w^2 \ln(1+\delta) - 2w\ln(1+\delta) - 1}{4(1+\delta)^{2w} \ln(1+\delta)} \sum_{i=1}^{q} \frac{c_i^2}{\beta_i}$$

10.2.3 DWC Under a PRW Policy

This section considers a PRW policy with period w for the q-component series system, and will derive the mathematical expressions for expected DWC and the cost variance.

Suppose that the PRW policy specifies that a consumer will get the amount of refund depending on the cost of the failed component instead of the total purchase price, then mathematically we can define the refund function, $K(\cdot)$, as follows: let $S_{ij}, 0 \le S_{ij} \le w$, be a system failure time, or equivalently, the jth failure time of component i, the warranty refund amount due to this failure is given by

$$K(S_{ij}) = c_i(1 - \frac{S_{ij}}{w})$$

Let $K'(S_{ij})$ be the ratio of the refund and the original cost c_i, then

$$K'(S_{ij}) = \left(1 - \frac{S_{ij}}{w}\right) \tag{10.50}$$

Again, let $H(\cdot)$ be the discounting function and define $Y_i'(w), \forall i, i \in \{1, 2, \cdots, q\}$ as follows:

$$Y_i'(w) = \begin{cases} \sum_{j=1}^{N_i(w)} c_i K'(S_{ij})H(S_{ij}) & \text{if } N_i(w) \in \{1, 2, \cdots\} \\ 0 & \text{if } N_i(w) = 0 \end{cases} \tag{10.51}$$

then the total discounted refund for a warranted series system can be written as

$$C(w) = \sum_{i=1}^{q} Y_i'(w) \tag{10.52}$$

Not surprisingly, if we let $G(S_{ij}) = K'(S_{ij})H(S_{ij})$, we can directly apply the results in Equations (10.36) and (10.37) after replacing the $H(\cdot)$ function by the $G(\cdot)$ function. Therefore, the case of determining the discounted value of the PRW policy is a special case of what we discussed in Section 10.2.2.

Equation (10.50) implicitly assumes that the refund function is linear in components' failure time (Blischke and Murthy 1994, p.295). However, our formulation does not necessarily require the linearity assumption. Actually it can be any continuous function. We also assume that each refund depends on component's cost instead of the system purchase price (or manufacturer's production cost), which is usually the refund base in practice. Nevertheless, theoretically speaking our formulation is more general and it can be modified easily when the purchase price is considered. In particular, for $q = 1$, the choice of base price has virtually no impact to our model since one can always interpret c1 as the purchase price. It should be noted that PRW is often used for non-repairable products such as automobile batteries, television picture tubes, and so forth (Blischke and Murthy 1994, p.169). However, such a policy could still be used for repairable products because: (1) consumers might be willing to share the repair cost with manufacturers for the benefits they has received after the purchase; (2) warranty issuers have the choice of offering longer warranties, which may be more attractive to buyers, provided that some of the cost is shared by consumers.

For illustration purposes, let us continue the Case III in Section 10.2.2 by considering the linear PRW policy with warranty period w. Since all new components' failure times are assumed to follow Weibull distributions, we have, $\forall i, i = 1, 2, ..., q$,

$$\text{(a) } \lambda_i(t) = \frac{\alpha_i}{\beta_i} t^{\alpha_i - 1}, \text{ (b) } \Lambda_i(w) = \frac{w_i^\alpha}{\beta_i}, \text{ (c) } f_{U_i}(u) = \frac{\alpha_i u^{\alpha_i - 1}}{w^{\alpha_i}}, \quad 0 \leq u \leq w.$$

If the continuous discounting is considered, then $G(U_i) = (1 - U_i / w)e^{-\delta U_i}$. For $\alpha_i = 3, \forall i, i = 1, 2, \cdots, q$, using Equations (10.36) and (10.37) (replacing $H(\cdot)$ by $G(\cdot)$), upon simplification, we obtain the following:

$$E[C_{d_1}(w)] = \frac{3(6 + 4w\delta + \delta^2 w^2 - 6e^{\delta w} + 2w\delta e^{\delta w}}{w\delta^4 w^{\delta w}} \sum_{i=1}^{q} \frac{c_i}{\beta_i}$$

$$V[C_{d_1}(w)] = \frac{-3(3 + 3w\delta + \delta^2 w^2 - 3e^{2\delta w} - \delta^2 w^2 e^{2\delta w} + 3w\delta e^{2\delta w}}{4w^2 \delta^5 w^{2\delta w}} \sum_{i=1}^{q} \frac{c_i^2}{\beta_i}$$

10.2.4 Numerical Examples

In the design phase of warranty programs, warranty program managers are often faced with the following two important questions:

(I) Warranty period determination (WPD) problem

Suppose that the budget c_0 for a warranty policy (either a FRW or PRW) is given. The warranty managers would like to determine the warranty period w^*, which is attractive to consumers, at the same time, the probability that the true warranty cost is over the budget is less or equal to α. Clearly it is desirable for manufacturers to offer the best possible warranty policy provided that the budget is followed properly. Therefore, w^*, the optimal warranty period, can be formulated as

$$w^* = \sup\{w : P[C(w) > c_0] \leq \alpha\}.$$

(II) Warranty reserve determination (WRD) problem

If due to competition, w, the parameter of the warranty period, is pre-determined, but it is of interest to find the required warranty reserve level c_{wr}^* such that the probability that the warranty reserve c_{wr} will be depleted is controlled. The corresponding mathematical expression for c_{wr}^* is

$$c_{wr}^* = \inf\{c_{wr} : P[c_{wr} > C(w)] \geq 1 - \alpha\}.$$

Mathematically, these two problems are like dual problems. The main difficulty lies in the complexity of the distribution of $C(w)$, which in general is unknown and difficult to obtain. However, with the information of the first and second centered moments of DWC per product, it is possible to utilize a normal approximation to solve them for the cases like single lot sales (Patankar and Worm 1981, p.142).

Let L be the size of a single lot sale, which is a relatively large number. The warranty policies under consideration are FRW or PRW. The total DWC (or the present value (PV) of the refund) for the lot sale is simply

$$TC = \sum_{i=1}^{L} C_i(w) \tag{10.53}$$

where $C_i(w)$ is the PV or the DWC of product i. Obviously, $C_i(w)$ s are $i.i.d.$. whose expectation $E[C(w)]$ and the variance $V[C(w)]$ are given in Section 10.2.2 and Section 10.2.3 for the FRW policy or the PRW policy respectively. By the central limit theory, as L is relatively large, the distribution of TC is approximately normal regardless the distribution of $C_i(w)$. Furthermore, from Equation (10.53) it is easy to obtain:

$$E[TC] = L \cdot E[C(w)] \tag{10.54}$$
$$V[TC\} = L \cdot V[C(w)] \tag{10.55}$$

Consequently, for the WRD problem of the lot sale, denote TC_{wr}^* as the required warranty reserve level, then

$$TC_{wr}^* = L \cdot E[C(w)] + z_{1-\alpha} \sqrt{L \cdot V[C(w)]} \tag{10.56}$$

If the warranty budget c_0 is given and it is desirable to determine the warranty period w, then this becomes the WPD problem. Again let w^* be the required warranty duration, then w^* is simply the solution for the following non-linear equation:

$$\frac{c_0 - L \cdot E[C(w)]}{\sqrt{L \cdot V[C(w)]}} = z_{1-\alpha} \tag{10.57}$$

where $z_{1-\alpha}$ is the $(1-\alpha)$ quantile of the standard normal distribution.

It is necessary to check the existence and the uniqueness of w^*. Let $\psi(w) = z_{1-\alpha} \sqrt{L \cdot V[C(w)]} + L \cdot E[C(w)] - c_0$; then it is sufficient to show that there exists a unique w^* such that $\psi(w^*) = 0$. Clearly $\psi(w)$ is continuous and monotonically increasing in w if both $E[C(w)]$ and $V[C(w)]$ are continuous and increasing functions of w, which are true in general. As $w \to 0$, $\psi(w) \to -c_0$, which is negative provided that $c_0 > 0$. Also it is obvious that there exists a positive number w^+ such that $\psi(w^+) > 0$. Therefore, a unique w^* exists in $(0, \infty)$ that can be expressed as

$$w^* = \psi^{-1}(0) \tag{10.58}$$

where $\psi^{-1}(0)$ is the inverse function of $\psi(\cdot)$ evaluated at the point 0. The existence of $\psi^{-1}(\cdot)$ is given by the fact that $\psi(\cdot)$ is a monotonically increasing function in $(0,\infty)$.

One can use Equation (10.58) directly for the purpose of determining w^*. In case the functional form of $\psi^{-1}(\cdot)$ is complicated and hard to obtain due to the complexity of $\psi(w)$, the Newton-Raphson method can be applied.

Consider a numerical example. Suppose the warranty policy under consideration is a FRPW with $w = 1$ and $L = 1000$. The product under warranty is a three-component series system for which all failures under warranty are minimally repaired. Other parameters are given in Table 10.3.

Table 10.3. Parameters for the 3-component minimally repaired series system

Component No.	1	2	3
c_i	$100	$150	$200
$\lambda_i(t)$	0.0611	0.0423t	0.0187t^2

If the continuous discounting method is used and the discounting rate δ is 5%, then according to Equations (10.38) and (10.39) and using the intermediate results in Cases I through III in Section 10.2.2, we have that $E[C_{d_1}(w)] = \$10.23$ and $V[C_{d_1}(w)] = 9.94$. Therefore, for the WRD problem, the required warranty reserve level TC^*_{wr} is $10461.25 for $\alpha = 0.01$.

For the WPD problem, if we are given that $c_0 = TC^*_{wr}$, then obviously the required warranty duration w^* is 1. However, in practice usually c_0 is different from TC^*_{wr} due to various budget constraints faced by warranty managers. If it is known that $c_0 = \$20,000$, by employing the Newton-Raphson method, we have that $w^* = 1.5037$. One should note that in this example both explicit forms exist for $E[C_{d_1}(w)]$ and $V[C_{d_1}(w)]$, which makes the computation of w^* relatively easy. However, even if no explicit form exists for either $E[C_{d_1}(w)]$ and $V[C_{d_1}(w)]$, Newton-Raphson method can still be applied to find w^*, for example, by considering polynomial approximations to $E[C_{d_1}(w)]$ and $V[C_{d_1}(w)]$ whenever necessary (Rustagi 1994, p.148).

10.2.5 Future Research

Section 10.2 has covered modeling of DWC for series systems with minimally repaired components. Both the expected value and the variance of DWC for FRW policies and PRW policies are considered. As shown in the applications, the results in this section can be used to determine the warranty duration or the required level of warranty reserve, two primary questions to be answered by warranty planning managers.

There may be some potential extensions to the warranty cost analysis of repairable complex systems: First is to consider imperfect repair (Pham and Wang 1996, p.146) instead of minimal repair. Second, one may consider non-instant repair as against to instant repair. Although instant repair can be justified when repair time is relatively short compared to warranty period, it might raise some concerns for multi-component systems since as the size of the system increases, the total repair time might be long enough that it has to be considered explicitly. Besides, in the automobile industry, additional warranty cost, such as transportation expense, may occur if a warranted vehicle cannot be repaired within one working day. Third, one may take into account of different system structure on warranty cost analysis. As shown in Section 10.1, various system structures could result in different warranty cost models. In this section, only series systems are considered. It is desirable to model DWC for more complicated systems such as series-parallel systems, parallel-series systems and k-out-of-n systems.

10.3 RLRFW Policies with Imperfect Repair

Numerous warranties have been studied in the past several decades. In general, one can divide them into three categories: free repair/replacement warranty (FRW), pro-rata warranty (PRW) and combination warranty that contains features of both FRW and PRW. In this section we introduce a repair-limit risk-free warranty (RLRFW) of fixed period w, based on Bai and Pham (2005). Different from ordinary FRW policies, this policy has a pre-determined limit m on the number of repairs. If there are more than m system failures within w, the failed product will be replaced instead of being repaired again. Such a policy is desirable for both manufacturers and consumers. For consumers, surely they will prefer such a policy to a simple free repair policy since there are chances that they could own another new product for free. From manufacturers' point of view, first of all, such a policy offers extra incentive for consumers to purchase their products. Second, if a single product has failed m times before w, this might have provided sufficient information that the particular product is indeed of low quality. So it could be economically sound for the manufacturer to simply provide replacements without wasting more time on repairs. In addition, such extra compensation for those unlucky consumers may effectively reduce the chance of high-cost lawsuits due to those products with 'proven' bad quality.

10.3.1 Introduction

As discussed in Chapter 3, many researchers have studied various repair limit problems, which can be categorized into three groups: repair-number limit problems, repair-time limit problems and repair-cost limit problems. Almost all researchers dealing with repair-limit problems assumed infinite horizon and performed the analysis based on asymptotic cost measures such as long-run average cost. This section focuses on finite horizon for the proposed repair-number limit warranty policy.

The main analytical tool used in this section is the truncated quasi-renewal processes. The concept and some properties of truncated quasi-renewal processes can be found in Chapter 4.

Another important issue in warranty cost analysis is the variability of warranty cost. It is often not sufficient for warranty managers to simply obtain the estimate of expected warranty cost. Additional information about variability of warranty cost is essential to evaluate risks involved in warranty programs.

The rest of this section is organized as follows. Section 10.3.2 provides the detailed analysis of the repair-limit risk-free warranty policy. Several special cases are discussed in Section 10.3.3. Section 10.3.4 presents sensitivity analyses for various policy parameters based on numerical examples.

In this section, we assume:

i)   All warranty service is instant.
ii) Repairs are imperfect and the failure process before the first replacement follows a truncated quasi-renewal process with parameters $\alpha (0 < \alpha < 1)$ and F.
iii) All warranty claims are executed and all claims are valid.
iv) Both repair cost c_a and replacement cost c_b are constant and $c_a < c_b$ to avoid triviality.

10.3.2 Analysis of Repair-limit Risk-free Warranties

In this section we discuss the warranty model of the repair-limit risk-free policy with warranty period w and repair limit m, based on Bai and Pham (2005). We assume that repairs are imperfect such that after each repair, the system is between the states "as good as new" and "as bad as old". Suppose that the failure process with the imperfect repair follows a quasi-renewal process with parameter α $(0 < \alpha < 1)$ and F (see Chapter 4). Since any warranted products will be repaired no more than m times according to this policy, the failure process before the first replacement is actually a quasi-renewal process truncated above m. As a reminder, by the definition of truncated quasi-renewal processes, the inter-failure times in the process are independent and have the distributions $F(\alpha^{1-n}x), n = 0,1,\cdots,m$.

One should note that for a repair-limit risk-free warranty policy, only Model II of truncated quasi-renewal processes is appropriate since the repair limit m has no impact on the probabilities of the number of repairs except when it is equal to m.

Let $N_a(w)$ and $N_b(w)$ be the number of repairs and the number of replacements within the warranty duration respectively. Since w is predetermined, we will suppress it later on for simplicity. Denote c_a as the repair cost per failure and c_b the replacement cost per unit, then for the warranty cost per product sold C, we have

$$E[C] = c_a E[N_a] + c_b E[N_b] \qquad (10.59)$$

and

$$V[C] = c_a^2 V[N_a] + c_b^2 V[N_b] + 2c_a c_b COV(N_a, N_b) \tag{10.60}$$

Note that N_a and N_b are correlated and $COV(N_a, N_b) \neq 0$. The relationship between them can be summarized as followings:

i) $N_a < m$ implies $N_b = 0$
ii) $N_b > 0$ implies $N_a = m$
iii) $N_b = 0$ implies $N_a \leq m$
iv) $N_a = m$ implies $N_b \geq 0$

Moments of N_a

Since under the warranty replacement instead of repair will be performed if there are more than m failures within w, it is obvious that N_a is a realization of a quasi-renewal process truncated above m. As mentioned at the beginning of Section 10.3.1, Model II of truncated quasi-renewal processes can be used in describing the behavior of N_a. Hence we have

$$E[N_a] = \sum_{i=1}^{m} G^{(i)}(w) \tag{10.61}$$

$$E[N_a^2] = \sum_{i=1}^{m} (2i-1) G^{(i)}(w) \tag{10.62}$$

The Pivot Point S_m

A pivot point is the time epoch that indicates the change of the type of warranty service. For the repair-limit risk-free policy, obviously S_m is the pivot point since after it any failed products will be replaced instead of being repaired again. Let $H(t_p)$ be the *cdf* of S_m, then

$$H(t_p) \equiv P[S_m \leq t_p]$$
$$= G^{(m)}(t_p), \qquad t_p \geq 0 \tag{10.63}$$

Moments of N_b

Suppose $S_m = t_p, 0 \leq t_p \leq w$, then starting from t_p, the system failure process becomes a delayed renewal process with the first failure time having the distribution $F_{m+1}(x)$ where $F_{m+1}(x) = F(\alpha^{-m} x)$ and all the following failure times are *i.i.d.* with distribution F. Conditioning on $S_m = t_p, t_p \leq w$, from the renewal theory, $E[N_b | S_m = t_p, t_p \leq w] = M_d(w - t_p)$ where $M_d(\cdot)$, the renewal function for the delayed renewal process, is given by

$$M_d(t) = \sum_{i=0}^{\infty} F_{m+1} * F^{(i)}(t)$$

where $F^{(i)}(\cdot)$ is the i-fold convolution of $F(\cdot)$ itself and $F^0(\cdot) = 1$ and

$$F_{m+1} * F^{(i)}(t) = \int_0^t F_{m+1}(t-x)dF^{(i)}(x), \quad i \ge 0$$

is the convolution of F_{m+1} and $F^{(i)}$. After un-conditioning on S_m,

$$E[N_b] = \int_0^w M_d(w-t_p)dH(t_p) \tag{10.64}$$

Similar technique can be used to obtain $E[N_b^2]$. Let $M_{d,2}(w-t_p) \equiv E[N_b^2 | S_m = t_p, t_p \le w]$ then

$$E[N_b^2] = \int_0^w M_{d,2}(w-t_p)dH(t_p) \tag{10.65}$$

where $M_{d,2}(\cdot)$ can obtained by

$$M_{d,2}(t) = \sum_{i=0}^{\infty}(2i+1)F_{m+1} * F^{(i)}(t).$$

Covariance of (N_a, N_b)

Next we determine covariance $COV(N_a, N_b)$. Since $COV(N_a, N_b) = E[N_a N_b] - E[N_a]E[N_b]$, it is sufficient to know $E[N_a N_b]$. Note

$$E[N_a N_b] = \sum_{n_b=0}^{\infty}\sum_{n_a=0}^{m} n_a n_b P[N_a = n_a, N_b = n_b]$$

$$= m\sum_{n_b=0}^{\infty} n_b p[m, n_b]$$

$$= m\sum_{n_b=1}^{\infty} n_b \int_0^w P[n_b | S_m = t_p]dH(t_p)$$

and for $t_p \le w$,

$$P[N_b = 0 | S_m = t_p] = 1 - F_{m+1}(w-t_p)$$

$$P[N_b = n_b | S_m = t_p] = F_{m+1} * G^{(n_b-1)}(w-t_p) - F_{m+1} * G^{(n_b)}(w-t_p), \quad n_b \ge 1$$

we obtain

$$E[N_a N_b] = m\sum_{n_b=1}^{\infty} n_b \int_0^w \left[F_{m+1} * G^{(n_b-1)}(w-t_p) - F_{m+1} * G^{(n_b)}(w-t_p)\right]dH(t_p) \tag{10.66}$$

Consequently,

$$COV[N_a, N_b] = m \sum_{n_b=1}^{\infty} n_b \int_0^w \left[F_{m+1} * G^{(n_b-1)}(w - t_p) - F_{m+1} * G^{(n_b)}(w - t_p) \right] dH(t_p)$$

$$- \int_0^w M_d(w - t_p) dH(t_p) \sum_{i=1}^m G^{(i)}(w) \tag{10.67}$$

Main Results: First and Second Moments of $C(w)$

Substituting Equations (10.61) and (10.64) into (10.59), we finally have the expected warranty cost per unit sold:

$$E[C] = c_a \sum_{i=1}^m G^{(i)}(w) + c_b \int_0^w M_d(w - t_p) dH(t_p) \tag{10.68}$$

The variance of the warranty cost per unit sold can be obtained through Equations (10.62), (10.65) and (10.67):

$$V[C] = c_a^2 \{ \sum_{i=1}^m (2i - 1) G^{(i)}(w) - [\sum_{i=1}^m G^{(i)}(w)]^2 \}$$

$$+ c_b^2 \{ \int_0^w M_{d,2}(w - t_p) dH(t_p) - [\int_0^w M_d(w - t_p) dH(t_p)]^2 \}$$

$$+ 2c_a c_b \{ m \sum_{n_b=1}^{\infty} n_b \int_0^w [F_{m+1} * G^{(n_b-1)}(w - t_p) - F_{m+1} * G^{(n_b)}(w - t_p)] dH(t_p)$$

$$- \int_0^w M_d(w - t_p) dH(t_p) \sum_{i=1}^m G^{(i)}(w) \} \tag{10.69}$$

10.3.3 Special Cases

We have derived $E(C)$ and $V(C)$, as shown in Equations (6.68) and (6.69). It is difficult to evaluate $E(C)$ and $V(C)$ analytically, and now we discuss some special cases.

Case I: Suppose for finite w, $m = 0$. In this case, no repair is allowed so all failed products within w will be replaced always. This implies that $F_i \sim F, \forall i, i \geq 1$. So our policy degenerates to the regular free replacement policy. As a result, Equation (10.68) becomes

$$E[C] = c_b M(w) \tag{10.70}$$

Equation (10.69) changes to

$$V[C] = c_b^2 \{ M_2(w) - [M(w)]^2 \} \tag{10.71}$$

where $M(t)$ and $M_2(w)$ are the first and the second moments of the number of renewals in a renewal process associated with F.

These are the well-known results for the FRW policy (see Blischke and Murthy 1994).

Case II: Consider $m = \infty$ and w is finite. Thus no change of warranty service will ever happen and all failed products within w will be repaired. Consequently, we have

$$E[C] = c_a \sum_{i=1}^{\infty} G^{(i)}(w) \qquad (10.72)$$

and

$$V[C] = c_a^2 \{ \sum_{i=1}^{\infty} (2i-1)G^{(i)}(w) - [\sum_{i=1}^{\infty} G^{(i)}(w)]^2 \} \qquad (10.73)$$

These results agree with the study in Wang and Pham (1996).

Case III: Suppose for finite positive integer valued m, w is large such that it can be treated as infinity. In this case, $E[C] \to \infty$ since it is strictly increasing in w. One may be interested in determining warranty cost per unit time (long-run average cost), another cost measure that is commonly used in the warranty and maintenance literature. It is not difficult to see that the long-run average cost $E[C']$ is given by

$$E[C'] = \frac{c_b}{\int_{-\infty}^{+\infty} x\,dF(x)} \qquad (10.74)$$

It is worth noting that this measure is only an approximation for the true warranty cost per unit time, and its accuracy heavily depends on the magnitude of w compared to the product life times.

Case IV: Assume F is a normal distribution with mean μ and variance σ^2 for finite positive integer-valued m and finite w, i.e., $F \sim N(\mu, \sigma^2)$. As a result, the interarrival failure times under the imperfect repairs are independent and also follow the normal distribution. In particular, it is easy to see that $F_i \sim N(\alpha^{i-1}\mu, \alpha^{2(i-1)}\sigma^2)$. Thus $G^{(i)} \sim N(\frac{1-\alpha^i}{1-\alpha}\mu, \frac{1-\alpha^{2i}}{1-\alpha}\sigma^2)$. The pivot point distribution $H(t_p)$ is given by

$$H(t_p) = \psi((t_p - \frac{1-\alpha^m}{1-\alpha}\mu) / [\sigma\sqrt{(1-\alpha^{2m})/(1-\alpha^2)}]) \qquad (10.75)$$

where $\psi(\cdot)$ is the *cdf* of the standard normal distribution.

To compute $M_d(w - t_p)$, we need to evaluate $F_{m+1} * F^{(i)}$, which obeys the distribution $N((\alpha^m + i)\mu, (\alpha^{2m} + i)\sigma^2)$. So

$$M_d(w-t_p)=\sum_{i=0}^{\infty}\psi(\frac{w-t_p-(\alpha^m+i)\mu}{\sigma\sqrt{\alpha^{2m}+i}} \qquad (10.76)$$

Similarly,

$$M_{d,2}(w-t_p)=\sum_{i=0}^{\infty}(2i+1)\psi(\frac{w-t_p-(\alpha^m+i)\mu}{\sigma\sqrt{\alpha^{2m}+i}}) \qquad (10.77)$$

It is also necessary to obtain $F_{m+1}*G^{n_b-1}$ and $F_{m+1}*G^{n_b}$. Clearly, they again follow the normal distribution with parameters

$$((\alpha^m+\frac{1-\alpha^{n_b-1}}{1-\alpha})\mu,(\alpha^{2m}+\frac{1-\alpha^{2(n_b-1)}}{1-\alpha^2})\sigma^2) \text{ and}$$

$$((\alpha^m+\frac{1-\alpha^{n_b}}{1-\alpha})\mu,(\alpha^{2m}+\frac{1-\alpha^{2n_b}}{1-\alpha^2})\sigma^2) \text{ respectively.}$$

To obtain the expected warranty cost, combining the previous results together, Equation (6.68) is simplified to

$$E[C]=c_a\sum_{i=1}^{m}\psi(\frac{w-\mu(1-\alpha^i)/(1-\alpha)}{\sigma\sqrt{(1-\alpha^{2i})/(1-\alpha^2)}})$$

$$+c_b\int_0^w\sum_{i=0}^{\infty}\psi(\frac{w-t_p-(\alpha^m+i)\mu}{\sigma\sqrt{\alpha^{2m}+i}}d\psi(\frac{t_p-\dfrac{1-\alpha^m}{1-\alpha}\mu}{\sigma\sqrt{(1-\sigma^{2m})/(1-\alpha^2)}}). \qquad (10.78)$$

The variance can be simplified in a similar way:

$$V[C]=c_a^2\{\sum_{i=1}^{m}(2i-1)\psi(\frac{w-\dfrac{1-\alpha^i}{1-\alpha}\mu}{\sigma\sqrt{\dfrac{1-\alpha^{2i}}{1-\alpha^2}}})-[\sum_{i=1}^{m}\psi(\frac{w-\dfrac{1-\alpha^i}{1-\alpha}\mu}{\sigma\sqrt{\dfrac{1-\alpha^{2i}}{1-\alpha^2}}})]^2\}$$

$$+c_b^2\{\int_0^w\sum_{i=0}^{\infty}(2i+1)\psi(\frac{w-t_p-(\alpha^m+i)\mu}{\sigma\sqrt{(\alpha^{2m}+i)}})d\psi(\frac{t_p-\dfrac{1-\alpha^m}{1-\alpha}\mu}{\sigma\sqrt{(1-\alpha^{2m})/(1-\alpha^2)}})$$

$$-[\int_0^w\sum_{i=0}^{\infty}\psi(\frac{w-t_p-(\alpha^m+i)\mu}{\sigma\sqrt{\alpha^{2m}+i}})d\psi(\frac{t_p-\dfrac{1-\alpha^m}{1-\alpha}\mu}{\sigma\sqrt{(1-\alpha^{2m})/(1-\alpha^2)}})]^2\}$$

$$+2c_ac_b\{m\sum_{n_b=1}^{\infty}n_b\int_0^w[\psi(\frac{w-t_p-(\alpha^m+\dfrac{1-\alpha^{n_b-1}}{1-\alpha})\mu}{\sigma\sqrt{\alpha^{2m}+\dfrac{1-\alpha^{2(n_b-1)}}{1-\alpha^2}}})$$

$$
\begin{aligned}
-\psi(\frac{w-t_p-(\alpha^m+\dfrac{1-\alpha^{n_b-1}}{1-\alpha})\mu}{\sigma\sqrt{\alpha^{2m}+\dfrac{1-\alpha^{2n_b}}{1-\alpha^2}i}})]d\psi(\frac{t_p-\dfrac{1-\alpha^m}{1-\alpha}\mu}{\sigma\sqrt{(1-\alpha^{2m})/(1-\alpha^2)}}) \\[2em]
-\int_0^w\sum_0^\infty\psi(\frac{w-t_p-(\alpha^m+i)\mu}{\sigma\sqrt{\alpha^{2m}+i}})d\psi(\frac{t_p-\dfrac{1-\alpha^m}{1-\alpha}\mu}{\sigma\sqrt{\dfrac{(1-\alpha^{2m})}{(1-\alpha^2)}}})\sum_{i=1}^m\psi(\frac{w-\dfrac{1-\alpha^i}{1-\alpha}\mu}{\sigma\sqrt{\dfrac{1-\alpha^{2i}}{1-\alpha}}})\}
\end{aligned}
$$

$$(10.79)$$

10.3.4 Numerical Examples and Sensitivity Analysis

Now, let us consider a numerical example. Suppose $F \sim N(4,1)$, $m=1$, $w=2.5$, $\alpha=.70$, $c_a=\$100$, and $c_b=\$5000$. Using Equation (6.80), we obtain that $E(C)=\$7.6726$, which only accounts for 0.15% of the unit production (replacement) cost. By looking into the components of the warranty cost, we find that the repair cost is the dominant source as it contributes 87.07% to $E(C)$. This is what one should expect since the probability of more than one failure within w is very small (0.02%), indicating that most of the time no replacement will ever happen under the warranty. The standard deviation (std) of the warranty cost is 75.9571, which indicates moderate risk contained in this warranty policy. When decomposing $V(C)$, not surprisingly we find that the dominant source of the variation in the warranty cost is from the replacement cost. The contributions in the variation from repair, replacement and the interaction between them are 10.81%, 85.99% and 3.21%, respectively.

Table 10.4. $E[C]$ and std(C) for $\alpha = 0.7$, $c_a = \$100$ and $c_b = \$5000$

w	E[C]				std(C)			
	m = 1	m = 2	m = 3	m = 4	m = 1	m = 2	m = 3	m = 4
1.0	0.1372	0.1351	0.1351	0.1351	4.9754	3.6785	3.6758	3.6757
1.5	0.6442	0.6217	0.6217	0.6217	13.5118	7.8895	7.8693	7.8691
2.0	2.4495	2.2799	2.2792	2.2792	33.6026	15.0592	14.9533	14.9525
2.5	7.6726	6.7069	6.7022	6.7022	75.9571	25.5623	25.0969	25.0921
3.0	20.3751	15.9878	15.9591	15.9587	156.9596	38.8223	36.9034	36.8782

It is of interest to know how the warranty cost measures $E(C)$ and std(C) change with regard to parameters m, w, α, and the replacement-repair cost ratio c_b/c_a. First vary m in {1, 2, 3, 4} and w in {1.0, 1.5, 2.0, 2.5, 3.0} while keeping other parameters unchanged. The corresponding reliability of a new warranted product evaluated at w are within the range of 84.13% to 99.87%.

From Table 10.4, it is clear that both $E(C)$ and std(C) are monotonically increasing in w. This is reasonable since the chance of failures happening increases as the warranty period becomes longer. As m, the repair limit increases from 1 to 4, both $E(C)$ and std(C) decreases, but the magnitude of decreasing becomes much smaller after $m = 2$. This suggests that $m = 2$ could be a good policy choice since the warranty cost and the risk associated with it are relatively small, while at the same time, smaller m tends to be more attractive to consumers.

To investigate how the effort of repair affects the warranty cost, we consider three different levels of α (higher α indicates better repair). All other parameters are kept the same and m is chosen as 2. As expected, as α goes up, both $E[C]$ and std(C) decrease, as indicated in Table 10.5. This implies that it is possible to reduce the warranty cost and the warranty cost risk by improving the quality of repair.

Table 10.5. $E[C]$ and std(C) under various levels of repair efforts

w	$E[C]$			std(C)		
	$a = 0.5$	$a = 0.7$	$a = 0.9$	$a = 0.5$	$a = 0.7$	$a = 0.9$
1	0.1356	0.1351	0.135	3.8738	3.6785	3.6738
1.5	0.6276	0.6217	0.6213	9.1006	7.8895	7.8626
2	2.324	2.2799	2.2766	19.9599	15.0592	14.9325
2.5	6.982	6.7069	6.6885	42.4468	25.5623	25.0368
3	17.4309	15.9878	15.8984	88.7695	38.8223	36.7474

Replacement-repair cost ratio is another important factor in determining the warranty cost. In Table 10.6, we report the results for various cost ratios for $w = 2.5$, $\alpha = 0.9$ and $m = 1$ while c_a is fixed at 100. It turns out that the cost ratio is positively related to both $E[C]$ and std(C), indicating that as replacement becomes more expensive compared to repair, the repair-limit warranty policy tends to be more costly with higher warranty cost risk.

Table 10.6. $E(C)$ and std(C) VS replacement-repair cost ratio

C_b/C_a	2	10	50	100	300
$E[C]$	6.6944	6.7493	7.0235	7.3662	8.7373
std[C]	25.0200	26.2891	48.308	86.4494	249.6575

10.3.5 Concluding Remarks

Section 10.3 has discussed repair-limit risk-free warranty policies and provided the first and second moments of the warranty cost per unit sold by means of truncated quasi-renewal processes. Warranty designers such as manufacturers are constantly in search of novel ideas to promote their products due to the more than ever fierce competition. This repair-limit risk-free warranty could be a good candidate for marketing purpose since it provides extra compensation to consumers suffering from low quality products with a relatively low cost.

The numerical examples assume that the failure time of a new product follows a normal distribution in order to simplify the computational efforts. More numerical studies are needed to include the cases when the failure time is modeled by non-negative distributions such as the Weibull distribution or Gamma distribution.

10.4 Optimal RRLRFW Policies with Minimal Repair

To gain some advantages in the highly competitive markets, manufacturers have to improve product quality continuously or to upgrade their products creatively. However, such strategies may be time consuming and very expensive. In comparison, a simple yet efficient alternative for the marketing purpose is to offer an attractive and comprehensive warranty. This leads to the question of designing and determining the optimal warranty policy. To make this problem more specific, this section focuses on a renewable repair-limit risk-free warranty (RRLRFW) for deteriorating repairable complex products:

Let $S_i, \forall i, i = 1,2,\cdots,m+1$ be the failure arrival times of a new product (or a replacement) under the warranty. For this product, any failure within warranty duration w will be repaired up to m times. If $S_{m+1} \leq w$, then a replacement will be provided. The warranty is then automatically renewed for another period of w. All the warranty service is free of charge to the consumer.

This RRLRFW is denoted as $\varphi(w,m)$, and studied by Bai (2004). This policy has several attractive features. First, it is a generalization of the ordinary renewable free repair/replacement warranty (Blischke and Murthy 1994, p.144), often offered for inexpensive products. Obviously, when $m = 0$, it degenerates to the latter. Second, this type of policies (when $m > 0$) provides consumers better warranty service than the conventional free repair policies because it offers a new replacement whenever more than m failures happened within a period of w, which may be a good indication that this particular product is of low quality or it does not worth being repaired any more. As a result, this policy could be applied for a wide range of products including expensive deteriorating systems. Third, the limit on the number of repairs presents producers the flexibility to control the warranty cost by choosing appropriate m in addition to the usual decision variable w.

To analyze the warranty cost of RRLRFW policies, it is necessary to model the repair impact on product reliability or the failure rate. Section 10.3 has presented a warranty cost model for non-renewable repair-limit risk-free policies based on an imperfect repair assumption. Although the imperfect repair model is flexible in a sense that the repair effort is represented by a parameter in (0, 1], it may not be easy for implementation due to the difficulty in estimating the parameter accurately. In contrast, minimal repair models have been studied extensively. This section assumes that repairs are minimal.

To determine the optimal RRLRFW policy, one has to assess quantitatively the benefit that a manufacturer might generate from a specific warranty. Usually there are two channels that producers could benefit from issuing a warranty: warranty pricing, which is an integrated part of product pricing strategy, and increase in sales or demand. In this section, we focus on the second channel since one of the main reasons of the existence of warranty is to promote sales instead of making extra profit per unit sold from warranty directly. Besides, many producers, as the so-called price takers, do not have the pricing power. For example, they may have to set the price equal or very close to that of the products with similar functionality of the competitors. The majority of the warranty literature focuses on warranty cost modeling and analysis for producers or consumers, and only a few study the demand side of warranty. Menezes (1992) posits a general deterministic demand function depending on price, warranty length, quality and many other economic factors, based on which the optimal warranty length was derived for profit-maximizing producers. Chun and Tang (1995) model consumers' acceptance of warranty through a general probability distribution of consumers' perception on product quality, which was represented by an exponential distribution. A linear form and a quadratic form of w, the warranty period, are employed by Thomas (1983, 1989) to model the warranty benefit directly. Alternatively, we propose a logistic regression model to estimate the demand of products with warranty.

The rest of this section is organized as follows: Section 10.4.1 gives a general optimization model for producers to determine the optimal warranty policy. The consumers are assumed to be homogeneous, thus the probability of accepting a warranted product is invariant among all consumers. A logistic regression model is used to determine the probability. For producers without the pricing power, their objective is to maximize the expected utility of the profit, generated from the products sold with warranty, by selecting appropriate parameters of the warranty based on the forecast of customers' demand. Section 10.4.2 discusses the warranty cost analysis of RRLRFW policies from manufacturers' perspective and some useful properties of the warranty cost per unit sold. These results combined with the optimization model are then used in Section 10.4.3 to determine the optimal RRLRFW through a numerical example. Section 10.4.4 discusses the limitations of the model and some future research directions. Following are assumptions in this section:

i)   Any warranty service is instantaneous.
ii)   All warranty claims are valid.
iii) All failures covered by warranty are claimed.
iv)   Consumers are homogeneous.
v)   Replacement products follow the same failure time distribution as of a new product.
vi)   Any repairs under warranty are minimal.

10.4.1 A General Optimization Model

In this section, we present a general decision model for manufacturers to determine the optimal warranty policy $\psi(\Theta^*)$, where $\Theta^* = \{\theta_1^*, \theta_2^*, \cdots, \theta_n^*\}$ is the optimal set of the parameters of ψ. One may use either ψ^* or Θ^* to represent the optimal policy. Let us suppose that the manufacturer has a utility function $U(\cdot)$, therefore his or her objective is to maximize the expected monetary utility of the profit, π, by choosing appropriate parameters of $\psi(\Theta)$.

10.4.1.1 Profit Per Unit Sold
Let's consider the situation that a typical producer, M, faces the problem of determining optimal parameters of a certain warranty policy $\phi(\Theta)$ for a certain product A, for which the market base, or the number of potential customers is B. In a competitive market with sufficient number of producers, producer M, as one of many other manufacturers in the market, may not have the pricing power. So, assume M is a price-taker, implying that the price of product A is exogenous. Due to technology constraints or the time constraint, it is possible that M cannot reduce the production cost in a short time, so it is reasonable to assume that the production cost is a constant. As a result, the profit margin per unit not accounting for the future warranty cost, denoted as ΔP, is a constant, which is assumed to be known by M.

The warranty cost per unit sold of product A under the $\psi(\Theta)$ can be represented by $C(\Theta)$, a *r.v.* depending on Θ. It should be noted that the quality of the product, represented by a distribution function F, surely will affect the warranty cost. However, it is not necessary to include F explicitly here as long as M has the complete knowledge of the quality. As a result, the profit per unit sold of product A is $\Delta P - C(\Theta)$, also a *r.v.*, due to the randomness of $C(\Theta)$.

10.4.1.2 Consumer's Responses
For homogeneous consumers in the customer base B, it is assumed that the probability p_c of purchasing product A solely depends on Θ, invariant among all the potential consumers. Therefore, p_c could be obtained by solving equation $h(p_c) = k(\Theta)$, where $h(\cdot)$ is the so-called link function and $k(\Theta)$ may be assumed to be a linear function of the elements in Θ, *i.e.*,

$$k(\Theta) = \beta_0 + \beta_1\theta_1 + \beta_2\theta_2 + \cdots + \beta_n\theta_n.$$

There are many choices for link function $h(\cdot)$. This section considers the commonly used logit link function $h(p_c) = \ln(p_c/(1-p_c))$ (see Stokes 2000). Thus,

$$\ln(\frac{p_c}{1-p_c}) = \beta_0 + \beta_1\theta_1 + \beta_2\theta_2 + \cdots + \beta_n\theta_n \qquad (10.80)$$

From Equation (10.80), it is easy to obtain that

$$p_c = \frac{\exp(\beta_0 + \beta_1\theta_1 + \beta_2\theta_2 + \cdots + \beta_n\theta_n)}{1 + \exp(\beta_0 + \beta_1\theta_1 + \beta_2\theta_2 + \cdots + \beta_n\theta_n)} \tag{10.81}$$

Note that β_0 could be interpreted as the baseline logarithm of the odds: the ratio of p_c to $(1 - p_c)$, for the products without warranty. Similarly, $\beta_i, i = 1,2,\cdots,n$ is the change in the logarithm of the odds due to the unit change of θ_i.

Let D be the demand of product A covered by $\psi(\Theta)$. It's easy to see that D follows a binomial distribution. The total units expected to sell under the $\psi(\Theta)$ is given by

$$E[D] = B\frac{\exp(\beta_0 + \beta_1\theta_1 + \beta_2\theta_2 + \cdots + \beta_n\theta_n)}{1 + \exp(\beta_0 + \beta_1\theta_1 + \beta_2\theta_2 + \cdots + \beta_n\theta_n)} \tag{10.82}$$

and the variance of D is

$$V[D] = B\frac{\exp(\beta_0 + \beta_1\theta_1 + \beta_2\theta_2 + \cdots + \beta_n\theta_n)}{[1 + \exp(\beta_0 + \beta_1\theta_1 + \beta_2\theta_2 + \cdots + \beta_n\theta_n)]^2} \tag{10.83}$$

If consumers are not homogeneous in a sense that they can be classified to several categories according to their personal profiles such as the risk attitude, the knowledge of the product, and the desire of the product, it is still possible to forecast the demand by fitting a logistic regression model for each category based on survey data or some historical data if available. However, such an extension is beyond the scope of this section.

10.4.1.3 A General Decision Model

The producer's profit π can be expressed as $\sum_{i=1}^{D}(\Delta P - C_i)$, where C_i is the future warranty cost to the producer due to the ith sale. Under the $\psi(\Theta)$, $C_i, i = 1,2,\cdots$, are i.i.d., following the same distribution as $C(\Theta)$. Assuming statistical independence between D and C_i, we then have

$$E[\pi] = B\frac{\exp(\beta_0 + \beta_1\theta_1 + \beta_2\theta_2 + \cdots + \beta_n\theta_n)}{1 + \exp(\beta_0 + \beta_1\theta_1 + \beta_2\theta_2 + \cdots + \beta_n\theta_n)}(\Delta P - E[C(\Theta)]) \tag{10.84}$$

So the problem of determining the optimal warranty policy $\psi(\Theta^*)$ can be formulated as

$$\underset{\theta_i, i=1,2,\cdots,n}{\text{Maximize}} \quad E[U(\pi)]$$

$$\text{Subject to} \quad \begin{cases} P[\sum_{i=1}^{D} C_i \geq R_0] \leq \varepsilon \\ \theta_i' \leq \theta_i \leq \theta_i'', \quad i = 1,2,\cdots,n \end{cases}$$

The first constraint is about the warranty budget R_0. Warranty managers usually require that the probability of the total warranty cost over the budget is controlled within ε, a pre-determined risk acceptance level. The second set of constraints defines the acceptable ranges of $\theta_i, i = 1,2,\cdots,n,$ which are usually affected by the reliability (quality) requirements and the competitors' strategy. Other constraints may also be included depending on individual applications.

If M is risk-neutral, implying that $U(\pi)$ is a linear function of π, $E[\pi]$ becomes the objective function, which is given in Equation (10.84). The constraints remain the same.

10.4.2 Cost Analysis of RRLRFW Policy

From the general optimization model illustrated in Section 10.4.1, it is clear that one of the main sources of the randomness is from the warranty cost per unit sold $C(\Theta)$. This section focuses on the RRLRFW with two parameters w and m for which $\Theta = \{w, m\}$, and will derive some useful statistical properties of $C(w, m)$ given minimal repairs. Such a situation is not rare for complex systems that are aging over time. Besides, manufacturers usually will not take extra effort to make a repaired system better than old unless they are required to do so. It is also assumed replacement products have the same failure time distribution as that of the original product.

Let T represent the warranty cycle of $\psi(w,m)$, starting from the date of a sale, ending at the warranty expiration date. Since $\psi(w,m)$ is renewable, T is actually an r.v. Denote $N_a(w,m)$ as the total number of renewals of the warranty, and let $t_1, t_2, \cdots, t_{N_a(w,m)}$ be the corresponding inter-arrival renewal times, then T can be expressed as

$$T = \begin{cases} w, & \text{for } N_a(w,m) = 0 \\ t_1 + t_2 + \cdots + t_{N_a(w,m)} + w, & \text{for } N_a(w,m) = 1,2,\cdots \end{cases}$$

The total warranty cost per item sold, or per cycle, $C(w,m)$, can be formulated as

$$C(w,m) = (c_a + c_f)N_a(w,m) + (c_b + c_f)N_b(w,m) \tag{10.85}$$

where $N_b(w,m)$ is the total number of minimal repairs within T. c_a and c_b are the replacement cost and repair cost per unit respectively. c_f represents the fixed cost per warranty service, which may include warranty managerial cost, handling cost, disposal cost, advertising cost, and so forth. c_a, c_b and c_f are assumed to be constant.

Define $N_b'(w,m)$ as the number of minimal repairs in the last warranty period w, then the relationship between $N_a(w,m)$ and $N_b(w,m)$ can be written as

$$N_b(w,m) = mN_a(w,m) + N_b'(w,m) \tag{10.86}$$

Using Equation (10.86), $C(w,m)$ can be re-written as

$$C(w,m) = [c_a + mc_b + (m+1)c_f]N_a(w,m) + (c_b + c_f)N'_b(w,m) \qquad (10.87)$$

To simplify the notion, we will suppress w and m later on whenever appropriate. The expectation of C, denoted as μ_ψ, is given by

$$\mu_\psi = [c_a + mc_b + (m+1)c_f]E[N_a] + (c_b + c_f)E[N'_b] \qquad (10.88)$$

The variance of C, σ^2_ψ, is

$$\sigma^2_\psi = [c_a + mc_b + (m+1)c_f]^2 V[N_a] + (c_b + c_f)^2 V[N'_b] \qquad (10.89)$$

It should be noted that Equation (10.89) holds since N'_b is independent of N_a, which can easily be proved through a conditioning argument.

We need the following lemma from Baxter (1982):

LEMMA 10.5 *Let* S_{m+1} *be the* $(m+1)^{th}$ *arrival failure time of a new or replacement product under the warranty, which is also the* $(m+1)^{th}$ *arrival time of the underlying NHPP associated with* Λ. *Let* G_{m+1} *and* g_{m+1} *be the cdf and pdf of* S_{m+1} *respectively. Then*

$$G_{m+1}(x) = 1 - e^{-\Lambda(x)} \sum_{i=0}^{m} \frac{[\Lambda(x)]^i}{i!} \qquad (10.90)$$

$$g_{m+1}(x) = \lambda(x)e^{-\Lambda(x)} \frac{[\Lambda(x)]^m}{m!} \qquad (10.91)$$

Next we derive some useful properties of N_a and N'_b. The following notation is used: $p_i^{N_a} = P[N_a = i]$; $p_i^{N_b} = P[N'_b = i]$; F is the *cdf* of the first failure time of a new product or a replacement. Accordingly, the accumulative failure rate function, Λ, is $\Lambda = -\ln(1-F)$, and the failure rate function is $\lambda = dF/(1-F)$.

PROPOSITION 10.3 *The first two centered moments of* N_a *are given by*

$$E[N_a] = \frac{G_{m+1}(w)}{1 - G_{m+1}(w)} \qquad (10.92)$$

$$V[N_a] = \frac{G_{m+1}(w)}{[1 - G_{m+1}(w)]^2} \qquad (10.93)$$

Proof. It is not difficult to see that under the $\psi(w,m)$, the warranty will not expire until the first time that $S_{m+1} > w$. Thus the total number of renewals, N_a, follows a geometric distribution of parameter $1 - G_{m+1}(w)$. Based on the properties of geometric distribution, the result then follows.   ◆

PROPOSITION 10.4 *The probability mass function (pmf) of N_b' is given by*

$$p_i^{N_b} = e^{-\Lambda(w)} \frac{(\Lambda(x))^i}{i!}, \quad \forall i, i = 0, 1, \cdots, m-1 \tag{10.94}$$

$$p_m^{N_b} = 1 - \sum_{i=0}^{m-1} e^{-\Lambda(w)} \frac{(\Lambda(x))^i}{i!} = G_m(w) \tag{10.95}$$

The expectation of N_b' is

$$E[N_b'] = m - e^{-\Lambda(w)} \sum_{i=0}^{m-1} (m-i) \frac{(\Lambda(x))^i}{i!} \tag{10.96}$$

and the variance of N_b' follows

$$V[N_b'] = m^2 - e^{-\Lambda(w)} \sum_{i=0}^{m-1} (m^2 - i^2) \frac{(\Lambda(w))^i}{i!} - (E[N_b'])^2 \tag{10.97}$$

Proof. Since any failed product under warranty will be minimally repaired up to *m* times, the failure process $\{N_a(t), t < 0\}$ forms an NHPP truncated above *m* with the accumulative failure rate function Λ. By realizing that the truncation does not make any change to the *pmf* of the underlying ordinary Poisson random variable when it is less or equal to $m-1$, we obtain Equation (10.94). Equation (10.95) holds due to the relationship that $P[N_b' = m] = P[S_m \leq w]$. Equations (10.96) and (10.97) can be verified easily using the *pmf* of N_b'.   ◆

Now it is ready to give the expressions of the first two centered moments of *C*.

PROPOSITION 10.5 *The first two moments of the warranty cost per unit sold are*

$$\mu_\psi = [c_a + mc_b + (m+1)c_f] \frac{G_{m+1}(w)}{1 - G_{m+1}(w)}$$

$$+ (c_b + c_f)(m - e^{-\Lambda(w)} \sum_{i=0}^{m-1} (m-i) \frac{(\Lambda(w))^i}{i!}) \tag{10.98}$$

$$\sigma_\psi^2 = [c_a + mc_b + (m+1)c_f]^2 \frac{G_{m+1}(w)}{(1 - G_{m+1}(w))^2}$$

$$+ (c_b + c_f)^2 \{(m^2 - e^{-\Lambda(w)} \sum_{i=0}^{m-1} (m^2 - i^2) \frac{[\Lambda(w)]^i}{i!}$$

$$- [m - e^{-\Lambda(w)} \sum_{i=0}^{m-1} (m-i) \frac{[\Lambda(w)]^i}{i!}]^2\} \tag{10.99}$$

Proof. The proof is straightforward, thus omitted here.

10.4.3 Optimal RRLRFW Policy

This sections uses the optimization model in Section 10.4.1 to determine the optimal RRLRFW policy. In particular, we assume that the seller is risk neutral; therefore, the optimal warranty policy is determined by maximizing the expected profit under some constraints. Using the results in Section 10.4.2, the model becomes

$$
\underset{(w,m)}{\textbf{Maximize}}\ B\frac{\exp(\beta_0 + \beta_1 w + \beta_2 m)}{1 + \exp(\beta_0 + \beta_1 w + \beta_2 m)}[\Delta P - [c_a + mc_b + (m+1)c_f]\frac{G_{m+1}(w)}{1 - G_{m+1}(w)}
$$

$$
+ (c_b + c_f)(m - e^{-\Lambda(w)}\sum_{i=0}^{m-1}(m-i)\frac{[\Lambda(w)]^i}{i!})]
$$

Subject to
$$
\begin{cases}
P[\sum_{i=1}^{D} C_i \ge R_0] \le \varepsilon \\[2mm]
w^l \le w \le w^u \\[2mm]
m \in \{1,2,\cdots\}
\end{cases}
$$

The above formulation is a non-linear optimization problem, which can be solved by non-linear optimization software, except that the first constraint requires the knowledge of the distribution of TC, defined as $TC = \sum_{i=1}^{D} C_i$. Next we simplify the constraint.

Since D follows a binomial distribution, and $C_i, i = 1,2,\cdots$, are *i.i.d.*, TC actually follows a compound binomial distribution, for which usually there is no close form expression and the computation is difficult in general. Several recursive algorithms exist to compute the exact distribution function (see Sundt 2002). This section considers a simple approximation of the compound binomial distribution by a normal distribution (Rolski *et al.* 1999) with parameters μ_{TC} and σ_{TC}^2, the mean and variance of TC respectively. It can easily be verified that

$$
\mu_{TC} = B\frac{\exp(\beta_0 + \beta_1 w + \beta_2 m)}{1 + \exp(\beta_0 + \beta_1 w + \beta_2 m)}\{[c_a + mc_b + (m+1)c_f]\frac{G_{m+1}(w)}{1 - G_{m+1}(w)}
$$

$$
+ (c_b + c_f)\{(m - e^{-\Lambda(w)}\sum_{i=0}^{m-1}(m-i)\frac{[\Lambda(w)]^i}{i!})\} \tag{10.100}
$$

$$
\sigma_{TC}^2 = B\frac{\exp(\beta_0 + \beta_1 w + \beta_2 m)}{1 + \exp(\beta_0 + \beta_1 w + \beta_2 m)}\{[c_a + mc_b + (m+1)c_f]^2\frac{G_{m+1}(w)}{[1 - G_{m+1}(w)]^2}
$$

$$
+ (c_b + c_f)^2\{m^2 - e^{-\Lambda(w)}\sum_{i=0}^{m-1}(m^2 - i^2)\frac{[\Lambda(w)]^i}{i!} - \frac{1}{1 + \exp(\beta_0 + \beta_1 w + \beta_2 m)}
$$

$$((c_a + mc_b + (m+1)c_f)\frac{G_{m+1}(w)}{1 - G_{m+1}(w)} + (c_b + c_f)(m - e^{-\Lambda(w)})$$

$$\times \sum_{i=0}^{m-1}(m - i)\frac{[\Lambda(w)]^i}{i!}))^2\} \tag{10.101}$$

Using Equations (10.100) and (10.101), and the normal approximation, the first constraint can be rewritten as

$$R_0 - \mu_{TC} - z_\alpha \sigma_{TC} \geq 0 \tag{10.102}$$

where z_α is the $(1 - \alpha)$ quantile of the standard normal distribution.

Now the optimization problem has been converted into a standard non-linear optimization problem. Let's see a numerical example.

Suppose a warranty manager, with a warranty budget $R_0 = \$4000$ and $\alpha = 10\%$, has decided to offer an RRLRFW. He (she) would like to consider $w \in [1,20]$ and $m \in \{1,2,3,4,5\}$. The objective is to determine the optimal warranty parameters w^* and m^* such that the expected profit is maximized while the constraints are satisfied. The customer base is estimated to be 10,000. Through the data from a market survey, he estimated that parameters β_0, β_1, and β_2 are − 0.4217, 0.0141 and − 0.2012 respectively. The profit margin per unit is known to be \$500. The reliability function of the product is given by $\exp(-t^{1.07}/247.16)$. The fixed warranty service cost per service c_f is \$50, the replacement cost c_a and repair cost c_b are \$10,000 and \$150 respectively.

The optimal RRLRF policy is found: $w^* = 9.5$ and $m^* = 2$, for which the expected total profit is \$1605,624, the expected warranty cost per unit sold is \$19.5, and the corresponding standard deviation is 330.61. It should be noted that the standard deviation is much higher than the expectation since the policy under consideration is renewable.

If there is no warranty budget constraint, then the optimal policy is given by $w^* = 13.2$ and $m^* = 2$. The corresponding expected profit is \$1,613,313.

10.4.4 Remarks

Among many warranty management problems, how to determine the optimal warranty policy that may help manufacturers to gain some advantage in the competitive market is a fundamental one. This section has provided a general optimization model for this problem. However, it should be noted that there are some limitations on this model. First, this optimization model deals with the demand of a warranted product through a logistic link function which only depends on the warranty parameters. Empirical study such as consumer surveys is needed to estimate the proposed link function. Second, the optimization model only considers the homogeneous consumers. The case of non-homogeneous consumers needs to be examined. Third, if the manufacturer is not a price taker, then the product price should be included as one of the decision variables. Fourth, only minimal repair is considered in this section. Other situations such as imperfect repair can be studied.

This section has presented some useful results that can be applied to determine the optimal RRLRFW policy. A numerical example is provided for illustration purposes. One natural extension is to apply this approach to other warranty policies such as a two-dimensional warranty or a combination warranty. Other generalizations of the RRLRFW policy may be considered. For example, Bai (2004) suggests use of the quasi renewal process to study the renewable repair-limit risk-free warranty.

Let $S_i, i = 1,2,\cdots,m+1$ be the failure arrival times of a new product (or a replacement) under the warranty, which has been renewed for j times, $j = 1,2,\ldots$ For this product, any failure within a period of w_j will be repaired up to m times. If $S_{m+1} \le w_j$, then a replacement will be provided. The warranty is then renewed for another period of w_{j+1}. The relationship between w_j and w_{j+1} is given by $w_{j+1} = \gamma w_j$, where $\gamma \in (0,1]$, and $w_0 = w$. All the warranty service is free of consumers.

Compared to an RRLRFW policy, the above policy is more general since it degenerates to the former when parameter $\gamma = 1$. It would be interesting to investigate the statistical properties of this warranty policy.

10.5 On Warranty Policies and their Comparison

Warranty managers usually have several choices among various warranty policies. This requires some basic measures as the criteria to make the comparison among these policies. Bai and Pham (2006b) and Mi (1999) discuss comparison of different warranties and some criteria, and following them we will compare various warranty policies in this section.

For a warranty policy, there are several measures available including EWC, expected DWC, monetary utility function and weighted objective function. EWC and expected DWC are more popular than others since they are easy to understand and can be estimated relatively easily. The key difference between them is that the latter considers the value of time, an important factor for the determination of warranty reserve.

Monetary utility function, $U(x)$, is a better candidate for the purpose of comparing warranty policies. The functional form of $U(x)$ reflects seller's risk attitude. If a seller is risk-neutral, then $U(x)$ is linear in x. This implies that maximizing $E[U(x)]$ is the same as maximizing $U(E[x])$. However, manufacturers may be risk-averse if they are concerned about the variations in profit or in warranty cost. For example, a particular seller may prefer a warranty with less warranty cost variation than another with much larger variation in warranty cost while the difference between the EWCs is small. If this is the case, then it can be shown that that the corresponding utility function is concave (Kreps 1990). The main difficulty of the utility theory approach is that utility functions are subjective.

Weighted objective functions could also be used in the comparison of warranties. One commonly used weighted objective function is $E[\pi(x)] - \rho V[\pi(x)]$, where ρ is a parameter representing the subjective relative importance of the risk (variation) against the expectation and $\pi(x)$ is the manufacturers' profit for a given warranty policy x. Interestingly, such an objective function coincides to a special case of the utility theory approach when the manufacturers' subjective utility function only depends on the first and the second moments of $\pi(x)$ (Markowitz 1959, p.126).

According to compensation methods specified in a warranty contract upon premature failures, there are three basic types of warranties: free replacement warranty (FRW), free repair warranty (FRPW), and pro-rata warranty (PRW). Combination warranty (CMW) contains both features of FRW/FRPW and PRW, which often has two warranty periods, a free repair/replacement period followed by a pro-rata period. Full-service warranty (FSW), also known as PM warranty, is a policy that may be offered for expensive deteriorating complex products such as automobiles. Under this type of policy, consumers not only receive free repairs upon premature failures, but also free (preventive) maintenance.

For non-repairable products, usually the failed products under warranty will be replaced free of charge to consumers. Such a policy is often referred as a free replacement warranty or a unlimited warranty. In practice, even if a product is technically repairable, sometimes it will be replaced upon failure since repair may not be economically sound. As a result, for inexpensive repairable products, warranty issuers could simply offer FRW policies. Consequently, those inexpensive repairable products can be treated as non-repairable. However, for repairable products, if the warranty terms specify that upon a valid warranty claim, the warranty issuer will repair the failed product to working condition free of charge to buyers, then such a policy is the so-called free repair warranty. In practice, it is not rare that a warranty contract specifies that the warranty issuer would repair or replace a defective product under certain conditions. This is the reason why most researchers do not treat FRW and FRPW separately. Nevertheless, it is necessary to differentiate these two types of policies based on the following reasoning: first, repair cost is usually much lower than replacement cost unless for inexpensive products; second, by clearly defining the compensation terms, warranty issuers may establish a better image among consumers, which can surely be helpful for the marketing purpose.

Under an FRW policy, since every failed product within T is replaced by a new one, it is reasonable to model all the subsequent failure times by a single probability distribution. However, under an FRPW, it is necessary to model the repair impact on failure times of a warranted product. If it is assumed that any repair is as-good-as-new (perfect repair), then from the modeling perspective, there is little difference between FRW and FRPW. For deteriorating complex systems, minimal repair is a commonly used assumption, as discussed in Chapter 2. In warranty literature, the majority of researchers consider repairs as either perfect or minimal. Little has been done on warranty cost analysis considering imperfect repair.

Both FRW and FRPW policies provide full coverage to consumers in case of product failures within T. In contrast, a PRW policy requires that buyers pay a proportion of the warranty service cost upon a failure within T in exchange for the warranty service such as repair or replacement, cash rebate or discount on purchasing a new product. The amount that a buyer should pay is usually an increasing function of the product age (duration after the sale). PRW policies are usually renewable, and offered for relatively inexpensive products like tires, batteries, and so forth.

Generally speaking, FRW and FRPW policies are in favor of buyers since manufacturers take all the responsibility of providing products that function properly during the whole warranty cycle (Blischke and Murthy 1994, p.221). In other words, it is the manufacturers that bear all the warranty cost risk. In contrast, for PRW policies manufacturers have the relative advantage with regard to the warranty cost risk. Although they do have to offer cash rebate or discounts to consumers if failures happen during T, they are usually better off no matter what consumers choose to do. If a consumer decides not to file a warranty claim, then the manufacturer saves himself the cash rebate or other types of warranty service. If instead a warranty claim is filed, the manufacturer might enjoy the increase in sales or at least the warranty service cost is shared by the consumer.

CMW can be used to balance the benefits between buyers and sellers, and is a policy that usually includes two warranty periods: a free repair/replacement period w_1 followed by a pro-rata period w_2. This type of warranty is not rare today because it has significant promotional value to sellers while at the same time it provides adequate control over the costs for both buyers and sellers (Blischke and Murthy 1996, p.12).

For deteriorating complex products, it is essential to perform PM to achieve satisfactory reliability performance. The burden of maintenance is usually on the consumers' side. Section 10.1 has discussed a renewable full-service warranty for multi-component systems under which the failed component(s) or subsystem(s) will be replaced, in addition, a PM action will be performed to reduce the chance of future product failures, both free of charge to consumers. Such a policy may be desirable for both consumers and manufacturers since consumers receive better warranty service compared to the traditional FRPW policies, at the same time manufacturers may enjoy cost savings due to the improved product reliability by the maintenance actions.

In the maintenance literature, many researchers studied maintenance policies set up in such a way that different maintenance actions may take place depending on whether or not some pre-specified limits are met. Three types of limits are usually considered: repair-number-limit, repair-time-limit, and repair-cost-limit, and these maintenance policies are summarized in Chapter 3. Similarly, three types of repair-limit warranties may be considered by manufacturers: repair-number-limit warranty (RNLW), repair-time-limit warranty (RTLW), and repair-cost-limit warranty (RCLW). Under a RNLW, the manufacturer agrees to repair a warranted product up to m times within a period of w. If there are more than m failures within w, the failed product shall be replaced instead of being repaired again. Section 10.3 has discussed this kind of policy under the imperfect repair assumption.

An RTLW policy specifies that within a warranty cycle T, any failures shall be repaired by the manufacturer, free of charge to consumers. If a warranty service cannot be completed within a certain time, then a penalty cost occurs to the manufacturer to compensate the inconvenience of the consumer. This policy was analyzed by Murthy and Asgharizadeh (1999) in the context of maintenance service operation.

An RCLW policy has a repair cost limit in addition to an ordinary FRPW policy, *i.e.*, upon each failure within the warranty cycle T, if the estimated repair cost is greater than a fixed number, then replacement instead of repair shall be provided to the consumer; otherwise, normal repair will be performed. This policy has been studied by Nguyen and Murthy (1989) and others.

Possible new warranty policies are those that combine various repair limits as well as other warranty characteristics such as renewing to define a new complex warranty. For example, it is possible to have a renewable repair-time-limit warranty for complex systems. Such combinations define a large set of new warranty policies that may appear in the market in the future.

In addition, most warranties in practice are one-attribute for which the warranty terms are based on either product age or product usage, but not both. Compared to one-attribute warranties, two-attribute warranties are more complex since the warranty obligation depends on both product age and product usage as well as the potential interaction between them. Two-attribute warranties are often seen in the automobile industry. For example, one automobile company, is currently offering 10 years/100,000 miles limited FRPW on the power train for their new car models in North America. Comparison and analysis of two-attribute warranties can be found in Murthy *et al.* (1995) and Singpurwalla and Wilson (1993).

11

Software Reliability, Cost, and Optimization Models

Practice over the years has shown that a software development process using software reliability models instead of traditional project management methods is efficient and effective to produce reliable software at low cost. This chapter models software reliability and debugging costs using the quasi-renewal process, and discusses optimal software testing and release policies, following Pham and Wang (2001). Several software reliability and cost models are presented in which successive error-free times are independent and increasing by a fraction, *i.e.*, they form an increasing quasi-renewal process. It is assumed that the cost of fixing a fault consists of deterministic and incremental random parts, and is increasing with the number of faults removed. The maximum likelihood estimates of parameters associated with these models are provided. Based on the valuable properties of quasi-renewal processes, the expected software testing and debugging cost, number of residual faults in the software, and mean error-free time upon testing are obtained. A class of related optimization problems are then contemplated, and optimum testing policies incorporating both reliability and cost measures are discussed. Finally, numerical examples are presented through a set of real testing, showing satisfactory results. The models in this chapter can also apply to modeling field reliability growth and maintenance cost.

11.1 Introduction

Research activities in software reliability have been conducted for the past several decades, and are still going on today because critical software applications are increasing in size and complexity. Since software is an interdisciplinary science, software reliability and cost models are developed from different perspectives towards software and with various applications. So far many software reliability models have been developed respectively by using nonhomogeneous Poisson processes, Markov processes, binary Markov processes, Bayesian statistics, classical statistics, input-domain-based methods, *etc.*, as shown in Musa *et al.* (1987), Xie (1991), Downs (1985), Goel (1985), Pham (2000), and Lyn (1996). However, there is still a great need to develop more practical and realistic models to estimate software reliability and testing costs, and to determine the desired reliability level before releasing it (Pham 2003b; Lyn 1996). The key modeling approaches and a critical analysis of underlying assumptions, limitations, and applicability of some previous software models during the software development cycle are discussed in Goel (1985). A quasi-renewal process is a new tool to facilitate modeling of both software reliability and testing costs (Pham and Wang 2001).

Software testing is an efficient and necessary way to remove faults in software products. However, exhaustive testing of all possible executable paths in a large program may be impractical. Debugging and testing reduce the error content but increase the development costs. In fact, after reaching a certain level of software refinement, further efforts to increase reliability will result in exponential increase in cost and debugging time (Pham 2000). Therefore, it is important to determine when to stop testing, or when to release the software to customers. One might consider what questions a software model should help answer for software developers and managers. The important questions are (Lyn 1996; Pham and Wang 2001):

i)   What would the failure rate of the software be if released now?

ii)   How many faults remain in the software? How many high severity faults? Fault location (subsystem)?

iii) How much more testing is needed to achieve software reliability targets? How should resources be scheduled to ensure the on-time and efficient delivery of a software product? Is the software product sufficiently reliable for release?

This chapter aims to present software reliability models which will help answer the above questions. Unlike most previous work, this chapter determines the optimal software release time based on two criteria: reliability of the released software and total software cost. In addition to the traditional software testing measures, software error-free time information upon testing is also provided in the models in this chapter. In fact, some unique properties of the quasi-renewal process ease modeling it. This chapter assumes the cost of fixing a fault during the software testing phase consists of deterministic and probabilistic parts, and grows as the number of faults removed increases. Obviously, this assumption is realistic because usually it may become difficult to fix a fault which is detected in the later testing phases. Besides, cost of testing per unit time is considered in this chapter, which is treated as a random variable. The second model in this chapter contemplates that in software there exist three types of faults which are classified in terms of failure effects and severities.

In Section 11.2, the quasi-renewal process is discussed with regard to its application in software reliability growth. Section 11.3 models software reliability and testing cost through the quasi-renewal process, and then investigates the optimal software testing policies by some numerical examples using a set of real testing data. Some concluding remarks are made in Section 11.4.

11.2 Use of Quasi-renewal Process in Software Reliability

In ordinary renewal process, the times between successive events are supposed to be independently and identically distributed (*i.i.d.*). As discussed in Chapter 4, a general renewal process, including ordinary renewal process as special case, is the quasi-renewal process. The quasi-renewal process is motivated by imperfect repair processes of hardware and in turn finds wide applications in modeling hardware maintenance as shown in Chapter 4. This chapter will use quasi-renewal process to model software reliability growth and testing costs.

Recall that Theorem 4.2 in Chapter 4 implies that after "renewal" the shape parameters of the inter-arrival times will not be changed. In reliability theory, shape parameters of lifetime of a hardware product tend to relate to its failure mechanism. Usually, a product will have the same shape parameters at different operating conditions if it possesses the same failure mechanism. Therefore, the use of a quasi-renewal process is generally justified in the maintenance process of a hardware system. The assumption that software debugging and testing or field use do not change the type of the error-free time distribution seems plausible. Note that the error-free times in the software during debugging phase or field use will have the same shape parameters, if modeled by a quasi-renewal process. In this sense a quasi-renewal process will be plausible to model the software reliability growth.

Recall from Section 4.1.1

$$\lim_{n \to \infty} \frac{E(X_1 + X_2 + \cdots + X_n)}{n} = \lim_{n \to \infty} \frac{\mu_1(1 - \alpha^n)}{(1 - \alpha)n} = \begin{cases} 0 & \text{when } \alpha < 1 \\ +\infty & \text{when } \alpha > 1 \end{cases}$$

Therefore, if the inter-arrival time represents the error-free time of a software system the average error-free time goes to infinite while its debugging process is going on forever. This conclusion seems reasonable because the faults in the software become generally less and less while it is subject to testing and debugging. When the debugging time is infinitely long, no faults in this software can be expected. Thus, the average error-free time and the error-free time is infinite as the debugging time goes to infinity. In practice, we can expect the error-free time of software upon testing is very large if the testing time is sufficiently long. In fact, Theorem 4.1 shows that if the first error-free time of software is DFR, then the successive error-free times are DFR. Therefore, in this case the failure rate of software can be expected to be smaller and smaller as faults in the software are being removed. The same arguments are true for software reliability growth during field use if faults found in field are also removed.

11.3 Software Reliability and Cost Modeling

If the inter-arrival time represents the error-free time (time between errors), a quasi-renewal process can be used to model reliability growth for software. Next we will utilize this quasi-renewal process to investigate software reliability and testing costs. Throughout this chapter we assume

- All faults of software are independent.
- All detected faults are removed immediately and no new faults are introduced.

11.3.1 Model 1

Suppose that all faults of software have the same chance of being detected. If the inter-arrival times of a quasi-renewal process represent the error-free times of software, the expected cumulative number of software faults in $[0, t)$ can be described by the quasi-renewal function $M(t)$ with parameter $\alpha > 1$. Denote by $\overline{M}(t)$ the number of remaining software faults at time t. It follows that

$$\overline{M}(t) = M(\tau) - M(t) \tag{11.1}$$

where $M(\tau)$ is the number of faults which can be detected through a long testing time τ, relative to t. We suggest taking $\tau \geq 6t$ in practice. In fact, this choice is somewhat arbitrary and the actual selection can also be determined through experience. However, any choice should make the difference between $M(\tau)$ and $M(\tau + \Delta)$ to be insignificant for any small value of Δ.

Assume that the cost of fixing software fault i is a random variable and consists of two parts – deterministic part c_0 and incremental random part $(i-1)W$:

$$c_i = c_0 + (i-1)W \qquad\qquad \forall i, \; i = 1,2,3,... \tag{11.2}$$

where c_0 is a constant and W is a random variable with mean c_v.

Note that the cost of fixing a fault is increasing as the number of faults removed is increasing. This is reasonable because it may become difficult to identify and fix a fault which occurs in the later testing phases. Then the expected total debugging cost in $[0, t)$ is given by

$$
\begin{aligned}
C_r(t) &= E\left[\sum_{i=1}^{N(t)} [c_0 + (i-1)W] \right] \\
&= \sum_{n=1}^{\infty} E\left[\sum_{i=1}^{N(t)} [c_0 + (i-1)W] \,\middle|\, N(t) = n \right] P(N(t) = n) \\
&= \sum_{n=1}^{\infty} E\left[\sum_{i=1}^{n} [c_0 + (i-1)W] \,\middle|\, N(t) = n \right] P(N(t) = n) \\
&= \sum_{n=1}^{\infty} \frac{n[2c_0 + (n-1)c_v]}{2} P(N(t) = n) \\
&= \frac{1}{2} \sum_{n=1}^{\infty} n[2c_0 - c_v + nc_v] P(N(t) = n) \\
&= \left(\frac{2c_0 - c_v}{2} \right) E[N(t)] + \frac{c_v}{2} \sum_{n=1}^{\infty} n^2 P(N(t) = n)
\end{aligned}
$$

$$= \left(\frac{2c_0 - c_v}{2}\right) M(t) + \frac{c_v}{2} E\left[N^2(t)\right]$$

$$= \left(\frac{2c_0 - c_v}{2}\right) M(t) + \frac{c_v}{2}\left[Var[N(t)] + M^2(t)\right] \qquad (11.3)$$

If in the above cost model we also consider the cost of testing per unit time and assume that it is a random variable V_1 with mean c_3, then the total expected testing and debugging cost up to time t is given by

$$C(t) = E[tV_1] + \left(\frac{2c_0 - c_v}{2}\right) M(t) + \frac{c_v}{2}\left[Var[N(t)] + M^2(t)\right]$$

$$= tc_3 + \left(\frac{2c_0 - c_v}{2}\right) M(t) + \frac{c_v}{2}\left[Var[N(t)] + M^2(t)\right] \qquad (11.4)$$

Now we determine the variance of $N(t)$. Pham and Wang (2001) prove

$$E[N^2(t)] = \sum_{n=0}^{\infty} n^2 P\{N(t) = n\}$$

$$= \sum_{n=0}^{\infty} n^2 \left[G^{(n)}(t) - G^{(n+1)}(t)\right]$$

$$= \sum_{n=1}^{\infty} (2n-1)G^{(n)}(t)$$

Therefore, the variance turns out to be

$$Var[N(t)] = E[N^2(t)] - E^2[N(t)]$$

$$= E[N^2(t)] - M^2(t)$$

$$= \sum_{n=1}^{\infty} (2n-1)G^{(n)}(t) - M^2(t) \qquad (11.5)$$

where $G^{(n)}(t)$ is the convolution of the inter-arrival times $F_1, F_2, ..., F_n$, defined in Section 11.2.

The cost of the expected testing and debugging up to time t is discussed above. Now we investigate the expected total software cost during its life cycle. Pham (1996) derives a software cost model with imperfect debugging, random life cycle and penalty cost using nonhomogeneous Poisson process. Similar to Pham (1996), we can derive the expected total software life-cycle cost. Let c_1 represent the cost of fixing a fault during testing phase, c_2 represent the cost of fixing a fault during operation phase, c_3 the cost of testing per unit time, T the software release time, T_d the scheduled delivery time, $g(t)$ the probability density function of the life-cycle length $(t > 0)$. Then it is easy to verify that the expected total software life-cycle cost is given by

$$C(T) = c_3 T + c_1 M(T) + \int_T^\infty c_2 [M(t) - M(T)] g(t) dt + I(T - T_d) \cdot c_p (T - T_d)$$

(11.6)

where $c_p(t)$ is a penalty cost for a delay of delivering software, and $I(t)$ is an indicator function, i.e.,

$$I(t) = \begin{cases} 1 & \text{if } t \geq 0 \\ 0 & \text{otherwise} \end{cases}$$

Usually, $M(t)$ and $Var[N(t)]$ contains some unknown parameters. Their estimation can be carried out by using the maximum likelihood or least squares method.

Denote by t_i the ith failure time since the software testing begins at time zero. Assume that $0 = t_0 < t_1 \cdots < t_n$. The likelihood function of the above software reliability model is, noting that Equation (11.1) in Section 11.2,

$$L(t_1, t_2, ..., t_n) = \prod_{i=1}^n f_1(t_1 \mid \Theta) f_2(t_2 \mid \Theta) \cdots f_n(t_n \mid \Theta)$$

$$= \prod_{i=1}^n f_1(t_1 \mid \Theta) \alpha^{-1} f_1(\alpha^{-1} t_2 \mid \Theta) \cdots \alpha^{1-n} f_1(\alpha^{1-n} t_n \mid \Theta)$$

$$= \alpha^{-n(n-1)/2} \prod_{i=1}^n f_1(t_1 \mid \Theta) f_1(\alpha^{-1} t_2 \mid \Theta) \cdots f_1(\alpha^{1-n} t_n \mid \Theta)$$

where Θ represents the parameter family including parameter α.

From the above likelihood function the parameters in $M(t)$ and $Var[N(t)]$ can be estimated by the maximum likelihood method. Let's take the normal distribution as an example.

Assume that the first failure time, X_1, of a new software system follows the normal distribution with mean μ and variance σ^2, that is,

$$f_1(x) = \frac{1}{\sigma \sqrt{2\pi}} e^{-(x-\mu)^2 / 2\sigma^2}$$

and that the testing process can be modeled by the quasi-renewal process. From testing data – failure times $\{t_1, t_2, ..., t_n\}$ we can easily estimate quasi-renewal process parameter α and normal distribution parameters μ and σ. Pham and Wang (2001) obtain the following MLEs of parameters α, μ, and σ, which can be obtained by solving following simultaneous equations:

$$
\left\{
\begin{aligned}
\hat{\mu} &= \frac{1}{n}\sum_{i=1}^{n}\hat{\alpha}^{1-i}t_i \\
n &= \frac{1}{2\hat{\sigma}^3}\sum_{i=1}^{n}(\hat{\alpha}^{1-i}t_i - \hat{\mu})^2 \\
\frac{n(n-1)}{2} &= \frac{1}{\hat{\sigma}^2}\sum_{i=1}^{n}\left(\hat{\alpha}^{1-i}t_i - \hat{\mu}\right)\frac{t_i}{\hat{\alpha}^{2i-1}}
\end{aligned}
\right.
$$

Now it is easy to compute the renewal function for the normal distribution. From Section 11.2, the renewal function is

$$
M(t) = \sum_{n=1}^{\infty} G^{(n)}(t)
$$

and

$$
G^{(n)}(t) = P\{SS_n \le t\}
$$

where random variable SS_n follows the normal cumulative distribution function with mean $\mu(1-\alpha^n)/(1-\alpha)$ and variance $\sigma^2(1-\alpha^{2n})/(1-\alpha)$.

Therefore, the renewal function is given by

$$
M(t) = \sum_{n=1}^{\infty} P\{SS_n \le t\} = \sum_{n=1}^{\infty} \Phi\left(\left[t - \frac{\mu(1-\alpha^n)}{1-\alpha}\right]\Big/\sqrt{\frac{\sigma^2(1-\alpha^{2n})}{1-\alpha^2}}\right)
$$

where $\Phi(\cdot)$ is the standard normal cumulative distribution function. Various types of approximations for the standard normal $\Phi(\cdot)$ have been developed and a simple approximation with high accuracy is by Zelen and Severo (1964):

$$
\Phi(x) \approx 1 - \left(0.4361836t - 0.1201676t^2 + 0.9372980t^3\right)\left(\sqrt{2\pi}\right)^{-1}\exp(-\tfrac{1}{2}x^2) \quad (11.7)
$$

where $t = (1+0.33267x)^{-1}$. The error in $\Phi(x)$, for $x \ge 0$, is less than 1×10^{-5}.

Note that the relationship $\Phi(x) = 1 - \Phi(-x)$. Thus, we can use this relationship and Equation (11.7) to approximate $\Phi(x)$ for $x < 0$.

11.3.2 Model 2

A software failure is one that occurs when the user perceives that the software ceases to deliver the expected result with respect to the specification input values. The user may need to identify the severity of failures, such as critical, major, or minor, depending on their impacts on systems. Severity levels may vary from one system to another, and from application to application (Pham 2000). Typically, the severity of software failure effects is classified into three categories (see, for example, Telcordia GR-1339-CORE 1997):

Type 1 fault (critical) – This category is for disastrous effects, such as loss of human life or permanent loss of property, for example, the effect of an erroneous medication prescription or an air-traffic controller error due to software failures. This type of fault may occur rarely in practice.

Type 2 fault (major) – This category is for serious failures of the software system where there is no physical injury to people or other systems. Included in this category might be erroneous purchase orders or the breakdown of a road vehicle. Usually, this type of fault occurs occasionally.

Type 3 fault (minor) – This category is reserved for those faults which lead to marginal inconveniences to a software system or its users. Examples might be a vending machine that momentarily cannot provide changes or a bank's computer system that is down when a consumer requests a balance (Pham 2000). Relatively, this may be a type of fault that occurs most in reality.

In Telcordia GR-1339-CORE 1997, the telecommunication system software reliability objectives for a given release are:

- The cumulative number of Critical software faults for each software release should be equal to 0.

- The cumulative number of Major faults for each software release should be less than or equal to 4.

- The cumulative number of Minor software faults for each software release should be less than or equal to 36.

Suppose that when a fault is detected it is a critical one with probability p_1, a major one with probability p_2, and a minor one with probability p_3 where $p_1 + p_2 + p_3 = 1$. When a critical, major, or minor fault is removed, the fault-free time will be independent of the previous ones and increased to a multiple, α_1, α_2, or α_3, of the immediate previous one, respectively, where parameters $\alpha_1 \geq \alpha_2 \geq \alpha_3 \geq 1$, or more generally $\alpha_1, \alpha_2, \alpha_3 \geq 1$. Thus, upon removal of the first fault, the *cdf* of the fault-free time X_2 is given by

$$F_2(t) = P\{X_2 \leq t\}$$

$$= \sum_{i=1}^{3} P\{X_2 \leq t \mid \text{first fault is type } i\} P\{\text{first fault is type } i\}$$

$$= \sum_{i=1}^{3} P\{\alpha_i Z_2 \leq t \mid \text{first fault is type } i\} P\{\text{first fault is type } i\}$$

$$= \sum_{i=1}^{3} F_1(\alpha_i^{-1} t) \cdot p_i$$

and the *pdf* and mean of X_2 are respectively,

$$f_2(t) = F_2'(t)$$

$$= \sum_{i=1}^{3} \alpha_i^{-1} f_1(\alpha_i^{-1} t) \cdot p_i$$

$$E(X_2) = \sum_{i=1}^{3} E\{X_2 \mid \text{first fault is type } i\} P\{\text{first fault is type } i\}$$

$$= \sum_{i=1}^{3} \alpha_i \mu p_i$$

$$= \mu \sum_{i=1}^{3} \alpha_i p_i$$

Similarly, upon removal of the second fault, the *cdf, pdf* and the expected fault-free time X_3 is given by

$$F_3(t) = P\{X_3 \leq t\}$$

$$= \sum_{i=1}^{3} F_2(\alpha_i^{-1} t) p_i$$

$$= \sum_{i=1}^{3} \sum_{j=1}^{3} F_1(\alpha_i^{-1} \alpha_j^{-1} t) p_i p_j$$

$$f_3(t) = F_3'(t)$$

$$= \sum_{i=1}^{3} \alpha_i^{-1} f_2(\alpha_i^{-1} t) p_i$$

$$= \sum_{i=1}^{3} \sum_{j=1}^{3} \alpha_i^{-1} \alpha_j^{-1} f_1(\alpha_i^{-1} \alpha_j^{-1} t) p_i p_j$$

$$E(X_3) = \sum_{j=1}^{3} E\{X_3 \mid \text{2nd fault is type } j\} P\{\text{2nd fault is type } j\}$$

$$= \sum_{j=1}^{3} \alpha_j \mu \sum_{i=1}^{3} \alpha_i p_i p_j$$

$$= \mu \sum_{j=1}^{3} \sum_{i=1}^{3} \alpha_i \alpha_j p_i p_j$$

By induction, we can obtain *cdf, pdf*, and the expected error-free time of the software upon removal of the kth fault:

$$F_{k+1}(t) = \sum_{i_1=1}^{3} \sum_{i_2=1}^{3} \cdots \sum_{i_k=1}^{3} F_1(\alpha_{i_1}^{-1} \alpha_{i_2}^{-1} \cdots \alpha_{i_k}^{-1} t) p_{i_1} p_{i_2} \cdots p_{i_k} \qquad (11.8)$$

$$f_{k+1}(t) = \sum_{i_1=1}^{3} \sum_{i_2=1}^{3} \cdots \sum_{i_k=1}^{3} \alpha_{i_1}^{-1} \alpha_{i_2}^{-1} \cdots \alpha_{i_k}^{-1} f_1(\alpha_{i_1}^{-1} \alpha_{i_2}^{-1} \cdots \alpha_{i_k}^{-1}) p_{i_1} p_{i_2} \cdots p_{i_k} \qquad (11.8a)$$

$$E(X_{k+1}) = \mu \sum_{i_1=1}^{3} \sum_{i_2=1}^{3} \cdots \sum_{i_k=1}^{3} \alpha_{i_1} \alpha_{i_2} \cdots \alpha_{i_k} p_{i_1} p_{i_2} \cdots p_{i_k} \qquad (11.8b)$$

From these *pdf*s and the failure times $\{t_1, t_2, ..., t_k\}$ we can estimate the parameters $\alpha_1, \alpha_2, \alpha_3, p_1, p_2, p_3$, and the error-free time distribution parameters by maximum likelihood method.

Suppose that the ith software fault may be of type j for $j = 1,2,3$ respectively. We assume that the cost of fixing this fault is a random variable c_{ij} and consists of two parts – deterministic part c_{0j} and incremental random part $(i-1)W_j$:

$$c_{ij} = c_{0j} + (i-1)W_j \qquad \forall i, i = 1,2,3,..., \forall j, j = 1,2,3$$

where c_{0j} is a constant and W_j is a random variable with mean c_{vj}.

Assume that the cost of testing per unit time is a random variable V_1 with mean c_3 and is independent of the error-free time. Then the expected total cost of fixing the first k faults during software testing phase is given by

$$C(t_k) = E \sum_{m=1}^{k} V_1 X_m + \sum_{i=1}^{k} \sum_{j=1}^{3} E[c_{ij} \mid \text{the } i^{th} \text{ fault is type } j] P\{\text{the } i^{th} \text{ fault is type } j\}$$

Pham and Wang (2001) show

$$C(t_k) = c_3 \sum_{m=1}^{k} E[X_m] + k \sum_{j=1}^{3} \left[c_{0j} + \frac{(k-1)}{2} c_{vj} \right] p_j \qquad (11.9)$$

where $E(X_m)$ is given by Equation (11.8b).

Now we consider a numerical example. Assume

$$\alpha_1 = 1.6 \qquad \alpha_2 = 1.4 \qquad \alpha_3 = 1.2$$
$$p_1 = 0.1 \qquad p_2 = 0.3 \qquad p_2 = 0.6$$
$$c_{01} = 30 \qquad c_{02} = 15 \qquad c_{03} = 5$$
$$c_{v1} = 1.4 \qquad c_{v2} = 1.2 \qquad c_{v3} = 0.8$$
$$c_3 = 0.5 \qquad \mu = 10 \text{ hrs}$$

The expected error-free time (execution time, hrs) in Equation (11.8b) and expected total cost in Equation (11.9) are computed and listed in Table 11.1. The above cost unit is staff-unit (Ehrlich *et al.* 1993). Table 11.1 shows that both the expected error-free time and the expected total cost are increasing as the number of faults removed is becoming large. However, increment of the expected total cost is faster in this example, and is not linearly proportional to that of the expected error-

Table 11.1. Expected error-free time and expected total cost

Number of errors removed k	Expected error-free time $E[X_{k+1}]$	Expected total cost $C(t_k)$
1	13.0	15.5
2	16.9	33.5
3	22.0	54.4
4	28.6	78.8
5	37.1	107.5
6	48.3	141.5
7	62.7	182.0
8	81.6	230.7
9	106.0	289.9
10	137.9	362.2
11	179.2	451.4
12	233.0	562.3
13	302.9	701.1
14	393.7	875.7
15	511.9	1096.8
16	665.4	1378.0
17	865.0	1736.8
18	1124.6	2196.5
19	1461.9	2786.9
20	1900.5	3547.0

free time. Therefore, this example demonstrates the fact that it may usually become difficult and costly to fix a fault which is detected in the later testing phases.

11.4 Optimization Models

In Sections 11.3.1 and 11.3.2 we have derived software reliability and testing cost measurements respectively. Now we discuss the optimal software testing policies. Usual criteria of optimization of software testing are based on testing cost indices only. However, to optimize cost measures alone is sometimes not sufficient; we may be required to consider both reliability measure and testing costs for optimization. In practice, two classes of optimal testing policies may be needed: optimal testing policies which minimize the testing cost while some reliability requirements are satisfied, or policies that maximize software reliability measure given testing cost is no more than some predetermined value. For example, from Equations (11.1) and (11.4) we can formulate the following optimization models in terms of the decision variable – software testing time t:

Minimize $C(t) = tc_3 + \left(\dfrac{2c_0 - c_v}{2}\right)M(t) + \dfrac{c_v}{2}\left\{Var[N(t)] + M^2(t)\right\}$

Subject to $\begin{cases} \overline{M}(t) = M(\tau) - M(t) \le N_r \\ t \ge 0 \end{cases}$

(11.10)

where constant N_r is the pre-determined requirement for number of the remaining faults in the software upon release. From the above model, the optimal testing time t^* can be achieved, which minimizes the expected total cost of testing and debugging, given that the number of remaining faults in the software upon release is no more than a constant N_r.

From Equations (11.8b) and (11.9) we can also establish the following optimization models in terms of the decision variable of testing stop number k:

Maximize $E(X_{k+1}) = \mu \sum\limits_{i_1=1}^{3} \sum\limits_{i_2=1}^{3} \cdots \sum\limits_{i_k=1}^{3} \alpha_{i_1} \alpha_{i_2} \cdots \alpha_{i_k} p_{i_1} p_{i_2} \cdots p_{i_k}$

Subject to $\begin{cases} C(t_k) = c_3 \sum\limits_{m=1}^{k} E[X_m] + \sum\limits_{j=1}^{3} \left[kc_{0j} + \dfrac{k(k-1)}{2} c_{vj} \right] p_j \le C_{r0} \\ \\ k \le a_0 \\ k = 1,2,3,\ldots \end{cases}$

(11.11)

where C_{r0} is the predetermined requirement for total testing-debugging cost, and a_0 is the initial number of errors in the software program and can be estimated by Halstead's software metric $\hat{B} = V/3000$ where V is defined in Pham (2000) or from the Goel-Okumoto model (Goel 1985).

From Equation (11.11), the optimal testing stop number k^* can be found, which maximizes the expected software error-free time upon release, given that the expected total cost of testing and debugging is no more than a constant C_{r0}.

The above two models can be solved by nonlinear programming software to obtain the optimal software release time or number.

Now a numerical example is used to illustrate the optimization model (11.11). A set of real testing data from Misra (1983) is shown in Table 11.2 and will be used in this example. We first estimate the number of initial errors in this software program. The Goel-Okumoto software model shows the following relationship between the expected number $m(t)$ of errors to be detected by time t and total number a_0 of faults that exist in a software before testing:

$$m(t) = a_0(1 - e^{-bt})$$

where b is a parameter representing the failure intensity of a fault.

Table 11.2. Failures in 1 hour (execution time) interval

Time (hours)	Number of failures	Cumulative failures
1	27	27
2	16	43
3	11	54
4	10	64
5	11	75
6	7	82
7	2	84
8	5	89
9	3	92
10	1	93
11	4	97
12	7	104
13	2	106
14	5	111
15	5	116
16	6	122
17	0	122
18	5	127
19	1	128
20	1	129
21	2	131
22	1	132
23	2	134
24	1	135
25	1	136

Using the maximum likelihood estimate method, we can obtain from Table 11.2 for the Goel-Okumoto model:

$$\hat{a}_0 = 143 \qquad\qquad \hat{b} = 0.1246$$

The cost coefficients are usually determined by empirical data, previous experiences, and software characteristics. Ehrlich *et al.* (1993) at AT&T studied some project data using the measure unit of staff-units and found that the ratio of the cost of removing an error during testing period and the testing cost per unit time is about 10~12. It is estimated that there are 370 CPU test-execution units during testing with 1.9 staff-units per CPU unit. Based on the above information, we assume in optimization model (11.11):

$$\alpha_1 = 1.020 \qquad\qquad \alpha_2 = 1.015 \qquad\qquad \alpha_3 = 1.010$$
$$p_1 = 0.1 \qquad\qquad\quad p_2 = 0.3 \qquad\qquad\quad p_2 = 0.6$$

$$c_{01} = 30 \qquad c_{02} = 15 \qquad c_{03} = 5$$
$$c_{v1} = 1.4 \qquad c_{v2} = 1.2 \qquad c_{v3} = 0.8$$
$$c_3 = 1.2 \qquad \mu = 0.1838 \text{ hrs} \qquad a_0 = 143$$
$$C_{r0} = 11,400$$

The cost unit above is staff-units, and the error-free time is in terms of execution time (hrs). By numerical method from optimization model (11.11), the optimal testing stop number k^* can be found to be 139, which results in the maximum expected software error-free time upon release of 1.46 hrs, given the expected total cost of testing and debugging is no more than 11400 staff-units. The corresponding total cost of testing and debugging is 11377.4 staff-units.

Next we see another example. Consider the software testing model in Section 11.3.1. Assume that the release time for the software is the time of detecting k faults. Then upon release the expected error-free time is

$$E(X_{k+1}) = \alpha^k E(Z_k) = \alpha^k \mu$$

The expected total testing-debugging cost until release is

$$C = E\left[\sum_{i=1}^{k} \left[c_0 + (i-1)W \right] \right]$$
$$= kc_0 + \frac{k(k-1)}{2} c_v$$

If the expected error-free time upon software release is required to be larger than a predetermined number L, then the following optimization model can be formulated:

$$\textbf{Minimize} \quad C = kc_0 + \frac{k(k-1)}{2} c_v$$

$$\textbf{Subject to} \quad \begin{cases} \alpha^k \mu \geq L \\ k \leq a_0 \\ k = 1,2,3,\ldots \end{cases}$$

$$(11.12)$$

where a_0 can be similarly estimated by Halstead's software metric or from the Goel-Okumoto model.

Now we assume that, in terms of execution time and staff-units,

$$c_0 = 13.5 \qquad c_v = 1 \qquad \alpha = 1.015$$
$$a_0 = 143 \qquad L = 1.4 \text{ hrs} \qquad \mu = 0.1838 \text{ hrs}$$

The optimal solution to above model is

$$k^* = 137$$

and the corresponding expected total cost is

$$C^* = 11,165.5 \text{ staff-units}$$

Therefore, we will obtain the minimum cost of 11,165.5 staff-units if we stop testing once the 137th fault is removed. The corresponding expected error-free time is 1.4 hrs. If the average error-free time requirement L is changed to 1.50 hrs from 1.40 hrs, the optimal release number is

$$k^* = 141$$

and the corresponding expected total cost and error-free time are respectively

$$C^* = 11,773.5 \text{ staff-units}$$

$$\mu^* = 1.50 \text{ hrs}$$

If we consider the cost of testing per unit time which is a random variable V_1 with mean c_3 and is independent of the error-free time, the optimization model (11.12) becomes

Minimize $\quad C = \dfrac{\mu(\alpha^k - 1)}{\alpha - 1} c_3 + kc_0 + \dfrac{k(k-1)}{2} c_v$

Subject to $\quad \begin{cases} \alpha^k \mu \geq L \\ k \leq B \\ k = 1,2,3,... \end{cases}$

$$(11.13)$$

Assume that $c_3 = 1.2$ staff-units per hour and the average error-free time requirement L is 1.40 hrs. The optimal release number, in terms of removed faults, is

$$k^* = 137$$

and the corresponding expected total cost and error-free time are respectively

$$C^* = 11,263.8 \text{ staff-units} \qquad \mu^* = 1.41 \text{ hrs}$$

The above numerical results based on the set of real testing data show that the models developed in this chapter work well in practice. Note that results from the three optimization models (Equations 11.11 through 11.13) are quite close.

In many cases a software is large-scale and consists of many software modules performing different functions. We can apply software reliability growth models to the module level to estimate how many faults are remaining in each module at different stages of software testing process, and prioritize future testing efforts.

11.5 Concluding Discussions

This chapter has discussed software reliability and cost modeling *via* quasi-renewal processes. From this chapter we can see that the quasi-renewal process is an effective tool to model software reliability and costs since measures and indices can be derived conveniently. Especially, software error-free time information upon testing can be obtained by using this modeling tool. In this chapter, we assume the cost of fixing a fault during software testing phase consists of deterministic and probabilistic parts, and it becomes larger as the number of faults removed is increasing. This assumption is justified by the fact that it may usually become difficult to fix a fault which is detected in the later testing phases. Besides, three types of faults in software are considered for Model 2. Testing and debugging costs are considered separately in this work. Obviously, all these assumptions and considerations make the proposed software models more realistic.

In software reliability and cost models introduced in this chapter, we note that most results – expected software testing and debugging cost, number of remaining faults in the software, and mean error-free time after testing – are in closed forms. The parameters associated with these models can be easily estimated through the maximum likelihood method and the likelihood equations can be solved by standard numerical methods. Unlike most other software reliability models, we combine reliability measures and testing cost measures of software in optimization problems and the optimal solutions to the optimization problems lead to optimal software testing policies with regards to both reliability measures and testing cost measures.

Software reliability models in this chapter can be used to estimate the residual faults upon release to know if software reliability objectives are met, for example, if telecommunication software can meet Telcordia software reliability objectives for a given release.

Generally, software reliability also grows in field use since faults found in the field may be removed. The models introduced in this chapter can also apply to modeling field reliability growth and maintenance cost.

12

Monte Carlo Reliability Simulation
of Complex Systems

To obtain the optimal maintenance policy for a complex system, we may need to evaluate system availability or MTBF first, as discussed in Section 1.5. The previous chapters have modeled reliability measures for some standard reliability architectures. However, in practice, many systems are complex systems. Generally there are four major difficulties in evaluating complex large-scale system reliability, availability and MTBF (MTTF): the system reliability structure may be very complicated; subsystems may follow different failure distributions; subsystems may have arbitrary failure and repair distributions for maintained systems; failure data of subsystems are sometimes not sufficient, sample size of life test or field population tends to be small. Therefore, it may be difficult and often impossible to obtain s-confidence limits of the reliability indices by classical statistics. It has been proven that Monte Carlo technique combined with Bayes method is a powerful tool to deal with this kind of complex systems. In this chapter, the typical existing Monte Carlo reliability, availability, and MTBF simulation procedures are analyzed together with variance reduction techniques and random variate generation algorithms. The advantages, drawbacks, accuracy and computer execution time of Monte Carlo simulation in evaluating reliability, availability and MTBF of complex networks are discussed. Some conclusions are summarized, and a general Monte Carlo reliability and MTTF assessment procedure is presented.

12.1 Introduction

Monte Carlo simulation methods are numerical methods which allow the solution to mathematical and technical problems by means of system probabilistic models and simulation of random variables. It was originated in the 1940s by mathematicians Newman and Ulam at an early development stage of nuclear technology. Scientists at the Los Alamos National Laboratory used it to model the random diffusion of neutrons. They gave it the name "Monte Carlo" after the city in Monaco and its many casinos. Today, its applications have been extended to many areas of science and technology. Monte Carlo simulation method is now recognized as playing an important role in system reliability, availability and MTTF (MTBF) assessment and optimal maintenance of large-scale complex networks. During the last 50 years a lot of efforts has been made in developing efficient Monte Carlo simulation methods and software programs for determining s-confidence bounds on system reliability, availability, and MTTF (MTBF). Using classical statistical methods, it is difficult and sometimes even impossible to obtain s-confidence intervals of system reliability, availability, and MTTF (MTBF) though it may be easy to find point estimates of complex system reliability. In fact, how to obtain s-confidence limits on them is not only a dilemma in engineering practice, but also one in statistical theory. By Monte Carlo simulation method, such analysis becomes relatively easy and at least possible, as fast computers are now readily available.

In the last several decades, a lot of Monte Carlo reliability, availability, and MTTF (MTBF) evaluation methods and software programs have been developed. Wang and Pham (1997) summarize the previous research in an overview paper. System reliability and MTTF estimation by Monte Carlo method began at least in 1960. Orkand (1960) presents his technical report on determining system reliability confidence limits from subsystem failure test data using Monte Carlo simulation method at the U.S. Picatinny Arsenal. Burnett and Wales (1961) discuss analytical and Monte Carlo techniques for obtaining confidence limits and the assumptions necessary for their use. They create the method for the case of components with exponentially distributed failures. Bernhoff (1963) in his thesis also studies the Monte Carlo reliability simulation at U.S. Air Force Institute of Technology. Moore (1965) develops a general Monte Carlo technique extending the Monte Carlo method to cases where the joint distribution of the estimators of the parameters of failure model is known. In fact, Moore and his graduate students Levy (1964, 1967), Lutton (1967), Lannon (1972), Snead (1978), Rice (1979), Putz (1979), Johnson (1980), MacDonald (1982), and others, have done much work in this field. Among them, Levy and Moore (1967) design a process to obtain system reliability s-confidence limits for a system composed of different subsystems whose failures follow the Normal, Lognormal, Gamma, or Weibull distributions and distribution parameters are supposed to be estimated by the maximum likelihood method based on life tests from a complete sample or from a censored sample where the distribution of the estimator is known.

Gilmore (1968) analyzes complex system MTBF using Monte Carlo simulation. Integrating Bayesian method with Monte Carlo simulation, Locks (1974a,b, 1978) proposes a Monte Carlo-Bayesian approach to determine reliability lower bounds and MTTFs of complex large-scale systems of any modular reliability structures. Massa develops a Monte Carlo reliability evaluation technique under binomial and exponential failure distributions. Rice and Moore (1983), Chao and Huwang (1987) investigate Monte Carlo reliability assessment for systems with binomial-failure subsystems. Kamat and Riley (1975) present a Monte Carlo system reliability estimation method in which its subsystems are allowed to conform to any different failure distributions. Later, Kamat and Franzmeier (1976) extend this procedure to determine reliability confidence intervals on systems including s-dependent subsystems and/or allowing repair of failed subsystems. Kumamoto *et al.* (1977, 1987), Fishman (1986a,b, 1987a,b,c, 1989), and Baca (1993) have done much research on Monte Carlo network reliability assessment and sampling plan, and investigated technique of variance reduction, a major problem for large Monte Carlo simulation. Availability and MTBF evaluation for reparable systems have also received certain attention (Kamat and Frazmeier 1976; Kim and Lee 1992; Kumamoto *et al.* 1980a; Moore *et al.* 1985). Today, Monte Carlo simulation technique for reliability, availability and MTBF assessment has been widely used in electric power systems, civil engineering, nuclear engineering, building industry, and probabilistic mechanics.

Monte Carlo reliability simulation methods of complex systems have become relatively mature, especially for non-repairable systems. In addition, fast computers can be available everywhere and using them to perform Monte Carlo simulation is very convenient. Note also that Monte Carlo reliability simulation research in some engineering fields, such as electric power systems, civil engineering, nuclear engineering, building industry, and probabilistic mechanics, still receive some attention. However, the focus of this chapter is general methodologies of Monte Carlo reliability and availability simulation for various reliability structures following Wang and Pham (1997), while its applications in individual fields will not be addressed in details.

12.2 Typical Monte Carlo Algorithms for Reliability

To study characteristics, accuracy and related problems of different Monte Carlo techniques, we have selected some typical ones, and next analyze and discuss them. First, we analyze a general approach by combining reliability flow graph representation, Boolean state representation and Monte Carlo simulation.

12.2.1 K-R Method

This Monte Carlo procedure, developed by Kamat and Riley (1975), is fairly general and can be applied to most systems with arbitrary system reliability structure and different subsystem failure distributions without modification. In this procedure, individual subsystems are assumed to be independent of each other and repair of failed subsystems are not allowed; the underlying life distribution is known for each subsystem and distribution parameters have been estimated. The key idea of this K-R method is:

(a) Find out all minimal tie-sets from system Reliability Block Diagram (RBD). Assume that we need to obtain system reliability interval estimates at some time point t.

(b) From the life distribution of each subsystem, a random failure time t_i is generated where i represents the ith subsystem, $0 < i < n$.

(c) Compare t_i with t for all subsystems. If $t_i > t$, this indicates that at time t subsystem i functions properly; if $t_i \leq t$, then subsystem i has failed.

(d) Determine whether the whole system is functioning or down according to the states of its subsystems at t from step (c). Check all subsystems in a minimal tie-set. If all of them are operational then the system operates properly at t. If one or more of them fail, then the tie-set is broken (failure) at t. Further, check next minimal tie-set until an unbroken one appears, which means that the system is operational at t. If all minimal tie-sets are broken then the system fails at t.

(e) Repeat steps (b), (c), (d) for, say, n times. Count failure and success numbers of the system respectively: $n_S(t)$ and $n_F(t)$. Note that

$$n = n_S(t) + n_F(t)$$

(f) The system reliability point estimate corresponding to t is given by

$$\hat{R}(t) = \frac{n_S(t)}{n_S(t) + n_F(t)}$$

Note that the simulation results are of binomial type. Based on the Normal approximation to the Binomial distribution, the $100(1-\gamma)\%$ confidence intervals of system reliability at time t are given by

$$[R_L(t), R_U(t)] = \hat{R}(t) \pm z_\gamma \frac{\hat{R}(t)\left(1 - \hat{R}(t)\right)}{\left[n_S(t) + n_F(t)\right]^{1/2}}$$

where z_γ is the double-side $100\gamma\%$ percentile of the standard Normal distribution with mean zero and variance 1.

An application example is given by Kamat and Riley (1975). The system reliability structure diagram in this example is shown in Figure 12.1 and lifetimes of all nine subsystems: a, b, c, d, e, f, g, h, i are assumed to follow the two-parameter Weibull distribution with survival function

$$sf(t; K_i, M_i) = \exp\left[-\frac{K_i}{M_i + 1} t^{M_i + 1}\right] \qquad t, M_i > 0, M_i > -1$$

From Figure 12.1, we can see that it is difficult to determine the reliability interval estimates of this system by using classical statistics. Per system reliability

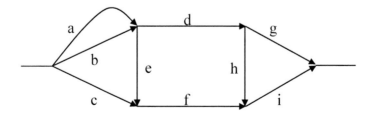

Figure 12.1. Reliability structure diagram

theory, the system's minimal tie-sets can be found to be

adg, bdg, adhi, bdhi, aefi, befi, cfi

The scale parameter K_i and shape parameter M_i values for all nine subsystems are listed in Table 12.1. Utilizing the K-R Monte Carlo algorithm, 1000 simulation replications are performed on the IBM360/65 computer by Kamat and Riley (1975). Table 12.2 summarizes the results of system reliability point estimates and 95% interval estimates at certain time points.

Since Kamat and Riley (1975) do not discuss the accuracy of their simulation results and we cannot derive exact 95% confidence intervals for this system by

Table 12.1. Weibull parameters for each component

Component no.	Scale parameter K	Shape parameter M
a	2.8	1.8
b	2.7	1.7
c	2.6	1.6
d	2.5	1.5
e	2.4	1.4
f	2.2	1.2
g	2.3	1.3
h	2.1	1.1
i	2	1

Table 12.2. Reliability simulation results – 95% confidence intervals

Time	Reliability point estimate	Upper 2.5% limit	Lower 2.5% limit
0	1	1	1
0.1	1	1	1
0.2	0.999	1	0.997
0.3	0.986	0.993	0.979
0.4	0.95	0.964	0.936
0.5	0.886	0.906	0.886
0.6	0.775	0.801	0.749
0.7	0.625	0.665	0.595
0.8	0.445	0.486	0.424

classical statistical methods, we are not able to draw conclusions on its accuracy. Note that the K-R method using the normal approximation to the binomial distribution will result in some error. However, from Table 12.2 we can see that the confidence intervals are generally very narrow and the point estimates are at the middle of them. Therefore, this method can basically be accepted in some engineering applications. Note that the s-confidence intervals obtained by the K-R method can be narrowed by increasing simulation replication number.

The simulation procedure are programmed in FORTRAN IV G level code. For the system structure in Figure 12.1, 16 seconds for execution are spent.

12.2.2 R-M Method

The drawback of the K-R approach is that all minimal tie-sets have to be determined in advance. Rice and Moore (1983) propose a special Monte Carlo method (R-M) dealing with the fail-pass failure. Using this technique, not only a lower confidence limit (LCL) but also the quantitative analysis of accuracy of LCLs can be determined. The R-M method can be applied to any complex system structure whose subsystem failures follow binomial distributions, especially to the systems with zero-failure subsystems. However, LCLs obtained by the R-M algorithm are somewhat larger than the exact LCLs.

The R-M simulation method is also based on the Normal approximation to the Binomial distribution: if the success probability of a binomial test is p, failure probability q, test number n, failure number f, then this binomial failure follows the Normal distribution with mean p and variance (pq/n). If a test has zero failure, that is, $f_i = 0$, then f_i can be replaced by the equivalent failure number f_i' given by Gatliffe (1976).

The key steps of the R-M process are

(a) Define the system and its reliability block diagram (RBD). Develop the algorithm to compute system reliability from its subsystems' reliabilities, $i.e.$, system reliability structure function.

(b) For each subsystem, determine its failure number f_i or equivalent failure number f_i'. To simplify, they are both represented by f_i^*.

(c) Calculate estimates:

$$p_i = 1 - \frac{f_i^*}{n_i} \qquad q_i = 1 - p_i \qquad \text{Asymptotic Variance} = \frac{p_i q_i}{n_i}$$

(d) For each subsystem, generate a random variable from $N(0,1)$, where $N(0,1)$ is the Normal distribution with mean zero and variance 1.

(e) Find the second estimate $p_i \sim \left(p_i, \frac{p_i q_i}{n_i}\right)$ by drawing an $r.v.$ from $N(0,1)$. Multiply by asymptotic standard variance and add it to p_i.

(f) Calculate system reliability R_s from subsystem reliabilities according to the algorithm created in step (a).

(g) Implement steps (d) – (f) many times for, say, 999 times.

(h) List these R_s values in order of non-decreasing magnitude.

(i) Determine the $100(1 - \gamma)\%$ percentile to obtain $100(1 - \gamma)\%$ LCLs of R_s .

The reliability LCLs of the system which consists of two and three subsystems in series have been computed by this algorithm and are compared with those by other methods. Table 12.3 lists these LCLs, where the other approaches are:

AN/MC   R-M algorithm (Rice and Moore 1983)
ML   Maximum-likelihood
LR   Likelihood ratio (Madansky 1965)
OPT   Optimal method (exact limits)
AO   Approximately optimum (Mann and Grubbs 1974)
MMLI   Modified maximum-likelihood (Easterling)

Table 12.3. Comparison of lower s-confidence limits on reliability by using three different Monte Carlo techniques

Num. comp.	CL	n	f_1	f_2	f_3	AN/MC	ML	LR	OPT	AO	MMLI
2	90%	10	1	1		0.655	0.655	0.629	0.607	0.606	0.585
			1	2		0.542	0.545	0.529	0.497	0.493	0.489
			2	2		0.458	0.456	0.451	0.445	0.43	0.441
			1	4		0.337	0.347	0.35	0.344	0.335	0.318
			2	3		0.372	0.373	0.375	0.354	0.353	0.362
		20	1	2		0.754	0.756	0.739	0.716	0.728	0.709
			2	2		0.7	0.701	0.687	0.683	0.678	0.669
			1	3		0.693	0.697	0.683	0.66	0.675	0.655
			2	3		0.646	0.647	0.638	0.622	0.628	0.619
			3	3		0.599	0.599	0.591	0.585	0.582	0.57
	95%	10	1	1		0.614	0.611	0.571	0.548	0.552	0.53
			1	2		0.495	0.495	0.473	0.443	0.435	0.436
			2	2		0.414	0.405	0.397	0.392	0.382	0.391
			1	4		0.29	0.292	0.301	0.298	0.293	0.271
			2	3		0.328	0.32	0.326	0.304	0.307	0.315
		20	1	2		0.724	0.728	0.7	0.677	0.693	0.671
			2	2		0.669	0.67	0.647	0.643	0.643	0.631
			1	3		0.663	0.665	0.643	0.62	0.639	0.616
			2	3		0.612	0.614	0.587	0.582	0.593	0.58
			3	3		0.565	0.551	0.544	0.548	0.548	0.532

Table 12.3. (continued)

3	90%	20	1	1	1	0.757	0.76	0.743	0.747	0.741	0.721
			1	1	2	0.704	0.704	0.69	0.693	0.689	0.669
			1	2	2	0.653	0.654	0.643	0.639	0.643	0.619
			1	2	3	0.608	0.605	0.596	0.595	0.597	0.587
		30	1	2	3	0.724	0.723	0.714	0.705	0.716	0.669
			1	1	1	0.833	0.835	0.822	0.825	0.82	0.803
			2	2	2	0.722	0.725	0.715	0.712	0.717	0.703
		50	1	2	4	0.806	0.805	0.798	0.789	0.8	0.788
			1	1	2	0.873	0.874	0.865	0.861	0.87	0.852
		100	1	1	2	0.935	0.936	0.931	0.929	0.932	0.923
			2	3	5	0.866	0.866	0.861	0.858	0.865	0.856
	95%	20	1	1	1	0.725	0.732	0.705	0.709	0.708	0.684
			1	1	2	0.667	0.673	0.651	0.644	657	0.631
			1	2	2	0.613	0.621	0.604	0.598	0.609	0.58
			1	2	3	0.567	0.571	0.557	0.544	0.554	0.549
		30	1	2	3	0.693	0.698	0.683	0.674	0.691	0.638
			1	1	1	0.81	0.816	0.794	0.796	0.698	0.775
			2	2	2	0.694	0.7	0.685	0.681	0.692	0.672
		50	1	2	4	0.784	0.788	0.776	0.767	0.781	0.766
			1	1	2	0.856	0.86	0.845	0.841	0.854	0.833
		100	1	1	2	0.926	0.929	0.92	0.918	0.923	0.913
			2	3	5	0.852	0.855	0.848	0.844	0.854	0.842

For an eight-subsystem series system with subsystem reliabilities respectively

$$0.95,\ 0.95,\ 0.90,\ 0.95,\ 0.85,\ 0.75,\ 0.95,\ 0.95$$

The 90% reliability LCL is 0.3665 using the R-M algorithm with Monte Carlo simulation replications of 1000. Among these 1000 LCLs, there exist 913 LCLs for which intervals (LCL, 1) contain the true reliability .42059 which are obtained by multiplying all subsystem reliabilities. In engineering, this accuracy can be acceptable in some cases.

12.2.3 C-H Method

Since a upper error exists for the R-M procedure, Chao and Huang (1987) have improved it. The numerical examples show that Chao and Huang's method (C-H) can reduce this error. The C-H algorithm is related to "bootstrap" method by Efron and replaces the (c) and (d) steps of the R-M process by the following (c') and (d'):

(c')   The success and failure probability estimates of subsystem i are given by

$$p_i = 1 - \frac{f_i + a}{n_i + a + b} \qquad q_i = 1 - p_i \qquad \text{where } a = 0.2 \text{ and } b = 0$$

The above equations are derived according to Bayes theorem using the Beta prior distribution. This choice of a and b are studied carefully to make LCL values by Monte Carlo simulation procedure close to the exact LCLs. Obviously, the above equation for p_i and q_i can apply to zero failure case.

(d') Generate a random variable $f_i^{\bullet}$ from the Binomial distribution with parameters n and p in (c') and compute

$$p_i^{\bullet} = 1 - \frac{f_i^{\bullet} + a}{n_i + a + b} \qquad\qquad q_i^{\bullet} = 1 - p_i^{\bullet}$$

The simulation results shows that the C-H approach can result in more exact LCLs than the R-M procedure and the ML methods, and LCLs given by the C-H procedure are close to those by other methods and the OPT (exact limits). Therefore, for binomial failure, we suggest the C-H method be used.

12.2.4 L-D-L Method

The L-D-L method was designed by Lin et al. (1988) and also used to analyze problems with binomial failure distribution. It increases failure information of subsystems using a priori failures from the Bayes method. The L-D-L method takes beta distribution $Beta(d_i, b_i)$ as a priori distribution where d_i values are determined in such a way that LCLs obtained can be made exact and d_i is the same for all subsystems, and $b_i = 1$ for all subsystems. Thus, the a priori distribution of each subsystem is $B(d,1)$. According to the Bayes theorem, its posterior distribution is $Beta(d + x_i, n_i + x_i + 1)$ where x_i and n_i are respectively failure and test numbers of subsystem i. Based on these obtained posterior distributions, the L-D-L algorithm is outlined as follows:

(a) Generate k random samples $r_1, r_2, ... r_k$ from $Beta(d + x_i, n_i + x_i + 1)$. Suppose that a system consists of k subsystems regardless of system reliability architecture.

(b) Calculate point estimate of system reliability
$$R_j = g(r_1, r_2 ... r_k)$$
where $g(r_1, r_2 ... r_k)$ are the structure function of the system.

(c) Repeat steps (a), (b) 10,000 times.

(d) Rank these R_j in ascending magnitude order.

(e) Find $100(1-\gamma)\%$ percentile $R_{1-\gamma}$ from step (d) and then $R_{1-\gamma}$ is the $100(1-\gamma)\%$ LCL which we need.

Consider a two-subsystem series system. The exact LCLs and LCLs by the L-D-L

Table 12.4. Comparison of exact LCLs and the ones by L-D-L

No. of subsystems		No. of failures		90% LCL		95% LCL		δ
n_1	n_2	f_1	f_2	L-D-L	Exact	L-D-L	Exact	
10	10	1	1	0.62	0.607	0.568	0.548	4.46
10	10	1	2	0.529	0.497	0.484	0.443	3.494
10	10	2	2	0.449	0.445	0.403	0.392	2.647
10	10	1	4	0.343	0.334	0.301	0.298	1.588
10	10	2	3	0.362	0.354	0.318	0.304	1.8
20	20	1	2	0.72	0.716	0.684	0.677	4.923
20	20	2	2	0.67	0.683	0.637	0.643	4.446
20	20	1	3	0.67	0.66	0.634	0.62	4.42
20	20	2	3	0.626	0.662	0.592	0.582	3.97
20	20	3	3	0.581	0.585	0.544	0.544	3.52

are listed in Table 12.4. We can see that they are close to each other and approximately equal.

12.2.5 L-D Method

Most Monte Carlo simulation procedures require their users to provide minimal tie-sets or cut-sets. Lin and Donaghey (1993) propose a new Monte Carlo procedure (L-D method) for system reliability. The advantages of this algorithm is that it first utilizes the Monte Carlo method to determine the minimal tie-sets by tracing through the system from the input components to the output components of a system modeled by the Reliability Block Diagram in a random manner, then uses the minimal tie-sets to simulate system failures, the minimal cut sets and system reliability at any time are determined again by the Monte Carlo approach. Therefore, it avoids providing minimal tie-sets prior to simulation. In addition, the system mean time to failure, system failure distribution and the cumulative failure rates can be obtained. The basic idea of this process is that using Monte Carlo simulation and the minimal tie-sets as the criteria for system failure, the system fails when all minimal tie-sets are broken. Components which have failed prior to the system failure constitute a cut set. The frequencies of the minimal cut sets are tallied during the simulation runs to show the distribution of the frequencies of the minimal cut sets (Lin and Donaghey 1993). For the single bridge system and a complex ten-subsystem network, the tie-sets given by this algorithm agree with the results generated by general reliability theory. The disadvantage of this method is that it does not give confidence interval estimates.

12.2.6 Other Methods for Non-repairable Systems

MacDonald (1982) creates a double Monte Carlo reliability evaluation procedure for a complex system which is composed of subsystems with Weibull failure distribution of three parameters. Putz (1979) presents a univariate Monte Carlo technique to approximate reliability confidence limits of systems with components characterized by the Weibull distribution. Lutton (1967) and Lannon (1972) study bivariate asymptotic Monte Carlo method by using the asymptotic s-normality of maximum likelihood estimates. Moore *et al.* (1980) compare these three Monte Carlo simulation methods: Double Monte Carlo, Univariate and Bivariate asymptotic Monte Carlo. For 3-subsystem series, parallel and series-parallel systems, a 5-subsystem complex system and a 25-subsystem network whose subsystems' failures follow respectively Weibull, Logistic and Gamma distributions. Conclusions by Moore *et al.* (1980) are that confidence bounds obtained by the bivariate asymptotic method are less than those by the double Monte Carlo which are less than those by the univariate approach. The percentage of times the confidence intervals covered the true system reliability is also compared with the desired confidence level and their results are that the bivariate method is conservative but more accurate than the univariate method and less sensitive to degradation due to high system reliability. The bivariate method is fast and accurate in some cases. The CPU times on the CDC6600 for the three methods (1000 Monte Carlo replications; 25-subsystem system) are respectively: Bivariate: 22 seconds; Double: 22 minutes; Univariate: 11 seconds. The double Monte Carlo uses much more computer time than the other two methods. Later, Depuy *et al.* (1982) modify the double Monte Carlo algorithm and compare it with the two asymptotic techniques. Chang *et al.* (2001) introduce VP (Variational Principle) technique over analog MC and use variationally processed Monte Carlo simulation for estimating system reliability.

Kumamoto *et al.* (1977) design a Monte Carlo method, KTI named after them, using variance-reducing technique which applies to fail-pass failure. Using their algorithm, an 18-subsystem complex network, as shown in Figure 12.2, is analyzed and network reliability upper and lower limits are determined. Later in 1987, using variance reduction technique, they propose another new Monte Carlo technique for evaluating the top-event probability of a coherent fault tree of complex systems which can have high reliability under the assumption that all the minimal cut sets are known. However, although the KTI algorithm with a smaller sample size and variance reduction is an improvement over direct Monte Carlo approach (Hammersley and Handscomb 1964, pp.51–52) and Mazumdar's importance sampling method (Mazumdar 1975), a general-purpose computer program is not available and some theoretical problems in statistics exist with this algorithm. Locks (1979) studies the KTI procedure and presents an alternative explanation of this procedure and discusses its usefulness compared to some alternative ones available.

Some other researchers, Su (1986), Fishman (1987, 1991), and Elperin *et al.* (1991), integrate the graph theory and network theory with Monte Carlo simulation to assess network reliability. Note that the research based on this idea will make Monte Carlo reliability simulation systematic and theoretical and thus is very useful.

In addition, Kim and Fard (1995) propose two types of discrete-event simulation models estimating reliability, mean time to failure and probability density function of time to failure for a complex system with general failure rates.

One advantage of this method is that these two discrete-event reliability models only require descriptive knowledge of network architecture rather than analytical network characteristics, such as cut or tie sets. Another advantage is that reliability modeling technique applied through this discrete-event simulation is modular and then its adaptability to large-scale complex systems is increased. They are programmed using SIMAN codes for both reliability modeling and evaluation. Actually before them, Prisker *et al.* (1989) suggest a sample system reliability modeling method using another discrete-event simulation language, SLAM. We believe that the discrete-event simulation for reliability is promising and suggest using it later.

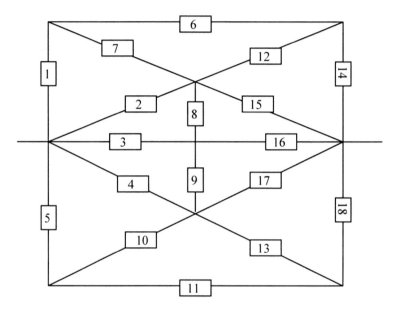

Figure 12.2. Reliability block diagram of a complex system

Worth mentioning is ESCAF – Electronic Simulator to Compute and Analyze Failure developed by Laviron *et al.* (1982) although it is a direct simulation method, not Monte Carlo. ESCAF can be employed to determine and list cut sets and minimal cut sets, or tie-sets and minimal tie-sets; to gauge the importance of each subsystem or event by constructing histograms; to compute system unreliability or unavailability for both *s*-coherent and non-coherent systems.

12.2.7 Monte Carlo Methods for Repairable Systems

Availability, MTBF and unreliability evaluation for a complex maintained system could be very complicated, sometimes even impossible to handle by classical probability, statistics, or Markov techniques. Using Monte Carlo method it is possible and practical to deal with this kind of problems. In fact, Monte Carlo simulation is the only practical technique to find *s*-confidence bounds for systems with general failure and repair distribution (Moore *et al.* 1985). In the past two decades, some researchers, for example, Kamat and Franzmeier (1976), Kumamto *et al.* (1980a,b), Moore *et al.* (1985), Kim *et al.* (1992), have succeeded in solving this class of problems by means of Monte Carlo techniques though less work has been done on reparable systems.

Moore (1985) designs a double Monte Carlo process (see also MacDonald, 1980) for determining confidence bounds on reliability and availability of a reparable system. The confidence limits obtained by this approach in two examples demonstrate a satisfactory accuracy for it. The key idea of this method is: the simulated component failure and repair times is used to estimate parameters of failure and repair distributions. Simulated values of parameters are obtained by generating sample failure and repair times of equal size to the original sample using as parameters the estimates from the simulated real data. The parameters are again estimated from the generated data using the same estimator to obtain simulated values. Then simulated values for reliability and availability can be obtained after inserting the simulated values in the equations for reliability and availability. The process is repeated for the desired number of Monte Carlo repetitions. These points are used to determine a cumulative distribution function of system reliability and availability estimates by plotting the order statistics at their median ranks. Then the $100(1 - \gamma)\%$ lower confidence bounds for reliability and availability are found.

Kumamto *et al.* (1980b) propose a state-transition Monte Carlo unreliability estimate method for evaluating large repairable systems which can be modeled by a stationary Markov transition diagram. Kumamto *et al.* (1980a) also create a dagger-sampling Monte Carlo approach for unavailability evaluation of systems which can be represented by a coherent fault tree. Generally, there are two kinds of Monte Carlo methods: direct and indirect simulation. The direct Monte Carlo is flexible but very wasteful of computer time, per Kumamto *et al.* (1980b). The above two methods are indirect Monte Carlo and thus reduce computation time. Along with reducing computation time, the dagger-sampling Monte Carlo combines success and failure states, and generates negatively-correlated state vectors of basic events. This negative correlation at the top-event level for fault tree applies to the corresponding states of the top event since the system structure function is a monotonously increasing one. The Monte Carlo estimator has a smaller variance than the direct Monte Carlo method because probabilistic fluctuations are canceled out by the correlation. A numerical example shows that the dagger-sampling could generate 100 trials in the computation time for one direct Monte Carlo trial. The disadvantage of the dagger-sampling Monte Carlo is that it only gives point estimate of system unavailability.

Besides Monte Carlo availability assessment, an MTBF estimate algorithm for a binomial coherent system is also presented by Kim *et al.* (1992), which is based on the assumptions that failed components are replaced with new ones at system failure ("as good as new" or perfect repair); if a minimal tie set fails, components therein (not included in any other minimal tie sets) cease to operate until repair of the system; replace time for any component is negligible; states of all components are *s*-independent. Note that Kamat and Franzmeier (1976) propose a Monte Carlo reliability determination method for systems which contain s-dependent components. In this method, the MTBF uses the special definition $t/E[N(t)]$, where $N(t)$ is the number of system failures in $[0, t]$ and $E[N(t)]$ is the expected value of $N(t)$. All minimal tie-sets of the system and lifetime distributions of all components are required in advance. If the component lifetime distributions are unknown, then a lower bound of system MTBF can be estimated by using known failure rates for each component. Comparing the simulation results against the theoretical ones given by Kim *et al.* (1992) indicates that this algorithm is successful. However, confidence limits on MTBF cannot be determined by this approach.

Monte Carlo simulation for maintained systems are more complicated than non-repairable systems. So far, there are no satisfactory Monte Carlo algorithms for evaluating MTBF and availability of general complex reparable systems. Some techniques can only give point estimates and some require much computer time. Most of them require determining all minimal tie-sets or path sets of systems in advance. Because many technical systems are repairable this seems to be a fruitful area on which later efforts can be concentrated. As shown in Chapter 2, researchers have developed many imperfect maintenance models for repairable systems, and Monte Carlo simulation of imperfect maintenance of repairable systems would be more realistic.

12.3 Variance Reduction and Random Number Generation

For a system with highly reliable components, the likelihood of observing a system failure during each Monte Carlo simulation is very low. In consequence there is a large variance in the estimation of the system reliability (Chang *et al.* 2001). Methods have been developed to reduce the variance associated with rare events. In fact, one of the difficulties in Monte Carlo simulation lies in the prohibitive computing time whenever a very rare event has to be shown. The time required for a single sample run could therefore be quite long. For most reliability problems, the computation time is a rapidly increasing function of the number of subsystems in the system and Monte Carlo simulation of reliability problems usually results in rare-event simulation. Hence, direct Monte Carlo methods are extremely wasteful of computer time. Variance Reducing Techniques (VRT) have a dual purpose: to reduce the length of a sample run and to increase accuracy using the same number of runs. The method of applying VRTs usually depends on the particular simulation model of interest. It is generally impossible to know beforehand how great a variance reduction might be realized or whether the variance will be reduced at all in comparison with straightforward simulation. However, preliminary runs could be made to compare the results of applying a VRT with those from straightforward simulation. It is worthwhile to note that some VRTs themselves will increase computing cost and this decrease in computational efficiency must be traded off against the potential gains in statistical efficiency, measured by the variances of the output random variables from a simulation.

There are several comprehensive surveys that provide useful ways of classifying VRTs and also contain extensive bibliographies, among which are Hammersley and Handscomb (1964), Wilson (1983) and Nelson (1987). Some commonly-used VRTs are control variates, antithetic variables, conditioning, stratified sampling and importance sampling. The control variate method attempts to take advantage of correlation between certain random variables to obtain a variance reduction. In the method of antithetic variables, the negative correlation is sought to reduce the variance of the output variable. Importance sampling reduces the variance of the output variable by increasing the frequency of rare and importance events.

Easton (1980) introduces a sequential destruction method which reduces the variance of the system reliability substantially and the amount of computation required. In evaluating system-failure probability, Kumamoto *et al.* (1980) show how to exploit sampling plans that induce negative correlation between replications (dagger-sampling Monte Carlo), and Kumamoto *et al.* (1987) proposes a new coverage Monte Carlo estimator with a smaller variance. Zio *et al.* (2004) further discuss dagger-sampling variance reduction in Monte Carlo reliability analysis to deal with components which may fail in more than one mode. Fishman (1986) compares the methodological features of four Monte Carlo sampling plans for estimating system reliability with particular emphasis on their statistical accuracy and variance reduction. Based on less prior information, Baca (1993) constructs a Monte Carlo procedure which yields estimators with smaller variance. The first application derives a variant of the sequential destruction method and the second application obtains the traditional importance sampling Monte Carlo method for static reliability problems.

Campioni *et al.* (2005) believe that, since for Monte Carlo reliability simulation it often happens that one has to deal with rare events, the use of a variance reduction technique is almost mandatory in order to have Monte Carlo efficient applications. The main issue associated with variance reduction techniques is related to the choice of the value of the biasing parameter. Actually, this task is typically left to the experience of the Monte Carlo user, who has to make many attempts before achieving an advantageous biasing. Campioni *et al.* (2005) provide a practical rule addressed to establish an *a priori* guidance for the choice of the optimal value of the biasing parameter. This rule, which has been obtained for a single component system, has the notable property of being valid for any multi-component system.

In addition, Chang *et al.* (2001) investigate use of VP (Variational Principle) for another variance reduction method.

Random sampling of operating or repair times is necessary for Monte Carlo simulation. There are many techniques for generating random numbers from continuous or discrete distribution. The commonly used techniques for simulating continuous random variables are the inverse transformation method, acceptance-rejection method, hazard rate method, composition method, and convolution method. The common used techniques for simulating discrete random variables are the inverse alias method developed by Walker (1977), and discrete-inverse-transformation method. The particular algorithms for generating random variates from several common occurring continuous distributions, *e.g.*, the Uniform, Exponential, Gamma, Weibull, Lognormal, Normal, *m*-Erlang, for generating random variates from some discrete distributions, *e.g.*, the Binomial, Geometric, Negative Binomial, Poisson, can be found in Law and Kelton (2000).

12.4 On Monte Carlo Reliability Simulation

Monte Carlo reliability simulation methods generate random failure times from each component's failure distribution. The overall system reliability is then obtained by simulating system operation and empirically calculating the reliability values for a series of time values. Through the use of computers, simulation has become a very popular analysis tool. Simulation is simple to apply and it can produce results that can be rather difficult to solve analytically. On the other hand, simulation methods also have certain drawbacks, not the least of which is that the results depend on the number of simulations, which results in a lack of repeatability.

Generally, there exist four major difficulties in evaluating complex large-scale system reliability, availability and MTBF:

(a) The system reliability structure may be very complex.

(b) Subsystems may follow different failure distributions.

(c) The failure data of subsystems are sometimes not sufficient. Test sample size or field population tends to be small.

(d) Subsystems may follow arbitrary failure and repair distribution for repairable systems.

Using Monte Carlo technique combined with Bayes method, these four major problems can be solved at the same time. Generally, Monte Carlo simulation methods have the following pros in evaluating system reliability and availability in summary:

(a) Monte Carlo simulation techniques can be used to analyze systems whose subsystems' lifetimes follow various distributions: Binomial, Exponential, Weibull, Lognormal, Gamma, Phase-type, *etc.* In fact, in engineering practice, all subsystems of a complex system may not follow a single failure distribution. Using classical statistics, it is difficult or impossible to determine the upper and lower bounds on various reliability measures for such systems. Therefore, Monte Carlo approach is a powerful tool to solve this kind of systems.

(b) Monte Carlo simulation methods can be applied to any network configuration and architecture: series, parallel, series-parallel, bridge, *k*-out-of-*n*, fault-tolerant, *etc.*, no matter how complex the network is, as long as we can determine the system reliability structure function in terms of subsystem reliabilities. For a system with several hundred or more subsystems in complex structures, it is difficult to obtain LCLs of its reliability measures by classical probability and statistics. However, at least in principle, Monte Carlo technique can be applied to evaluate system reliability measures and confidence limits for this kind of systems.

(c) Combining Bayes method, Monte Carlo procedure can easily integrate *a priori* information into its simulation modules. In practice, some systems tend to possess high reliability and/or be subject to reliability tests with small sample sizes or have small field population. For reliability assessment from small testing sample sizes or field population, this Monte Carlo-Bayes method is relatively effective because Bayes method can enlarge failure information of subsystems by using expert experience on similar subsystems' failures (Martz and Waller 1982).

(d) Monte Carlo method can be used to evaluate availability and MTBF of a reparable system. It is noted that so far there are no effective probabilistic and statistical methods for repairable systems with arbitrary time to failure and arbitrary time to repair. Markov chains are usually applied to exponential failures. In addition, Monte Carlo method can simulate availability and MTBF of maintained systems with imperfect maintenance.

(e) After Monte Carlo approaches are programmed in some programming languages, users can easily obtain system reliability and availability interval estimates by inputting the related data into computers. It is not necessary for them to be familiar with Monte Carlo methods applied and the software programs they are using.

(f) Modern computers have made actual Monte Carlo simulation time on computers shorter and shorter for most applications, and it is also convenient to implement Monte Carlo simulation on personal computers which are available almost everywhere now.

(g) By using Monte Carlo simulation, it is easy to find the system reliability and availability confidence interval and point estimates as well as their distributions at any time point t.

(h) The accuracy of the results by some Monte Carlo methods can be estimated.

(i) Discrete-event reliability simulation is robust in the modeling of complex system reliability structure and subsystem failure/repair functions.

The drawbacks of Monte Carlo simulation for evaluating system reliability and availability are:

(a) To get Monte Carlo simulation results with high accuracy, the number of simulation operation may becomes very large and computer time will then be increased.

(b) Confidence bounds obtained by some Monte Carlo methods are not exact or narrow enough.

(c) The significant digit number of confidence limits by Monte Carlo is small.

Sreider (1960) states that the error E from Monte Carlo methods is less than some value d generally, where d is approximately equal to $1/\sqrt{n}$:

$$E = |S - A| < d = 1/\sqrt{n}$$

where S is a simulation result, A is the true value, and n the number of simulation replications. According to this relationship, the error of Monte Carlo simulation technique decreases as the number of simulation replications increases. However, once E is smaller than certain value, it will no longer decrease basically. That is, it is impossible that E becomes zero. If special techniques for reducing E from Monte Carlo method are not used, this error have a maximum of 0.001 – 0.1 (Singh *et al.* 1993).

Simulation can be used for analyzing any system. However, the accuracy of the results depends on the number of iterations and the complexity of the system. To achieve the desired level of accuracy, the number of simulations can be determined. Analytical methods based on advanced algorithms are in general quicker and produce more accurate results than simulation. Therefore, whenever possible, it is better to use analytical methods. However, if analytical results are not possible or prone to round-off errors, then simulation should be used.

Most Monte Carlo availability assessment, and MTBF estimate algorithms assume that failed components are replaced with new ones at failures, *i.e.*, repair is perfect. As pointed out in Chapter 1, imperfect repair is more general and realistic, so allowing imperfect repair would be future direction for Monte Carlo availability and MTBF simulation.

12.5 Commercial Monte Carlo Reliability Simulation Tools

Worth mentioning is the simulation language. Some existing Monte Carlo programs have used the FORTRAN, BASIC, C, *etc.*, which may limit the adaptability and the scope of analysis. Kim and Fard (1995) state that many existing Monte Carlo simulation programs use a high level language which require considerable programming and development effort. Consequently, the result is a customized application-specific program usually limited in modeling flexibility and capability.

An alternative approach to improve the efficiency and flexibility in reliability modeling and assessment is to use discrete-event simulation language, for example, SIMAN, SLAM, or GPSS.

The good news is that today quite a few commercial general-purpose Monte Carlo reliability and availability simulation programs have appeared, for example, by Relex, Isograph, SoHaR, ReliaSoft, *etc.* Details on those general-purpose programs can be found in respective internet websites. An example is AvSim+ developed by Isograph. AvSim+ is a Windows-based availability and reliability simulation program capable of analyzing complex and dependent systems. AvSim+ allows users to construct fault tree or network diagrams (reliability block diagrams) using drag and drop facilities. Historical data (times to failure and times to repair) is automatically analyzed using the built-in Weibull Analysis facility and connected directly through to component failure models. This allows users to update their historical data records and almost immediately see the effects on predicted system performance. The AvSim+ Monte Carlo simulator engine enables one to model complex redundancies, common failures and component dependencies which cannot be modeled using standard analytical techniques. Complex dependencies include spares requirements, labor availability, operational phases, and standby arrangements. AvSim+ can also model ageing and effectiveness of planned maintenance, and determine optimal maintenance intervals.

12.6 A General Monte Carlo Reliability Procedure

In Section 12.4, we state the four key difficulties in determining complex system reliability and MTTF interval estimates. Although Monte Carlo-Bayes integrated method can solve these four problems at the same time, so far there are none to fulfill this goal satisfactorily among the existing Monte Carlo algorithms. Examining the typical existing Monte Carlo procedures we can see that the S-R procedure utilizes no *a priori* information and its accuracy cannot be determined. L-D-L, R-M, and C-H algorithms can only apply to binomial failure distribution. The double Monte Carlo technique (MacDonald 1982) can only deal with Weibull distribution and does not make use of *a priori* failure information in its simulation model.

Wang and Pham (1997) propose a general Monte Carlo reliability simulation procedure which can apply to any complex systems with arbitrary failure distributions of their subsystems and employs Bayes method to increase failure information.

Assume that subsystems of a system follow different failure distributions. For each kind of failure distribution, using Bayes theorem, we can derive its posterior distribution from its *a priori* distribution based on engineering experiences of experts and engineers and related testing or field reliability data. Methods for determining posterior distribution have been discussed extensively for various distributions and can been found in Martz and Waller (1982). Based on the posterior failure distributions of all subsystems which possess more failure information, we can use Monte Carlo technique to obtain system reliability and MTTF confidence bounds. Here we can utilize a special Monte Carlo method similar to the S-R algorithm.

Different from the S-R method, the proposed method generates a random sample T_i from the posterior distribution of subsystem i for all subsystems. The other steps are the same until we obtain system reliability interval estimates. Then it is easy to obtain system MTTF interval estimates. The procedure for obtaining MTTF intervals is to generate a random sample T_i from the posterior distribution of subsystem i for all subsystems first, and then to find the minimum T_i from among the T_is generated that make tie-set j unbroken (success) for all tie-sets. Take the maximum of all T_is as system MTTF q_k. Repeating these steps m times, m system MTTFs $q_1, q_2, ..., q_n$ are obtained. From them we can finally determine confidence bounds on system MTTF.

The minimal tie-sets can be obtained by using Monte Carlo technique in advance. In fact, we can employ the idea of an existing program "MINCUT" developed by Lin *et al.* (1993). For details see Lin *et al.* (1993).

Checking the frequency that the intervals (LCL, 1) cover the true reliability or MTTF we can determine the accuracy of this proposed method.

Similar algorithms can be used for Monte Carlo availability and MTBF simulation of maintained systems with imperfect maintenance under different maintenance policies. Various imperfect maintenance situations and maintenance policies can be found in Chapters 1 and 3.

Appendix

Elements of Reliability and Probability

The fundamental definitions of statistical reliability must depend on concepts from probability theory. This appendix describes the concepts of system reliability, examines common distribution functions useful in reliability and maintenance engineering and stochastic processes including Markov process, Poisson process, renewal process, and nonhomogeneous Poisson process. In general, a system may be required to perform various functions, each of which may have a different reliability.

A.1 Reliability Measures

The reliability definitions given in the literature vary among different practitioners as well as researchers. The generally accepted definition is as follows.

Definition A.1 *Reliability is the probability of success or the probability that the system will perform its intended function under specified design limits.*

More specific, reliability is the probability that a product or part will operate properly for a specified period of time (design life) under the design operating conditions (such as temperature, volt, *etc.*) without failure. Mathematically, reliability $R(t)$ is the probability that a system will be successful in the interval from time 0 to time t:

$$R(t) = P(T > t) \qquad t \geq 0$$

where T is a random variable denoting the time-to-failure or failure time.

If the time-to-failure random variable T has a density function $f(t)$, then

$$R(t) = \int_{t}^{\infty} f(s)\,ds$$

System Mean Time to Failure

Suppose that the reliability function for a system is given by $R(t)$. The expected failure time during which a component is expected to perform successfully, or the system mean time to failure (MTTF), is given by

$$MTTF = \int_0^\infty t f(t) dt$$

or equivalently,

$$MTTF = \int_0^\infty R(t) dt$$

Thus, MTTF is the definite integral evaluation of the reliability function. In general, if $\lambda(t)$ is defined as the failure rate function, then, by definition, MTTF is not equal to $1/\lambda(t)$.

Failure Rate Function

The hazard function is defined as the limit of the failure rate as the interval approaches zero. Thus, the hazard function $h(t)$ is the instantaneous failure rate, and is defined by

$$
\begin{aligned}
h(t) &= \lim_{\Delta t \to 0} \frac{R(t) - R(t + \Delta t)}{\Delta t R(t)} \\
&= \frac{1}{R(t)} \left(-\frac{d}{dt} R(t) \right) \\
&= \frac{f(t)}{R(t)}
\end{aligned}
$$

The quantity $h(t)dt$ represents the probability that a device of age t will fail in the small interval of time t to $(t + dt)$. The importance of the hazard function is that it indicates the change in the failure rate over the life of a population of components by plotting their hazard functions on a single axis. For example, two designs may provide the same reliability at a specific point in time, but the failure rates up to this point in time can differ.

The hazard function or hazard rate or failure rate function is the ratio of the probability density function (*pdf*) to the reliability function.

A.2 Common Probability Distribution Functions

This section presents some of the common distribution functions and several hazard models that have applications in reliability and maintenance. For each distribution, we will give its distribution form, reliability function, mean, variance, and other useful properties. This appendix is, by no means, comprehensive in its coverage of statistical distributions.

A.2.1 Discrete Random Variable Distributions

Binomial Distribution

The binomial distribution is one of the most widely used discrete random variable distributions in reliability and quality inspection. It has applications in reliability engineering, *e.g.*, when one is dealing with a situation in which an event is either a success or a failure.

The binomial distribution can be used to model a random variable X which represents the number of successes (or failures) in n independent trials (these are referred to as Bernoulli trials), with the probability of success (or failure) being p in each trial. The pdf of the distribution is given by

$$P(X = x) = \binom{n}{x} p^x (1-p)^{n-x} \quad x = 0, 1, 2, ..., n$$

$$\binom{n}{x} = \frac{n!}{x!(n-x)!}$$

where n = number of trials; x = number of successes; p = single trial probability of success.

Poisson Distribution

Although the Poisson distribution can be used in a manner similar to the binomial distribution, it is used to deal with events in which the sample size is unknown. A Poisson random variable is a discrete random variable distribution with probability density function given by

$$P(X = x) = \frac{\lambda^x e^{-\lambda}}{x!} \quad \text{for } x = 0, 1, 2, ...$$

where λ = constant failure rate; x = is the number of events. In other words, $P(X = x)$ is the probability of exactly x failures occur.

Geometric Distribution

Consider a sequence of independent trials, each having the same probability for success, say p. Let N be a random variable that counts for the number of trials until the first success. This distribution is called the geometric distribution. It has a *pdf* given by

$$P(N = n) = p(1-p)^{n-1} \quad n = 1, 2, ...$$

The expected value and variance are, respectively

$$E(N) = \frac{1}{p}$$

and

$$V(N) = \frac{1-p}{p^2}$$

Hypergeometric Distribution

A discrete distribution that arises in sampling, for example, is the hypergeometric distribution. It has a pdf given by

$$f(x) = \frac{\binom{k}{x}\binom{N-k}{n-x}}{\binom{N}{n}}, \quad x = 0, 1, 2, ..., n$$

Typically, N will be the number of units in a finite population; n will be the number of samples drawn without replacement from N; k will be the number of failures in the population; and x will be the number of failures in the sample.

A.2.2 Continuous Random Variable Distributions

Exponential Distribution

Exponential distribution plays an essential role in reliability engineering because it has a constant failure rate. This distribution has been used to model the lifetime of electronic and electrical components and systems. This distribution is appropriate when a used component that has not failed is as good as a new component – a rather restrictive assumption. The *pdf* and reliability functions are given by, respectively,

$$f(t) = \frac{1}{\theta} e^{-\frac{t}{\theta}} = \lambda e^{-\lambda t}, \quad t \geq 0$$

$$R(t) = e^{-\frac{t}{\theta}} = e^{-\lambda t}, \quad t \geq 0$$

where $\theta = 1/\lambda > 0$ is an MTTFs parameter and $\lambda \geq 0$ is a constant failure rate.

The hazard function or failure rate for the exponential density function is constant, *i.e.*,

$$h(t) = \frac{f(t)}{R(t)} = \frac{1}{\theta} = \lambda$$

It should be noted that the exponential distribution is the only continuous distribution satisfying

$$P\{T \geq t\} = P\{T \geq t + s \mid T \geq s\} \quad \text{for } t > 0, s > 0$$

Uniform Distribution

Let us denote X be a random variable having a uniform distribution on the interval (a, b) where $a < b$. The *pdf* is given by

$$f(x) = \begin{cases} \dfrac{1}{b-a} & a \leq x \leq b \\ 0 & \text{otherwise} \end{cases}$$

The expected value and variance are, respectively,

$$E(X) = \frac{a+b}{2}$$

and

$$V(X) = \frac{(b-a)^2}{12}$$

Normal Distribution
Normal distribution plays an important role in classical statistics owing to the *Central Limit Theorem*. In reliability engineering, the normal distribution primarily applies to measurements of product susceptibility and external stress. The *pdf* of the normal random variable is given by

$$f(t) = \frac{1}{\sigma\sqrt{2\pi}}\exp(-\frac{(t-\mu)^2}{2\sigma^2}) \qquad -\infty < t < \infty$$

where μ is the mean value and σ is the standard deviation.

Log Normal Distribution
The log normal lifetime distribution is a very flexible model that can empirically fit many types of failure data. The log normal density function is given by

$$f(t) = \frac{1}{\sigma t\sqrt{2\pi}}\exp(-\frac{(\ln t - \mu)^2}{2\sigma^2}) \qquad -\infty < t < \infty, \quad \sigma > 0$$

where μ and σ are parameters such that $-\infty < \mu < \infty$, and $\sigma > 0$. Note that μ and σ are not the mean and standard deviations of the distribution.

Mathematically, if a random variable X is defined as $X = \ln T$, then X is normally distributed with a mean of μ and a variance of σ^2. That is,

$$E(X) = E(\ln T) = \mu$$

and

$$V(X) = V(\ln T) = \sigma^2$$

The cumulative distribution function for the log normal is

$$F(t) = \int_0^t \frac{1}{\sigma s\sqrt{2\pi}} e^{-\frac{1}{2}(\frac{\ln s - \mu}{\sigma})^2} ds$$

and this can be related to the standard normal deviate Z by

$$F(t) = P[T \le t] = P(\ln T \le \ln t)$$

$$= P\left[Z \le \frac{\ln t - \mu}{\sigma}\right]$$

Therefore, the reliability function is given by

$$R(t) = P\left[Z > \frac{\ln t - \mu}{\sigma}\right]$$

and the hazard function would be

$$h(t) = \frac{f(t)}{R(t)} = \frac{\Phi\left(\frac{\ln t - \mu}{\sigma}\right)}{\sigma t R(t)}$$

where Φ is a *cdf* of standard normal density.

Weibull Distribution

The exponential distribution is often limited in applicability owing to the memoryless property. The Weibull distribution (Weibull 1951) is a generalization of the exponential distribution and is commonly used to represent fatigue life, ball bearing life, and vacuum tube life. The three-parameters probability density function is

$$f(t) = \frac{\beta(t-\gamma)^{\beta-1}}{\theta^\beta} e^{-(\frac{t-\gamma}{\theta})^\beta} \quad t \ge \gamma \ge 0$$

where θ and β are known as the scale and shape parameters, respectively, and γ is known as the location parameter. These parameters are always positive. By using different parameters, this distribution can follow the exponential distribution, the normal distribution, *etc.* It is clear that, for $t \ge \gamma$, the reliability function $R(t)$ is

$$R(t) = e^{-(\frac{t-\gamma}{\theta})^\beta} \quad \text{for } t > \gamma > 0, \beta > 0, \theta > 0$$

hence,

$$h(t) = \frac{\beta(t-\gamma)^{\beta-1}}{\theta^\beta} \quad t > \gamma > 0, \beta > 0, \theta > 0$$

It can be shown that the hazard function is decreasing for $\beta < 1$, increasing for $\beta > 1$, and constant when $\beta = 1$.

Gamma Distribution

The gamma distribution can be used as a failure probability function for components whose distribution is skewed. The failure density function for a gamma distribution is

$$f(t) = \frac{t^{\alpha-1}}{\beta^{\alpha}\Gamma(\alpha)} e^{-\frac{t}{\beta}} \qquad t \geq 0, \ \alpha, \ \beta > 0$$

where α is the shape parameter and β is the scale parameter. In this expression, $\Gamma(\alpha)$ is the gamma function, which is defined as

$$\Gamma(\alpha) = \int_0^{\infty} t^{\alpha-1} e^{-t} dt \qquad \text{for } \alpha > 0$$

Hence, the gamma reliability function is given by

$$R(t) = \int_t^{\infty} \frac{1}{\beta^{\alpha}\Gamma(\alpha)} s^{\alpha-1} e^{-\frac{s}{\beta}} ds$$

If α is an integer, it can be shown by successive integration by parts that

$$R(t) = e^{-\frac{t}{\beta}} \sum_{i=0}^{\alpha-1} \frac{\left(\frac{t}{\beta}\right)^i}{i!}$$

A common use of the gamma lifetime model occurs in Bayesian reliability applications.

Beta Distribution
The two-parameter beta density function, $f(t)$, is given by

$$f(t) = \frac{\Gamma(\alpha+\beta)}{\Gamma(\alpha)\Gamma(\beta)} t^{\alpha-1} (1-t)^{\beta-1} \qquad 0 < t < 1, \ \alpha > 0, \ \beta > 0$$

where α and β are the distribution parameters. This two-parameter beta distribution has commonly used in many reliability engineering applications and also an important role in the theory of statistics. Note that the beta-distributed random variable takes on values in the interval (0, 1), so the beta distribution is a natural model when the random variable represents a probability. The mean and variance of the beta distribution are, respectively, given by

$$E(T) = \frac{\alpha}{\alpha+\beta}$$

and

$$V(T) = \frac{\alpha\beta}{(\alpha+\beta+1)(\alpha+\beta)^2}$$

Pareto Distribution
The Pareto distribution was originally developed to model income in a population. Phenomena such as city population size, stock price fluctuations, and personal

incomes have distributions with very long right tails. The probability density function of the Pareto distribution is given by

$$f(t) = \frac{\alpha k^{\alpha}}{t^{\alpha+1}} \qquad k \le t \le \infty$$

Rayleigh Model
The Rayleigh model is a flexible lifetime model that can apply to many degradation process failure modes. The Rayleigh probability density function is

$$f(t) = \frac{t}{\sigma^2} \exp\left[\frac{-t^2}{2\sigma^2}\right]$$

Vtub-shaped Hazard Rate Distribution
Pham (2002) recently developed a two-parameter lifetime distribution with a Vtub-shaped hazard rate, also known as *Loglog distribution* or Pham distribution.

Note that the loglog distribution with Vtub-shaped and Weibull distribution with bathtub-shaped failure rates are not the same. As for the bathtub-shaped, after the infant mortality period, the useful life of the system begins. During its useful life, the system fails as a constant rate. This period is then followed by a wear out period during which the system starts slowly increases with the on set of wear out. For the Vtub-shaped, after the infant mortality period, the system starts to experience at a relatively low increasing rate, but not constant, and then increasingly more failures due to aging.

The probability density function of the distribution is (Pham 2002)

$$f(t) = \alpha \ \ln a \ t^{\alpha-1} \ a^{t^{\alpha}} \ e^{1-a^{t^{\alpha}}} \qquad \forall t > 0, \ a > 0, \ \alpha > 0$$

The loglog reliability function is given by

$$R(t) = e^{1-a^{t^{\alpha}}}$$

The corresponding failure rate of the loglog distribution is given by

$$h(t) = \alpha \cdot \ln a \cdot t^{\alpha-1} \cdot a^{t^{\alpha}}$$

A.3 Stochastic Processes Concepts

Stochastic processes are used for the description of a systems operation over time. There are two main types of stochastic processes: continuous and discrete. The complex continuous process is a process describing a system transition from state to state. The simplest process that is discussed here is a Markov process. Given the current state of the process, its future behavior does not depend on the past.

A.3.1 Markov Processes

Definition A.2 *Let* $t_0 < t_1 < ... < t_n$. *If*

$$P[X(t_n) = A_n \mid X(t_{n-1}) = A_{n-1}, X(t_{n-2}) = A_{n-2}, ..., X(t_0) = A_0]$$
$$= P[X(t_n) = A_n \mid X(t_{n-1}) = A_{n-1}]$$

then the process is called a Markov process.
Given the present state of the process, its future behavior does not depend on past information of the process.

The essential characteristic of a Markov process is that it is a process that has no memory; its future is determined by the present and not the past. If, in addition to having no memory, the process is such that it depends only on the difference $(t + dt) - t = dt$ and not the value of t, *i.e.*, $P[X(t + dt) = j \mid X(t) = i]$ is independent of t, then the process is Markov with stationary transition probabilities or homogeneous in time. This is the same property noted in exponential event times, and referring back to the graphical representation of $X(t)$, the times between state changes would in fact be exponential if the process has stationary transition probabilities.

Thus, a Markov process which is time homogeneous can be described processes where events have exponential occurrence times. The random variable of the process is $X(t)$, the state variable rather than the time to failure as in the exponential failure density.

A.3.2 Counting Processes

Among discrete stochastic processes, counting processes in reliability engineering are widely used to describe the appearance of events in time, *e.g.*, failures, number of perfect repairs, *etc*. The simplest counting process is a Poisson process. The Poisson process plays a special role to many applications in reliability (Pham 2000). A well-known counting process is the so-called renewal process. This process is described as a sequence of events, the intervals between which are independent and identically distributed random variables. In reliability theory, this type of mathematical model is used to describe the number of occurrences of an event in the time interval. In this section, we also discuss the quasi-renewal process and the non-homogeneous Poisson process.

A non-negative, integer-valued stochastic process, $N(t)$, is called a counting process if $N(t)$ represents the total number of occurrences of the event in the time interval [0, t] and satisfies these two properties:

i)   If $t_1 < t_2$, then $N(t_1) \le N(t_2)$

ii) If $t_1 < t_2$, then $N(t_2) - N(t_1)$ is the number of occurrences of the event in the interval $[t_1, t_2]$

For example, if $N(t)$ equals the number of persons who have entered a restaurant at or prior to time t, then $N(t)$ is a counting process in which an event occurs whenever a person enters the restaurant.

A.3.3 Poisson Processes

One of the most important counting processes is the Poisson process.

Definition A.3 *A counting process, N(t), is said to be a Poisson process with intensity* λ *if*

   i)   *The failure process, N(t), has stationary independent increments*
   ii)   *The number of failures in any time interval of length s has a Poisson distribution with mean* λs*, that is,*

$$P\{N(t+s) - N(t) = n\} = \frac{\exp(-\lambda s) \cdot (\lambda s)^n}{n!} \qquad n = 1, 2, ...$$

   iii) *Tthe initial condition is N(0) = 0*

   This model is also called a homogeneous Poisson process, indicating that the failure rate λ does not depend on time t. In other words, the number of failures occurring during the time interval $(t, t + s]$ does not depend on the current time t but only the length of time interval s. A counting process is said to possess independent increments if the number of events in disjoint time intervals are independent.

   For a stochastic process with independent increments, the auto-covariance function is

$$Cov[X(t_1), X(t_2)] = \begin{cases} Var[N(t_1 + s) - N(t_2)] & \text{for } 0 < t_2 - t_1 < s \\ 0 & \text{otherwise} \end{cases}$$

where

$$X(t) = N(t+s) - N(t).$$

If $X(t)$ is Poisson distributed, then the variance of the Poisson distribution is

$$Cov[X(t_1), X(t_2)] = \begin{cases} \lambda[s - (t_2 - t_1)] & \text{for } 0 < t_2 - t_1 < s \\ 0 & \text{otherwise} \end{cases}$$

This result shows that the Poisson increment process is covariance stationary.

A.3.4 Renewal Processes

A renewal process is a more general case of the Poisson process in which the inter-arrival times of the process or the time between failures do not necessarily follow the exponential distribution. For convenience, we will call the occurrence of an event a renewal, the inter-arrival time the renewal period, and the waiting time the renewal time.

Definition A.4 *A counting process N(t) that represents the total number of occurrences of an event in the time interval (0, t] is called a renewal process, if the* *time between failures are independent and identically distributed random variables.*

The probability that there are exactly n failures occurring by time t can be written as

$$P\{N(t) = n\} = P\{N(t) \geq n\} - P\{N(t) > n\}$$

Note that the times between the failures are $T_1, T_2, \ldots, T_n$ so the failures occurring at time W_k are

$$W_k = \sum_{i=1}^{k} T_i$$

and

$$T_k = W_k - W_{k-1}$$

Thus,

$$
\begin{aligned}
P\{N(t) = n\} &= P\{N(t) \geq n\} - P\{N(t) > n\} \\
&= P\{W_n \leq t\} - P\{W_{n+1} \leq t\} \\
&= F_n(t) - F_{n+1}(t)
\end{aligned}
$$

where $F_n(t)$ is the cumulative distribution function for the time of the nth failure and $n = 0,1,2, \ldots$

Renewal reward theory is often used in maintenance modeling. Consider a renewal process $\{N(t), t \geq 0\}$ with interarrival times $D_n, \forall n, n \geq 1$. Suppose that each time a renewal occurs we receive a reward. Denote by R_n, the reward earned at the time of the nth renewal. Assume further that $R_n, \forall n, n \geq 1$ are independently and identically distributed, and they may (and usually will) depend on D_n, the duration of the nth renewal interval. If we let

$$R(t) = \sum_{n=1}^{N(t)} R_n$$

then $R(t)$ represents the total reward earned by time t. Let

$$E(R) = E(R_n), \quad E(D) = E(D_n)$$

Following Ross (1972), we have the following theorem:

Theorem A.1 *(renewal reward). If $E(R) < \infty$ and $E(D) < \infty$, then with probability 1,*

$$\frac{R(t)}{t} \rightarrow \frac{E(R)}{E(D)} \qquad \text{as } t \rightarrow \infty$$

A.3.5 Non-homogeneous Poisson Processes

The non-homogeneous Poisson process model (NHPP) that represents the number of failures experienced up to time t is a non-homogeneous Poisson process $\{N(t), t \geq 0\}$. The main issue in the NHPP model is to determine an appropriate mean value function to denote the expected number of failures experienced up to a certain time.

Note that in a renewal process, the exponential assumption for the inter-arrival time between failures is relaxed, and in the NHPP, the stationary assumption is relaxed.

The NHPP model is based on the following assumptions:

- The failure process has an independent increment, *i.e.*, the number of failures during the time interval $(t, t + s)$ depends on the current time t and the length of time interval s, and does not depend on the past history of the process.
- The failure rate of the process is given by

$$P\{\text{exactly one failure in } (t, t + \Delta t)\} = P\{N(t + \Delta t) - N(t) = 1\}$$
$$= \lambda(t)\Delta t + o(\Delta t)$$

  where $\lambda(t)$ is the intensity function.
- During a small interval Δt, the probability of more than one failure is negligible, that is,

$$P\{\text{two or more failure in } (t, t + \Delta t)\} = o(\Delta t)$$

- The initial condition is $N(0) = 0$.

On the basis of these assumptions, the probability that exactly n failures occurring during the time interval $(0, t)$ for the NHPP is given by

$$\Pr\{N(t) = n\} = \frac{[m(t)]^n}{n!} e^{-m(t)} \qquad n = 0, 1, 2, ...,$$

where $m(t) = E[N(t)] = \int_0^t \lambda(s)ds$ and $\lambda(t)$ is the intensity function. It can be easily shown that the mean value function $m(t)$ is non-decreasing.

Reliability Function

The reliability $R(t)$, defined as the probability that there are no failures in the time interval $(0, t)$, is given by

$$R(t) = P\{N(t) = 0\}$$
$$= e^{-m(t)}$$

In general, the reliability $R(x|t)$, the probability that there are no failures in the interval $(t, t + x)$, is given by

$$R(x \mid t) = P\{N(t+x) - N(t) = 0\}$$
$$= e^{-[m(t+x)-m(t)]}$$

and its density is given by

$$f(x) = \lambda(t+x)e^{-[m(t+x)-m(t)]}$$

where

$$\lambda(x) = \frac{\partial}{\partial x}[m(x)]$$

The variance of the NHPP can be obtained as follows:

$$Var[N(t)] = \int_0^t \lambda(s)ds$$

and the auto-correlation function is given by

$$Cor[s] = E[N(t)]E[N(t+s) - N(t)] + E[N^2(t)]$$
$$= \int_0^t \lambda(s)ds \int_0^{t+s} \lambda(s)ds + \int_0^t \lambda(s)ds$$
$$= \int_0^t \lambda(s)ds \left[1 + \int_0^{t+s} \lambda(s)ds\right]$$

References

Abdel-Hameed M (1987a) An imperfect maintenance model with block replacements. *Applied Stochastic Models and Data Analysis* 3:63-72

Abdel-Hameed M (1987b) Inspection and Maintenance Policies of Devices subject to deterioration. *Advance in Applied Probability* 10:917-931

Abdel-Hameed M (1995) Correction to: "Inspection and maintenance policies of devices subject to deterioration. *Advance in Applied Probability* 27/2:584

Albin SL, Chao S (1992) Preventive replacement in systems with dependent components. *IEEE Transactions on Reliability* 41/2:230-238

Alexopoulos C, Fishman SG (1988) Stochastic flow networks: How component criticality changes with component reliability. *Winter Simulation Conference Proceedings*, San Diego, CA, USA, 12-14 Dec 1988

Allan RN, Bhuiyan MR (1994) Application of sequential simulation to the reliability assessment of bulk power systems. *Proceedings of the 29th Universities Power Engineering Conference*, Part 2 (of 2), Galway, Irel, 1994, pp 763-766

Altiok T (1996) *Performance Analysis of Manufacturing Systems*. Springer

Amato HN, Anderson EE (1976) Determination of warranty reserves: an extension. *Management Science* 22/12:854-862

Archibald TW, Dekker R(1996) Modified block-replacement for multiple-component systems. *IEEE Transactions on Reliability* 45/1, 75-83

Arjas E, Norros I (1989) Change of life distribution via a hazard transformation: an inequality with application to minimal repair. *Mathematics of Operations Research* 14/2:355-361

Arjas E, Norros I (1990) Should minimal repair depend on information? In: Block HW, Sampson AR, Savits TH (eds) *Proceedings of the Symposium on Dependence in Probability and Statistics*. (held in Somerset, Pennsylvania, August 1-5, 1987; *IMS Lecture Notes Monograph Ser.*, 16) Inst. Math. Statist., Hayward, CA.

Arsham H (1989) On the inverse problem in Monte Carlo experiments. *Inverse Problems* 5/6:927-934

Asher H, Feingold H (1984) *Repairable Systems Reliability*. Marcel Dekker, New York

Assaf D, Shanthikumar JG (1987) Optimal group maintenance policies with continuous and periodic inspections. *Management Science* 33:1440-1450

Aven T (1983) Optimal replacement under a minimal repair strategy - a general failure model. *Advances in Applied Probability* 15/1:198-211

Aven T (1985) Determination/estimation of an optimal replacement interval under minimal repair. *Optimization: A Journal of Mathematical Programming and Operations Research* 16/5:743-754

Aven T, Jensen U (1999) *Stochastic Models in Reliability*. Springer-Verlag, New York

Baca A (1993) Examples of Monte Carlo methods in reliability estimation 'based on reduction of prior information. *IEEE Transactions on Reliability* 42/4:645-649

Bai J (2004) On the study of warranties for repairable complex systems. PhD Dissertation, Rutgers University, USA

Bae SI, Ichikawa M (1993) Attempt to unify reliability and confidence level in reliability-based design (The case of 2-parameter Weibull distribution). *Nippon Kikai Gakkai Ronbunshu, A Hen/Transactions of the Japan Society of Mechanical Engineers*, Part A, 59/558:478-482

Bai J, Pham H (2004) Discounted warranty cost for minimally repaired series systems. *IEEE Transactions on Reliability* 53/1:37-42

Bai J, Pham H (2005) Repair-limit risk-free warranty policies with imperfect repair. *IEEE Trans. on Systems, Man, and Cybernetics* (Part A) 35/6:765-772

Bai J, Pham H (2006a) Cost analysis on renewable full-service warranties for multi-component systems. *European Journal of Operational Research* 168/2: 492-508

Bai J, Pham H (2006b) Promotional warranty policies: analysis and perspectives. In Pham H (ed) *Springer Handbook of Engineering Statistics*. Springer, London.

Balaban HS, Singpurwalla ND (1984) Stochastic properties of a sequence of interfailure times under minimal repair and under revival. In: Abdel Hameed MS, Cinlar E, Quinn J (eds) *Reliability Theory and Models*. Academic Press, Orlando, Fla., pp 65-80

Balachandran KR, Maschmeyer RA, Livingstone JL (1981) Product warranty period: A Markovian approach to estimation and analysis of repair and replacement costs. *The Accounting Review* 1:115-124

Balasubramanya KS, Vasudevan S, Rao PK (1985) State transition Monte Carlo simulation of electronic systems. IREECON, International (Convention Digest) (Institution of Radio and Electronics Engineers Australia) 2:730-732

Barlow RE, Hunter LC (1960) Optimum preventive maintenance policies. *Operations Research* 8:90-100

Barlow RE, Proshan F (1965) *Mathematical Theory of Reliability.* John Wiley & Sons, New York

Barlow RE, Proshan F (1975) *Statistical Theory of Reliability and Life Testing.* Holt, Renehart & Winston, New York

Baxter LA. (1982) Reliability applications of the relevation transform. *Naval Research Logistics Quarterly* 29:323-330

Beichelt F (1976) A general preventive maintenance policy. *Mathematische Operationsforschung und Statistik Series, Statistics* 7:927-932

Beichelt F (1978) A new approach to repair limit replacement policies. *Transactions of the Eighth Prague Conference on Information Theory, Statistical Decision Functions, Random Processes*, Prague, vol.C, pp 31-37

Beichelt F (1981a) A generalized block-replacement policy. *IEEE Transactions on Reliability* R-30/2:171-173

Beichelt F (1981b) Replacement policies based on system age and maintenance cost limits. *Mathematische Operationsforschung und Statistik Series, Statistics* 12/4:621-627

Beichelt F, Fischer K (1980) General failure model applied to preventive maintenance policies. *IEEE Transactions on Reliability* R-29/1:39-41

Bererhoff OA (1963) Confidence limits for system reliability based on component test data. AD-42845, available from National Technical Information Service (NTIS), Department of Commerce, Springfield, VA22161, USA

Berg M (1976a) A proof of optimality for age replacement policies. *Journal of Applied Probability* 13:751-759

Berg M (1976b) Optimal replacement policies for two-unit machines with increasing running costs - I. *Stochastic Processes and Applications* 5:89-106

Berg M (1978) General trigger-off replacement procedures for two-unit systems. *Naval Research Logistics* 25:15-29.

Berg M, Epstein B (1976) A modified block replacement policy. *Naval research Logistics* 23:15-24

Berg M, Epstein B (1978) Comparison of age, block, and failure replacement. *IEEE Transactions on Reliability* R-27/1:25-29

Bergman B (1978) Optimal replacement under a general failure model. *Advances in Applied Probability* 10/2:431-451

Bergman B (1980) On the optimality of stationary replacement strategies. *Journal of Applied Probability* 17:178-186

Bertoldi O, Rivoiro A, Salvaderi L (1994) New trends in power system planning and related effect on the reliability evaluation. *Reliability Engineering & System Safety* 46/1:49-61

Bhat KS, Gururajan M (1993) A two-unit cold standing system with imperfect repair and excessive availability period. *Microelectronics and Reliability* 33/4:509-514

Bhattacharjee MC (1987) New results for the Brown-Proschan model of imperfect repair. *Journal of Statistical Planning and Inference* 16:305-316

Bianu A, Frant S, Gurevich V (1995) Generation system reliability model with Monte Carlo simulation. *Proceedings of the 18th Convention of Electrical and Electronics Engineers in Israel*, pp 3.1.4/1-5

Billinton R, Li W (1991a) Composite system reliability assessment using a Monte Carlo approach. *Third International Conference on Probabilistic Methods Applied to Electric Power Systems*, London, UK, 1991 Jul 3-5

Billinton R, Li W (1991b) Hybrid approach for reliability evaluation of composite generation and transmission systems using Monte-Carlo simulation and enumeration technique. *IEE Proceedings, Part C: Generation, Transmission and Distribution* 138/3:233-241

Billinton R, Li W (1992) A Monte Carlo method for multi-area generation system reliability assessment. *IEEE Transactions on Power Systems* 7/4:1487-1492

Billinton R, Lian G (1993) Monte Carlo approach to substation reliability evaluation. *IEE Proceedings, Part C: Generation, Transmission and Distribution* 140/2:147-152

Blacer Y, Sahin I (1986) Replacement costs under warranty: cost moments and time variability. *Operations Research* 34/4:554-559

Blischke WR (1990) Mathematical models for analysis of warranty policies. *Math. Comput. Model* 13:1-16

Blischke WR, Murthy DNP (1993) Product warranty management-I: A taxonomy for warranty policies. *European Journal of Operational Research* 62:127-148

Blischke WR, Murthy DNP (1994) *Warranty Cost Analysis*. Marcel Dekker

Blischke WR, Murthy DNP (eds) (1996) *Product Warranty Handbook*. Marcel Dekker

Blischke WR, Scheuer EM (1975) Calculating the cost of warranty policies as a function of estimated life distributions. *Naval Research Logistics Quarterly* 28:193-205

Bloch-Mercier S (2002) A preventive maintenance policy with sequential checking procedure for a Markov deteriorating system. *European Journal of Operational Research* 147:548-576

Block HW, Borges WS, Savits TH (1985) Age dependent minimal repair. *Journal of Applied Probability* 22:370-385

Block HW, Borges WS, Savits TH (1988) A general age replacement model with minimal repair. *Naval Research Logistics, An International Journal* 35/5:365-372

Block HW, Langberg NA, Savits TH (1990) Comparisons for maintenance policies involving complete and minimal repair. In: Block HW, Sampson AR, Savits

TH (eds) *Proceedings of the Symposium on Dependence in Probability and Statistics.* (held in Somerset, Pennsylvania, August 1-5, 1987. *IMS Lecture Notes Monograph Ser.* 16) Inst. Math. Statist., Hayward, CA.

Block HW, Langberg NA, Savits TH (1993) Repair replacement policies. *Journal of Applied Probability* 30/1:194-206

Blumenthal S, Greenwood JA, Herbach LH (1976) A comparison of the bad as old and superimposed renewal models. *Management Science* 23/3:280-285

Boehm F, Hald UP, Lewis EE (1988) Parts renewal in continuous-time Monte Carlo reliability simulation. *Proceedings of the Annual Reliability and Maintainability Symposium,* pp 345-349

Boland PJ (1982) Periodic replacement when increasing minimal repair costs vary with time. *Naval Research Logistics* 29:541-546

Boland PJ, El-Neweihi E (1998) Statistical and information based (physical) minimal repair for *k* out of *n* systems. *Journal of Applied Probability* 35/3:731-740

Boland PJ, Proschan F (1982) Periodic replacement with increasing minimal repair costs at failure. *Operations Research* 30:1183-1189

Boland PJ, El-Neweihi E, Proschan F (1991) Stochastic order for inspection and repair policies. *The Annals of Applied Probability* 1/2:207-218

Brennan JR (1994) *Warranties: Planning, Analysis and Implementation.* McGraw-Hill, New York

Bris R, Chatelet E, Yalaoui F (2003) New method to minimize the preventive maintenance cost of series-parallel systems. *Reliability Engineering and System Safety* 82:247-255

Brown M, Proschan F (1982) Imperfect maintenance. In: IMS Lecture Notes-Monograph Ser. 2: *Survival analysis.* Inst. Math. Statist., Hayward, Calif., pp 179-188

Brown M, Proschan F(1983) Imperfect repair. *Journal of Applied Probability* 20:851-859

Buehler RJ (1957) Confidence intervals for the product of two binomial parameters. *J. American Statistical Assoc.* 52:482-493

Bustamante AS (1988) Monte Carlo methods in Reliability Engineering. In: Amendola A, Bustamantre AS (eds) *Proceedings of the ISPRA.* Kluwer Academic Publishers

Campioni L, Scardovelli R, Vestrucci P (2005) Biased Monte Carlo optimization: the basic approach. *Reliability Engineering & System Safety* 87/3: 387-94

Canfield RV (1986) Cost optimization of periodic preventive maintenance. *IEEE Transactions on Reliability* R-35/1:78-81

Carter LL, Miles TL, Binney SE (1993) Quantifying the reliability of uncertainty predictions in Monte Carlo fast reactor physics calculations. *Nuclear Science and Engineering* 113/4:324-338

Cha JH, Kim JJ (2001) On availability of Bayesian imperfect repair model. *Statistics & Probability Letters* 53/2:181-187

Chan PKW, Downs T (1978) Two criteria for preventive maintenance. *IEEE Transactions on Reliability* R-27:272-273

Chan JK, Shaw L (1993) Modeling repairable systems with failure rates that depend on age & maintenance. *IEEE Transactions on Reliability* 42:566-570

Chang M, Parks GT, Lewins JD (2001) Estimation of system reliability by variationally processed Monte Carlo simulation. In: Pham H (ed) *Recent Advances in Reliability and Quality Engineering.* World Scientific, New Jersey, pp 93-122

Chao A, Huwang LC (1987) A modified Monte Carlo technique for confidence limits of system reliability using pass-fail data. *IEEE Trans. reliability* R-36:109-112

Chaudhuri D, Sahu KC (1977) Preventive maintenance intervals for optimal reliability of deteriorating system. *IEEE Transactions on Reliability* R-26:371-372

Chelbi A, Ait-Kadi D (1999) An optimal inspection strategy for randomly failing equipment. *Reliability Engineering and System Safety* 63:127-131

Chen J, Yao DD, Zheng S (1988) Quality control for products supplied with warranty. *Operations Research*, 46/1:107-115

Chen M, Feldman RM (1997) Optimal replacement policies with minimal repair and age-dependent costs. *European Journal of Operational Research* 98/1:75-84

Cho ID, Parlar M (1991) A survey of maintenance models for multi-unit systems. *European Journal of Operational Research* 51:1-23

Chukova S, Dimitrov B (1996) Warranty analysis for complex systems. In: Blischke WR, Murthy DNP (eds) *Product Warranty Handbook, Chapter 22.* Marcel Dekker, pp 543-584

Chun YH (1992) Optimal number of periodic preventive maintenance operations under warranty. *Reliability Engineering & System Safety* 37/3:223-225

Chun YH (1992) Optimal number of periodic preventive maintenance operations under warranty. *Reliability Engineering and System Safety* 37:223-225

Chun YH, Tang K (1995) Determining the optimal warranty price based on the producer's and customers' risk preferences. *European Journal of Operational Research* 85:97-110

Chun YH, Tang K (1999) Cost analysis of two-attribute warranty policies based on the product usage rate. *IEEE Transactions on Engineering Management* 46/2:201-209

Cinlar E (1975) *Introduction to Stochastic Processes.* Prentice-Hall, Englewood Cliffs, NJ

Cox DR (1962) *Renewal Theory.* Methuen, London

Cox DR (1972) Regression models and life tables (with discussion). *Royal Statistical Society B* 34:187

Crawford RH, Rao SS (1987) Reliability analysis of function generating mechanisms through Monte Carlo simulation. *Advances in Design Automation, Volume Two: Robotics, Mechanisms, and Machine Systems.* Boston, MA, USA, 1987 Sep 27-30

Csenki A (1989) Improved Monte Carlo method in structural reliability. *Reliability Engineering & System Safety* 24/3:275-292

Dagpunar JS (1996) Maintenance model with opportunities and interrupt replacement options. *Journal of the Operational Research Society* 47/11:1406-1409

Dagpunar JS (1999) New approach for solving repair limit problems. *European Journal of Operational Research* 113/1:137-146

Dagpunar JS, Jack N (1992) Optimal repair-cost limit for a consumer following expiry of a warranty. *IMA Journal of Mathematics Applied in Business and Industry* 4:155-161

Dagpunar JS, Jack N (1994) Preventative maintenance strategy for equipment under warranty. *Microelectronics and Reliability* 34/6:1089-1093

Davani D (1994) Parametric what-if analysis in MTTF: a single-run Monte-Carlo-based approach. *Microelectronics & Reliability* 34/2:275

DeCroix GA (1999) Optimal warranties, reliabilities and prices for durable goods in an oligopoly. *European Journal of Operational Research* 112:554-569

DeGroot MH (1984) *Probability and Statistics, 2nd edn.* Addison-Wesley

Dekker R (1996) Applications of maintenance optimization models: a review and analysis. *Reliability Engineering & System Safety* 51/3:229-240

Dekker R, Roelvink IFK (1995) Marginal cost criteria for preventive replacement of a group of components. *European Journal of Operational Research* 84/2:467-480

Dekker R, Wilderman RE, van der Duyn Schouten FA (1997) A review of multi-component maintenance models with economic dependence. *Math. Methods Oper. Res.* 45/3:411-435

Depuy M, Hobbs JR, Moore AH, Johnson JW (1982) Accuracy of univariate, bivariate, and a 'modified double Monte Carlo' technique for finding lower confidence limits of system reliability. *IEEE Trans Reliability* R31:474-477

Devooght J, Dubus A, Smidts C (1990) Suboptimal inspection policies for imperfectly observed realistic systems. *European Journal of Operational Research* 45/2-3:203-218

Dey DK, Lee TM (1992) Bayes computation for life testing and reliability estimation. *IEEE Transactions on Reliability* 41/4:621-626

Dias JR (1990) Some approximate inspection policies for a system with imperfect inspections. *RAIRO Recherche Operationnelle* 24/2:191-199

Dieulle L, Berenguer C, Gralland A, Roussignol M (2003) Sequential condition-based maintenance scheduling for a deteriorating system. *European Journal of Operational Research* 150:451-461

Djamaludin I, Murthy DNP (1994) Quality control through lot sizing for items sold with warranty. *International Journal of Production Economics* 33:97-107

Djamaludin I, Murthy DNP (2001) Warranty and preventive maintenance. *International Journal of Reliability, Quality and Safety Engineering* 8:89-107

Dohi T, Matsushima N, Kaio N, Osaki S (1997) Nonparametric repair-limit replacement policies with imperfect repair. *European Journal of Operational Research* 96/2:260-273

Dohi T, Kaio N, Osaki S (1998) On the optimal ordering policies in maintenance theory - survey and applications. *Appl. Stochastic Models Data Anal.* 14/4:309-321

Dohi T, Kaio N, Osaki S (2000) A graphical method to repair-cost limit replacement policies with imperfect repair. *Mathematical and Computer Modelling* 31:99-106

Doshay I (1971) System availability and service simulation; on-line program using Monte-Carlo model. *IEEE Trans. on Reliability* R20:142-147

Downs T (1985) An approach to the modeling of software testing with some applications. *IEEE Trans. Software Engineering* se-11/4:356-363

Doyen L, Gaudoin O (2004) Classes of imperfect repair models based on reduction of failure intensity or virtual age. *Reliability Engineering and System Safety* 84/1:45-56

Drinkwater RW, Hastings NVJ (1967) An economic replacement model. *Oper. Res. Quart.* 18:121-138

Dubi A, Gandini A, Goldfeld A, Righini R (1989) New multipurpose Monte-carlo code for reliability analysis of complex systems: Application to a FBR decay heat removal system. *Reliability '89*, Part 1, London, UK, 1989 Jun 14-16 (Publ by Inst of Quality Assurance, London), pp 2B/6/1-2B/614

Dubi A, Gandini A, Goldfeld A, Righini R, Simonot H (1991) Analysis of non-markovian systems by a Monte-Carlo method. *Annals of Nuclear Energy* 18/3:125-130

Easton MC, Wong CK (1980) Sequential destruction method for Monte Carlo evaluation of system reliability. *IEEE Trans. Reliability* R-29:27-32

Ebrahimi N (1986) Two new replacement policies. *IEEE Transactions on Reliability* 42/1:141-145

Efton B (1979) Bootstrap method: another look at the jackknife. *Annals of Statistics* 7:1-26

Ehrlich W, Prasanna B, Stampfel J, Wu J (1993) Determining the cost of a stop-testing decision. *IEEE Trans. Software Engineering* 19:33-42

Elperin T, Gertsbakh I, Lomonosov M (1991) Estimation of network reliability using graph evolution models. *IEEE Transactions on Reliability* 40/5:572-581

Emons W (1988) Warranties, moral hazard, and the lemons problem. *Journal of Economic Theory* 46:16-33

Emons W (1989) On the limitation of warranty duration. *Journal of Industrial Economics* 37:287-301

Faddy MJ (1995) Phase-type distributions for failure times. *Mathematical and Computer Modeling* 22:63-70

Faddy MJ, Wilson RJ (2000) Compartmental modeling of equipment subject to partial repair. *Mathematical and Computer Modeling* 31:115-120

Feller W (1966) *An Introduction to Probability Theory and Its Applications, vol.II.* John Wiley and Sons, New York

Feller W (1968) *An Introduction to Probability Theory and Its Applications, vol.I,* 3rd ed. John Wiley and Sons, New York

Finkelstein MS (1992) Some notes on two types of minimal repair. *Advances in Applied Probability* 24/1:226--228

Finkelstein MS (1997) Imperfect repair models for systems subject to shocks. *Applied Stochastic Models & Data Analysis* 13/3-4:385-390

Fishman GS (1986a) A comparison of four Monte Carlo methods for estimating the probability of *s-t* connectedness. *IEEE Trans. Reliability* R-35:145-154

Fishman GS (1986b) A Monte Carlo sampling plan for estimating network reliability. *Operations Research* 34:581-592

Fishman GS (1987a) A Monte Carlo sampling plan for estimating reliability parameters and related functions. *Networks* 17:169-186

Fishman GS (1987b) Distribution of maximum flow with applications to multistate reliability systems. *Operations Research* 35/4:607-618

Fishman GS (1987c) Monte Carlo estimation of function variation. *Winter Simulation Conference Proceedings 1987*, Atlanta, GA, USA, 1987 Dec 14-16 (Available from IEEE Service Cent (Cat n 87CH2512-2), Piscataway, NJ, USA), pp 347-350

Fishman GS (1989) Estimating the s-t reliability function using importance and stratified sampling. *Operations Research* 37/3:462-473

Flynn J (1988) Optimal replacement policies for a multicomponent reliability system. *Operations Research Letters* 7/4:167-172

Fontenot RA, Proschan F (1984) Some imperfect maintenance models. In: Abdel Hameed MS, Cinlar E, Quinn J (eds) *Reliability Theory and Models*. Academic press, Orlando, Fla.

Frees EW (1986) Warranty analysis and renewal function estimation. *Naval Research Logistics Quarterly* 33:361-372

Frees EW (1988) Estimating the cost of a warranty. *Journal of Business and Economic Statistics* 6/1:79-86

Frenk H, Dekker R, Kleijn M (1997) A unified treatment of single component replacement models, Stochastic models of reliability. *Mathematical Methods of Operations Research* 45/3:437-454

Gasmi S, Love CE, Kahle W (2003) A General Repair, Proportional-Hazards, Framework to Model Complex Repairable Systems. *IEEE Trans. on Reliability* 52/1:26-32

Gatliffe TR (1976) Accuracy analysis for a lower confidence limit procedure for system reliability. AD-A 031817, available from US NTIS

Geist RM, Smotherman MK (1989) Ultrahigh reliability estimates through simulation. *Annual Reliability and Maintainability Symposium - 1989 Proceedings*, Atlanta, GA, USA, 1989 Jan 24-26. Available from IEEE Service Cent (cat n 89CH2580-9), Piscataway, NJ, USA., pp 350-355

Gertsbakh IB (1977) *Models of Preventive Maintenance.* North-Holland, Amsterdam

Gertsbakh IB (1989) Optimal dynamic opportunistic replacement with random resupply of spare parts. *Communications in Statistics, Stochastic Models* 5/2:315-326

Ghajar R, Billinton R (1988) Monte Carlo simulation model for the adequacy evaluation of generating systems. *Reliability Engineering & System Safety* 20/3: 173-186

Glickman TS, Berger PD (1976) Optimal price and protection period decisions for a product under warranty. *Management Science* 22:1381-1390

Goel AL (1985) Software reliability models: assumptions, limitations, and applicability. *IEEE Trans. Software Engineering* se-11/12:1411-1423

Goel LR, Taiga VK (1993) A two unit series system with correlated failures and repairs. *Microelectronics and Reliability* 33/14:2165-9

Goel LR, Shrivastava P, Gupta R (1992) Two unit cold standby system with correlated failures and repairs. *International Journal of Systems Science* 23/3:379-391

Goel LR, Gupta R, Tyagi PK (1993) Cost benefit analysis of a complex system with correlated failures and repairs. *Microelectronics and Reliability* 33/15: 2281-4

Goel LR, Mumtaz SZ, Gupta R (1996) A 2-Unit Duplicating Standby System with Correlated Failure Repair Replacement Times. Microelectronics and Reliability 36/4:517-523

Grall A, Berenguer C, Dieulle L (2002a) A condition-based maintenance policy for stochastically deteriorating systems. *Reliability Engineering and System Safety* 76:167-180

Grall A, Dieulle L, Berenguer C, Roussignol M (2002b) Continuous-time predictive-maintenance scheduling for a deteriorating system. *IEEE Trans. on Reliability* 51/2:141-150

Gubbala N, Singh C (1995) Models and considerations for parallel implementation of Monte Carlo simulation methods for power system reliability evaluation. *IEEE Transactions on Power Systems* 10/2:779-787

Guo R, Love CE (1992) Statistical analysis of an age model for imperfectly repaired systems. *Quality and Reliability Engineering International* 8:133-146

Gupta R (1999) Profit analysis of a system with mutual changeover of units and correlated failures and repairs. *Journal of Quality in Maintenance Engineering* 5/2:128-140

Gupta RC, Kirmani SNUA (1988) Closure and momotonicity properties of non-homogeneous Poisson processes and record values. *Probability in the Engineering and Informational Science* 2:475-484

Gupta RC, Kirmani SNUA (1989) On predicting repair times in a minimal repair process. *Communications in Statistics, Simulation and Computation* 18/4:1359-1368

Gupta S (1984) Replacement policies involving idle time and minimal repair under Markov renewal processes. *Journal of the Indian Statistical Association* 22:53-62

Gupta YP, Chand S (1993) Strategies of replacement under Markov renewal process. *International Journal of Information and Management Sciences* 4/1:41-50

Gururajan M, Stanley ADJ (1988) A complex two unit system with priority and imperfect repair. *IAPQR Transactions, Journal of the Indian Association for Productivity, Quality and Reliability* 13/1:65-70

Gutjahr WJ (1995) Optimal test distributions for software failure cost estimation. *IEEE Trans. Software Engineering* 21/3:219-228

Hammersley JM, Handscomb DC (1964) *Monte Carlo Method*. Methuen and Co. Ltd, London

Haringa GE, Jordan GA, Garver LL (1991) Application of Monte Carlo simulation to multi-area reliability evaluations. *IEEE Computer Applications in Power* 4/1:21-25

Hegde GG, Kubat P (1989) Diagnosic design: A product support strategy. *European Journal of Operational Research* 38:35-43

Heidergott B (1999) Optimization of a single-component maintenance system: A smoothed perturbation analysis approach. *European Journal of Operational Research* 119:181-190

Helvic BE (1980) Periodic maintenance, on the effect of imperfectness. *10th Int. Symp. Fault-tolerant Computing*, pp 204-206

Henley J, Kumamoto H (1981) Reliability Engineering and Risk Assessment. Prentice-Hall

Heyman D, Sobel MJ (1982) *Stochastic Models in Operations Research* (vol. I). McGraw-Hill

Heyman D, Sobel MJ (1984) *Stochastic Models in Operations Research* (vol. II). McGraw-Hill

Hill VL, Beall CW, Blischke WR (1998) A simulation model for warranty analysis. *International Journal of Production Economics* 16:463-491

Hollander M, Presnell B, Sethuraman J (1992) Nonparametric methods for imperfect repair models. *The Annals of Statistics* 20/2:879-887

Holmberg K, Folkeson A (eds) (1991) *Operational Reliability and Systematic Maintenance*. Elsevier Applied Science, London

Hopp WJ, Wu SC (1988) Multiaction maintenance under Markovian deterioration and incomplete state information. *Naval Research Logistics, An International Journal* 35/5:447-462

Hosseini MM, Kerr RM, Randall RB (2000) An inspection model with minimal and major maintenance for a system with deterioration and Poisson failures. *IEEE Trans. Reliability* 49/1:88-98

Huang JS, Okogbaa OG (1996) A heuristic replacement scheduling approach for multi-unit systems with economic dependency. *International Journal of reliability, Quality, and safety Engineering* 3/1:1-10

Hudes ES (1979) Availability theory for systems whose components are subjected to various shut-off rules. *Ph.D. dissertation*, Department of Statistics, University of California, Berkeley, USA

Hussain AZMO, Murthy DNP (1998) Warranty and redundancy design with uncertain quality. *IIE Transactions* 30:1191-1199

Ingle AD, Siewiorek DP (1977) Reliability models for multiprocessor systems with and without periodic maintenance. *7th Int. Symp. Fault-Tolerant Computing*, pp 3-9

Isaacson D, Reid S, Brennan J (1991) Warranty cost-risk analysis. *Proceedings of the Annual Reliability and Maintainability Symposium*, pp 332-339

Iskandar BP, Sandoh H (1999) An opportunity-based age replacement policy considering warranty. *International Journal of Reliability, Quality and Safety Engineering* 6:229-236

Iyer S (1992) Availability results for imperfect repair. *Sankhya: the Indian Journal of statistics* 54/2:249-259.

Ja S, Kulkarni V, Mitra A, Partaker G (2001) A renewable minimal-repair warranty policy with time-dependent costs. *IEEE Transactions on Reliability* 50/4:346-352

Ja S, Kulkarni V, Mitra A, Partaker G (2002) Warranty reserves for non-stationary sales processes. *Naval Research Logistics* 49/5:499-513

Jack N (1991) Repair replacement modeling over finite time horizons. *Journal of the Operational Research Society* 42/9:759-766

Jack N, Dagpunar JS (1994) An optimal imperfect maintenance policy over a warranty period. *Microelectronics and Reliability* 34:529-534

Jacobsen SE, Arunkumar S (1973) Investment in series and parallel systems to maximize expected life. *Management Science* 19:1023-1028

Jardine AKS, Buzacott JA (1985) Equipment reliability and maintenance. *European Journal of Operational Research* 19:285-296

Jayabalan V, Chaudhuri D (1992a) Optimal maintenance and replacement policy for a deteriorating system with increased mean downtime. *Naval Research Logistics* 39:67-78.

Jayabalan V, Chaudhuri D (1992b) Cost optimization of maintenance scheduling for a system with assured reliability. *IEEE Transactions on Reliability* R-41/1:21-26

Jayabalan V, Chaudhuri D (1992c) Sequential imperfect preventive maintenance policies: A case study. *Microelectronics and Reliability* 32/9:1223-1229

Jayabalan V, Chaudhuri D (1992d) Optimal maintenance - Replacement policy under imperfect maintenance. *Reliability Engineering & System Safety* 36/2:165-169.

Jayabalan V, Chaudhuri D (1992e) Heuristic approach for finite time maintenance policy. *International Journal of Production Economics* 27/3:251-256

Jayabalan V, Chaudhuri D (1995) Replacement policies: a near optimal algorithm. *IIE Transactions* 27:784-788

Jensen U (1995) Stochastic models of reliability and maintenance: an overview. In: Ozekici S (ed) *Reliability and maintenance of complex systems.* (Proceedings of the NATO Advanced Study Institute on Current Issues and Challenges in the Reliability and Maintenance of Complex Systems, held in Kemer-Antalya, Turkey, 12- 22 June 1995) Springer-Verlag, Berlin., pp 3-36

Jia X, Christer AH (2002) A periodic testing model for a preparedness system with a defective state. *IMA Journal of Management Mathematics* 13/1:39-49

Jiang X, Cheng K, Makis V (1998) On the optimality of repair-cost-limit policies. *Journal of Applied Probability* 35/4:936-949

Jin G, Chen L, Dong J (1993) Monte Carlo finite element method of structure reliability analysis. *Reliability Engineering & System Safety* 40/1:77-83

Johnson NI, Kotz S (1970) *Distributions in Statistics: Continuous Univariate Distributions*-1. Houghton Mifflin Co, Boston

Johnson R, Wichern D (2002) *Applied Multivariate Statistical Analysis, 5th edn.* Prentice Hall

Jorion P (2000) *Value-at-Risk: The New Benchmark for Managing Financial Risk.* McGraw-Hill

Kaio N, Osaki S (1982) Optimum repair limit policies with time constraint. *International Journal of Systems Science* 13:1345-1982

Kalbfleisch JD, Lawless JF, Robinson JA.(1991) Methods for the analysis and prediction of warranty claims. *Technometrics* 33:273-285

Kamat SJ, Riley MW (1975) Determination of reliability using event-based Monte Carlo simulation. *IEEE Transactions on reliability* R-24/1:73-75

Kamat SJ, Franzmeier WE (1976) Determination of reliability using event-based Monte Carlo simulation II. *IEEE Transactions on reliability* R-25/4:254-255

Kaminskiy M, Krivtsov V (2000) G-renewal process as a model for statistical warranty claim prediction. *Proceedings Annual Reliability and Maintainability Symposium*, pp 276-280

Kao EPC (1998) Computing the phase-type renewal and related functions. *Technometrics* 30/1:87-93

Kao EPC, Smith MS (1992) On excess, current and total life distributions of phase-type renewal processes. *Naval Research Logistics* 39:789-799

Kao EPC, Smith MS (1996) Computational approximations of renewal process relating to a warranty problem: the case of phase-type lifetimes. *European Journal of Operational Research* 90:156-170

Kapur PK, Garg RB, Butani NL (1989) Some replacement policies with minimal repairs and repair cost limit. *International Journal of Systems Science* 20/2:267-279

Karpinski J (1986) Multistate system under an inspection and repair. *IEEE Transactions on Reliability* R-35/1:76-77

Kay E (1976) The effectiveness of preventive maintenance. *Int. J. Prod. Res.* 14:329-344

Kececioglu D, Jiang S (1990) Error band estimation on Monte Carlo simulations. *Proceedings of the 36th Annual Technical Meeting of the Institute of Environmental Sciences*, New Orleans, LA, USA, 1990 Apr 23-27

Keller AZ (1987) Monte Carlo simulation in reliability. In: Colombo AG, Keller AZ (eds) *Proceedings of the ISPRA*. D.Reidel Publishing Co.

Khalil ZS (1985) Availability of series systems with various shut-off rules. *IEEE Transaction Reliability* 34:187-189

Kijimma M (1989) Some results for repairable systems with general repair. *Journal of Applied Probability* 26:89-102

Kijima M (1992) Replacement policies of a shock model with imperfect preventive maintenance. *European Journal of Operational Research* 57:100-110

Kijima M, Nakagawa T (1991) Accumulative damage shock model with imperfect preventive maintenance. *Naval research Logistics* 38:145-156

Kijima M, Nakagawa T (1992) Replacement policies of a shock model with imperfect preventive maintenance. *European Journal of Operations Research* 57:100-110

Kijima M, Morimura H, Suzuki Y (1988) Periodical replacement problem without assuming minimal repair. *European Journal of Operational Research* 37/2:194-203

Kim J, Fard N (1995) Discrete-event simulation of network reliability and Markovian models. *International Journal of Modeling and Simulation*

Kim C, Lee HK (1992) A Monte Carlo simulation algorithm for finding MTBF. *IEEE Transactions on Reliability* 41/2:193-195

Kim HG, Rao BM (2000) Expected warranty cost of two-attribute free-replacement warranties based on a bivariate exponential distribution. *Computers and Industrial Engineering* 38:425-434

Kirmani SNUA, Gupta RC (1989) On repair age and residual repair life in the minimal repair process. *Probability in the Engineering and Informational Science* 3:381-391

Kirmani SNUA, Gupta RC (1992) Some moment inequalities for the minimal repair process. *Probability in the Engineering and Informational Science* 6:245-255

Klein JP, Moeschberger ML (1997) *Survival Analysis: Techniques for Censored and Truncated Data*. Springer

Klutke GA, Yang YJ (2002) The availability of inspected systems subjected to shocks and graceful degradation. *IEEE Trans. on Reliability* 44:371-374

Kochar SC (1996) Some results on interarrival times of nonhomogeneous Poisson processes. *Probability in the Engineering and Informational Science* 10:75-85

Koshimae H, Tanaka H, Osaki S (1994) Some remarks on MTBF's for non-homogeneous Poisson processes. *IEICE Trans. on Fundamentals of Electronics, Communications and Computer sciences* E77-A/1:144-149

Koshimae H, Dohi T, Kaio N, Osaki S (1996) Graphical/statistical approach to repair limit replacement problem. *Journal of the Operations Research Society of Japan* 39:230-246

Kreps DM (1990) *A Course in Microeconomic Theory*. Princeton University Press, Princeton, NJ

Kulkarni VG (1995) *Modeling and Analysis of Stochastic Systems*. Chapman and Hall

Kumamoto H, Tanaka T, Inoue K (1977) Efficient evaluation of system reliability reliability by Monte Carlo method. *IEEE Trans. Reliability* R-26:311-315

Kumamoto H, Kazuo T, Koichi I, Henley EJ (1980a) State transition Monte Carlo for evaluating large, repairable systems. *IEEE Trans. Reliability* R-29:376-380

Kumamoto H, Tanaka T, Inoue K, Henley EJ (1980b) Dagger sampling Monte Carlo for system unavailability evaluation. *IEEE Trans. Reliability* R-29:122-125

Kumamoto H, Tanaka T, Inoue K (1987) A new Monte Carlo methods for evaluating system-failure probability. *IEEE Trans. Reliability* R-36:63-69

Kvam PH, Singh H, Whitaker LR (2002) Estimating distributions with increasing failure rate in an imperfect repair model. *Lifetime data analysis* 8/1: 53-67

Lam Y (1988) A note on the optimal replacement problem. *Adv. Appl. Prob.* 20:479-482

Lam Y (1996) Analysis of a two-component series system with a geometric process model. *Naval Research Logistics* 43:491-502

Lam Y, Lam PKW (2001) An extended warranty policy with options open to consumers. *European Journal of Operational Research* 131:514-529

Lam CT, Yeh RH (1994a) Optimal maintenance-policies for deteriorating systems under various maintenance strategies. *IEEE Trans. on Reliability* 43/3:423-430

Lam CT, Yeh RH (1994b) Optimal replacement policies for multistate deteriorating systems. *Naval Research Logistics* 41:303-315

Lannon RG (1972) A Monte Carlo technique for approximating system reliability confidence limits using the Weibull distribution. AD-743633, Available from US NTIS

Laprie JC, Costes A, Landrault C (1981) Parametric analysis of 2-unit redundant computer systems with corrective and preventive maintenance. *IEEE Trans. Reliability* R-30/2:139-144

Laviron A, Carnino A, Manaranche JC, (1982) ESCAF – a new and cheap system for complex reliability analysis and computation. IEEE Trans. Reliability R-3:339-341

Law AM, Kelton WD (2000) Simulation modeling and analysis. McGraw-Hill

Levy LL, Moore AH (1967) A Monte Carlo Technique for obtaining system reliability confidence limits from component test data. *IEEE Transaction on reliability* R-16:69-72

Lewis EE, Boehm F, Kirsch C, Kelkhoff BP (1989) Monte Carlo simulation of complex system mission reliability. *1989 Winter Simulation Conference Proceedings – WSC '89*, Washington, DC, USA, 1989 Dec 4-6

Li HJ, Shaked M (2003) Imperfect repair models with preventive maintenance. *Journal of Applied Probability* 40/4:1043-1059

Li W, Pham H (2005a) An inspection-maintenance model for systems with multiple competing processes. *IEEE Trans. on Reliability* 54/2:318-327

Li W, Pham H (2005b) Reliability modeling of multi-state degraded systems with multiple multi-competing failures and random shocks. *IEEE Trans. on Reliability* 54/2:297-303

Lie CH, Chun YH (1986) An algorithm for preventive maintenance policy. *IEEE Trans. Reliability* R-35/1:71-75

Lie CH, Hwang CL, Tillman FA (1977) Availability of maintained systems, A state-of-the-art survey. *AIIE Transactions* 9:247-259

Lim JH, Lu KL, Park DH (1998) Bayesian imperfect repair model. *Communications in Statistics-Theory and Methods* 27/ 4:965-984

Lin CT, Duran BS, Lewis TO (1988) Estimating lower confidence limits on system reliability using a Monte Carlo technique on binomial data. *Microelectronics and Reliability* 28/3:487-493

Lin JY, Donaghey CE (1993) Monte Carlo simulation to determine minimal cut sets and system reliability. *Proceedings of the 1993 Annual Reliability and Maintainability Symposium*, Atlanta, GA, USA.

Liu XG, MakisV, Jardine AKS (1995) A replacement model with overhauls and repairs. *Naval Research Logistics* 42:1063-1079

Locks MO (1974a) Monte Carlo Bayesian system reliability and MTBF - confidence assessment. AD-A057068, Available from US NTIS

Locks MO (1974b) Monte Carlo Bayesian system reliability - and MTBF - confidence assessment II. AD-A025820, Available from US NTIS

Locks MO (1979) Evaluating of the KTI Monte Carlo method for system reliability calculations. *IEEE Trans. Reliability* R-28:369-372

Love CE, Guo R (1993) An application of a bathtub failure model to imperfectly repaired systems data. *Quality and Reliability Engineering International* 9:127-134

Love CE, Guo R (1996) Utilizing Weibull failure rates in repair limit analysis for equipment replacement/preventive maintenance decisions. *Journal of the Operational Research Society* 47/11:1366-1376

Love CE, Rodger A, Blazenko G (1982) Repair limit policies for vehicle replacement. *INFOR* 20:226-236

Luby MG (1983) Monte-Carlo methods for estimating system reliability. Ph.D. Dissertation in computer science, University of California, Berkeley, USA

Lutton SC (1967) A Monte Carlo technique for approximating system reliability confidence limits from component failure data. MS thesis (GRE/MA/67-9), Air Force Institute of Technology, Wright-Patterson AFB, Ohio, USA

Lutz MA, Padmanabhan V (1998) Warranties, extended warranties, and product quality. *International Journal of Industrial Organization* 16:463-493

Lyu M (ed) (1996) *Handbook of software reliability engineering.* McGraw-Hill.

MacDonald MR (1982) A Monte Carlo technique suitable for obtaining complex space system reliability confidence limits from component test data with three unknown parameters. MS thesis, Air force Institute of Technology, Wright-Patterson AFB, Ohio, USA

Madansky A (1965) Approximate confidence limits for the reliability of series and parallel systems. *Technometrics* 7:495-506

Mahajan V, Muller E, Wind Y (2000) New product diffusion models. *Kluwer Academic*

Makis V, Jardine AKS (1991) Optimal replacement of a system with imperfect repair. *Microelectronics and Reliability* 31/2-3:381-388

Makis V, Jardine AKS (1992) Optimal replacement policy for a general model with imperfect repair. *Journal of the Operational Research Society* 43/2:111-120

Makis V, Jardine AKS (1993) A note on optimal replacement policy under general repair. *European Journal of Operational Research* 69:75-82

Malik MAK (1979) Reliable preventive maintenance policy. *AIIE Transactions* 11/3:221-228

Mamer JW (1969) Determination of warranty reserves. *Management Science* 15/10:542-549

Mamer JW (1982) Costs analysis of pro rata and free-replacement warranties. *Naval Research Logistics Quarterly* 29:345-356

Mamer JW (1987) Discounted and per unit costs of product warranty. *Management Science* 33/7:916-930

Mann NR, Grubbs FE (1974) Approximately optimum confidence bounds for system reliability based on component test data. *Technometrics* 16:335-347

Marcellus R, Projboot B (1996) *Design of Warranty Policies to Balance Consumer and Producer Risks and Benefit.* In: Blischke WR, Murthy DNP (eds) *Product Warranty Handbook.* Marcel Dekker, pp 483-510

Markowitz H (1959) *Portfolio Selection.* Yale University Press, USA

Marseguerra M, Zio E (1993) Nonlinear Monte Carlo reliability analysis with biasing towards top event. *Reliability Engineering & System Safety* 40/1:31-42

Marshall CW (1981) Design trade-offs in availability warranties. *Proceedings of the Annual Reliability and Maintenance Symposium*, pp 95-100

Martz HF, Waller R (1982) *Bayesian Reliability Analysis.* Wiley

Matteis AD, Pagnutti S (1995) Controlling correlations in parallel Monte Carlo. *Parallel Computing* 21/1:73-84

Mazumdar M (1975) Importance sampling in reliability estimation. *Reliability and Fault Tree Analysis.* SIAM, Philadelphia, pp153-163

McCall JJ (1963) Operating characteristics of opportunistic replacement and inspection policies. *Management Science* 10/1:85-97

McCall JJ (1965) Maintenance policies for stochastically failing equipment: A survey. *Management Science* 11/5:493-524

McGuire EP (1980) *Industrial Product Warranties: Policies and Practices.* The Conference Board Inc., New York

Mello JCO, Pereira MVF, Leite da Silva AM (1994) Evaluation of reliability worth in composite systems based on pseudo-sequential Monte Carlo simulation. *IEEE Transactions on Power Systems* 9/3:1318-1326

Melo ACG, Oliveira GC, Morozowski M, Pereira MVF (1991) Hybrid algorithm for Monte Carlo/enumeration based composite reliability evaluation. *Third International Conference on Probabilistic Methods Applied to Electric Power Systems*, London, 1991 Jul 3-5

Menezes M (1992) An approach for determination of warranty length. *International Journal of Research in Marketing* 9:177-195

Menipaz E (1978) Optimization of stochastic maintenance policies. *European Journal of Operational Research* 2/2:97-106

Menzefricke U (1992) On the variance of total warranty claims. *Comm. Statist. Theory and Methods* 21/3:779-790

Mi J (1996) Warranty and burn-in. *Naval Research Logistics* 44:199-210

Mi J (1999) Comparisons of renewable warranties. *Naval Research Logistics* 46:91-106

Misra PN (1983) Software reliability analysis. *IBM Systems Journal* 22:262-270

Mitra A, Patankar JG (1993) An integrated multicriteria model for warranty cost estimation and production. *IEEE Transactions on Engineering Management* EM-40/3:300-311

Monga A, Zuo MJ, Toogood R (1997) Reliability based design of systems considering preventive maintenance and minimal repair. *International Journal of Reliability, Quality and Safety Engineering* 4/1:55-71

Moore AH (1965) Extension of Monte Carlo technique for obtaining system reliability confidence limits from component test data. *Proc. National Aerospace Electronics Conf*, 1965 May, pp 459-463

Moore AH, Harter HL, Snead RC (1980) A comparison of Monte Carlo technique for obtaining system reliability confidence limits from component test data. *IEEE Trans Reliability* R-29:327-332

Moore AH, Hobbs JR, Hasaballa MSB (1985) A Monte Carlo method for determining confidence bounds on reliability and availability of maintained systems. *IEEE Trans. reliability* R-34:497-498

Morimura H (1970) On some preventive maintenance policies for IFR. *Journal of the Operations Research Society of Japan* 12/3:94-124

Morimura H, Makabe H (1963a) A new policy for preventive maintenance. *Journal of the Operations Research Society of Japan* 5:110-124

Morimura H, Makabe H (1963b) On some preventive maintenance policies. *Journal of the Operations Research Society of Japan* 6:17-43

Morse PM (1958) *Queues, Inventories, and Maintenance*. Wiley, New York

Moskowitz H, Chun YH (1994) A Poisson regression model for two-attribute warranty policies. *Naval Research Logistics* 41:355-376

Murthy DNP (1991a) A note on minimal repair. *IEEE Transactions on Reliability* 40/2:245-246

Murthy DNP (1991b) A usage dependent model for warranty costing. *European Journal of Operational Research* 57/1:89-99

Murthy DNP (1992a) Product warranty management - I: A review of mathematical models. *European Journal of Operational Research* 62:127-148

Murthy DNP (1992b) Product warranty management - II: A review of mathematical models. *European Journal of Operational Research* 62:261-281

Murthy DNP (1992c) Product warranty management - III: A review of mathematical models. *European Journal of Operational Research* 62:1-34

Murthy DNP, Asgharizadeh E (1999) Optimal decision making in a maintenance service operation. *European Journal of Operational Research* 116:259-273

Murthy DNP, Djamaludin I (2002) New product warranty: A literature review. *International Journal of Production Economics* 79:231-260

Murthy DNP, Hussain AZMO (1993) Warranty and optimal redundancy design. *Engineering Optimization* 23:301-314

Murthy DNP, Iskandar BP, Wilson RJ (1995) Two dimensional failure-free warranty policies: Two-dimensional point process models. *Operations Research* 43/2:356-366

Murthy DNP, Jack N (2003) *Warranty and Maintenance.* In: Pham H (ed) *Handbook of Reliability Engineering.* Springer, London

Murthy DNP, Nguyen DG (1981) Optimal age policy with imperfect preventive maintenance. *IEEE Trans. Reliability* R-30:80-81

Murthy DNP, Wilson RJ (1994) Parameter estimation in multi-component systems with failure interaction. *Applied Stochastic Models and Data Analysis* 10:47-60

Muth EJ (1977) An optimal decision rule for repair *vs* replacement. *IEEE Transactions on Reliability* R-26/3:179-181

Musa JD, Iannino A, Okumoto K (1987) *Software Reliability: Measurement, Prediction, Application.* McGraw-Hill, New York

Nakagawa T (1979a) Optimum policies when preventive maintenance is imperfect. *IEEE Transactions on Reliability* R-28/4:331-332

Nakagawa T (1979b) Imperfect preventive maintenance. *IEEE Transactions on Reliability* R-28/5:402

Nakagawa T (1980) A summary of imperfect maintenance policies with minimal repair. *RAIRO, Recherche Operationnelle* 14:249-255

Nakagawa T (1981) A summary of periodic replacement with minimal repair at failure. *Journal of the Operations Research Society of Japan* 24:213-228

Nakagawa T (1984a) Periodic inspection policy with preventive maintenance. *Naval Research Logistics Quarterly* 31:33-40

Nakagawa T (1984b) Optimal policy of continuous and discrete replacement with minimal repair at failure. *Naval Research Logistics Quarterly* 31/4, 543-550.

Nakagawa T (1985) Optimization problems in *k*-out-of-*n* systems. *IEEE Transactions on Reliability* R-34:248-250

Nakagawa T (1986) Periodic and sequential preventive maintenance policies. *Journal of Applied Probability* 23/2:536-542

Nakagawa T (1988) Sequential imperfect preventive maintenance policies. *IEEE Trans. Relia.* 37/3:295-298

Nakagawa T, Kowada M (1983) Analysis of a system with minimal repair and its application to replacement policy. *European Journal of Operational Research* 12:176-182

Nakagawa T, Murthy DNP (1993) Optimal replacement policies for a two-unit system with failure interactions. *RAIRO: Recherche Operationnelle* 27/4:427-438

Nakagawa T, Yasui,K. (1987) Optimum policies for a system with imperfect maintenance. *IEEE Trans. Reliability* R-36/5:631-633.

NAPS document No. 03476-A; 3 pages in this Supplement. Order NAPS document No. 03476, 19 pages. ASIS-NAPS; Microfiche Publications, P.O.Box 3513, Grand Central Station, New York, NY 10017, USA.

Natvig B (1990) On information-based minimal repair and the reduction in remaining system lifetime due to the failure of a specific module. *Journal of Applied Probability* 27/2:365-375

Nelson BL (1987) Variance reduction for the simulation practitioners. *Proc. 1987 Winter Simulation Conference*, Atlanta, pp 43-51

Nelson SC, Haire MJ, Schryver JC (1992) Network-simulation modeling of interactions between maintenance and process systems. *Proceedings of the 1992 Annual Reliability and Maintainability Symposium*, Las Vegas, NV, USA,1992 Jan 21-23, pp 150-156

Neuts MF (1978) Renewal processes of phase type. *Naval Logistics Quarterly* 25:445-454

Neuts M (1981) *Matrix-Geometric Solutions in Stochastic Process.* John Hopkins University press, USA

Nguyen DG, Murthy DNG (1981a) Optimal repair limit replacement policies with imperfect repair. *Journal of Operational Research Society* 32:409-416

Nguyen DG, Murthy DNG (1981b) Optimal maintenance policy with imperfect preventive maintenance. *IEEE Trans. Reliability* R-30/5:496-497.

Nguyen DG, Murthy DNG (1981c) Optimal preventive maintenance policies for repairable systems. *Operations Research* 39:1181-1194

Nguyen DG, Murthy DNP (1984a) A general model for estimating warranty costs for repairable products. *IIE Transactions* 16:379-386

Nguyen DG, Murthy DNP (1984b) Cost analysis of warranty policies. *Naval Research Logistics Quarterly* 31:525-541

Nguyen DG, Murthy DNP (1988) Optimal reliability allocation for products sold under warranty. *Engineering Optimization* 13:35-45

Nguyen DG, Murthy DNP (1989) Optimal replacement-repair strategy for servicing products sold under warranty. *European Journal of Operational Research* 39/2:206-212

Nicol DM, Palumbo DL (1995) Reliability analysis of complex models using SURE bounds. *IEEE Transactions on Reliability* 44/1:46-53

Nicola VF (1990) Fast simulation of dependability models with general failure, repair and maintenance processes. *Proceedings of the 20th International Symposium on Fault-tolerant Computing*, New Castle Upon Tyne, England, June 1990, pp 491-498

Ohi F (1989) Notes on imperfect repair, Keikaku sugaku to sono kanren bunya. In: *Mathematical programming and its related fields*. Proceedings of a symposium held at the Research Institute for Mathematical Sciences, Kyoto University, Kyoto, December 8-10, 1988, Kokyuroku No. 680, pp146-154

Ohnishi M, Kawai H, Mine H (1986) An optimal inspection and replacement policy for a deteriorating system. *Journal of Applied Probability* 23:973-988

Ohnishi M, Ibaraki T, Liu CG, Mine H (1987) Adaptive (*t*, *T*)-minimal repair and replacement policy when failure distribution includes unknown parameter. In: *Reliability theory and applications* (Shanghai, Xi'an, Beijing, 1987), World Sci. Publishing, Singapore, pp 304-313

Okuda S, Yonezawa M (1995) Structural reliability analysis based on importance sampling simulation defined in failure region. *Zairyo/Journal of the Society of Materials Science of Japan* 44/500:517-522

Okumoto K, Elsayed EA (1983) An optimum group maintenance policy. *Naval Research Logistics Quarterly* 30:667-674

Oliveira GC, Pereira MVF, Cunha SHF (1989) Technique for reducing computational effort in Monte-Carlo based composite reliability evaluation. *IEEE Transactions on Power Systems* 4/4:1309-1315

Osaki S, Nakagawa T (1976) Bibliography for reliability and availability of stochastic systems. *IEEE Transactions on Reliability* R-25:284-287

Osaki S, Yamada S, Hishitani J (1989) Availability theory for two-unit nonindependent series systems subject to shut-off rules. *Reliability Engineering & System Safety* 25/1:33-42

Ozekici S (1988) Optimal periodic replacement of multicomponent reliability systems. *Operations Research* 36/4:542-552

Ozekici S (ed)(1996) *Reliability and maintenance of complex systems*. (NATO ASI series, vol. 154) Springer-Verlag, Berlin

Padmanabhan V, Rao RC (1993) Warranty policy and extended service contracts: theory and an application to automobiles. *Marketing Science* 12/397-117

Pan ZJ, Tai YC (1988) Variance importance of system components by Monte Carlo. *IEEE Transactions on Reliability* 37/4:421-423

Pandey M, Uddin B, Ferdous J (1992) Reliability estimation of an s-out-of-k system with non-identical component strength: The Weibull case. *Reliability Engineering & System Safety* 36/2:109-116

Papadopoulos AS (1993) Hierarchical confidence bounds for the exponential failure model. *Microelectronics and Reliability* 33/5:719-727

Patankar JG, Mitra A (1995) Effect of warranty execution on warranty reserve costs. *Management Science* 4:395-400

Patankar JG, Worm GH (1981) Prediction intervals for warranty reserves and cash flows. *Management Science* 27:237-241

Patton AD, Blackstone JH, Balu NJ (1988) Monte Carlo simulation approach to the reliability modeling of generating systems recognizing operating considerations. *IEEE Transactions on Power Systems* 3/3:1174-1180

Pereira MVF, Maceira MEP, Oliveira GC, Pinto LMVG (1992) Combining analytical models and Monte Carlo techniques in probabilistic power system analysis. *IEEE Transactions on Power Systems* 7/1:265-272

Pham H (1992) Reliability analysis of a high voltage system with dependent failures and imperfect coverage. *Reliability Engineering and System Safety* 37/1:25-28.

Pham H (1996) A software cost model with imperfect debugging, random life cycle and penalty cost. *International Journal of Systems Science* 5:455-463

Pham H (2000) *Software Reliability*. Springer-Verlag, Singapore

Pham H (2002) A Vtub-shaped hazard rate function with applications to system safety. *International Journal of Reliability and Applications* 3/1:1-16

Pham H (2003a) Commentary: Steady-state series-system availability. *IEEE Trans on Reliability* 52/3:146-147

Pham H (2003b) Software reliability and cost models: perspectives, comparison and practice. *European Journal of Operational Research* 149: 475-489

Pham H (ed)(2003c) *Handbook of Reliability Engineering*. Springer-Verlag, London

Pham H, Wang HZ (1996) Imperfect maintenance. *European Journal of Operational Research* 94:425-438

Pham H, Wang HZ (2000) Optimal (τ, T) opportunistic maintenance of a k-out-of-n:G system with imperfect PM and partial failure. *Naval Research Logistics* 47:223-239

Pham H, Wang HZ (2001) A quasi-renewal process for software reliability and testing costs. *IEEE Transactions on Systems, Man and Cybernetic, Part A: Systems and Humans* 31:623-631

Pham H, Xie M (2002) A generalized surveillance model with applications to systems safety. *IEEE Trans. on Systems, Man and Cybernetics*, Part C 32:485-492

Pham H, Suprasad A, Misra RB (1996) Reliability and MTTF prediction of k-out-of-n complex systems with components subjected to multiple stages of degradation. *International Journal of Systems Science* 27/10:995-1000

Pham H, Suprasad A, Misra RB (1997) Availability and mean life time prediction of multi-stage degraded system with partial repairs. *Reliability Engineering and System Safety* 56:169-173

Phelps RI (1981) Replacement policies under minimal repair. *Operational Research Society Journal* 32/7:549-554

Pierskalla WP, Voelker JA (1976) A survey of maintenance models: the control and surveillance of deteriorating systems. *Naval Research Logistics Quarterly* 23:353-388

Pignal PI (1987) Analysis of a communication system with imperfect repair. *Microelectronics and Reliability* 27/1:165-169

Pijnenburg M, Ravichandran N, Regterschot G (1993) Stochastic analysis of a dependent parallel system. *European Journal of Operational Research* 68/1:90-104

Polatoglu H, Sahin I (1998) Probability distribution of cost, revenue and profit over a warranty cycle. *European Journal of Operational Research* 108:170-183

Popova E, Wilson JG (1999) Group replacement policies for parallel systems whose components have phase distributed failure times. *Annals of operations research* 91:163-190

Prasad MS, Rattihalli SR (1987) Optimum repair limit replacement policy when the lifetime depends on the number of repairs. *Opsearch, The Journal of the Operational Research Society of India* 24/3:155-162

Presnell B, Hollander M, Sethuraman J (1994) Testing the minimal repair assumption in an imperfect repair model. *Journal of the American Statistical Association* 89/425:289-297

Pulat S, Leemis L (1989) Network reliability and availability analysis to minimize downtime costs for communication networks. *Microelectronics and Reliability* 29/1:37-48

Purohit SG (1994) Testing for the minimal repair model versus additional damage at failures. *Communications in Statistics, Simulation and Computation* 23/1:89-107

Putz RB (1979) A Univariate Monte Carlo technique to approximate reliability confidence limits of systems with components characterized by the Weibull distribution. MS thesis (GOR/MA/79D-7), Air Force Institute of Technology. Available from US NTIS

Ramachandran V, Sankaranarayanan V (1990) Dynamic redundancy allocation using Monte-Carlo optimization. *Microelectronics and Reliability* 30/6:1131-1136

Rander R, Jorgenson DW (1963) Opportunistic replacement of a single part in the presence of several monitored parts. *Management Science* 10/1:70-83

Rander MC, Kumar A, Tuteja RK (1993) Analysis of a two unit cold standby system with imperfect assistant repairman, perfect master repairman and inspection after repair by assistant repairman. *Journal of the Indian Association for Productivity, Quality and Reliability* 18/1:41-53

Rangan A (1994) Time for failure free tests of systems subject to shocks and imperfect repair. *Opsearch* 31/3:228-236

Rangan A, Grace RE (1989) Optimal replacement policies for a deteriorating system with imperfect maintenance. *Advances in Applied Probability* 21/4:949-951

Rao BM (1995) Algorithms for the free replacement warranty with phase-type lifetime distributions. *IIE Transactions* 27:348-357

Rardin RL (1998) *Optimization in Operations Research*. Prentice Hall.

Resende LIP (1988) Computing network reliability using exact and Monte-Carlo method. Ph.D. Dissertation, Dept. of Industrial Engineering and Operations Research, University of California, Berkeley, USA

Rice RE, Moore AH (1983) A Monte Carlo technique for estimating lower confidence limits on system reliability using pass-fail data. *IEEE Trans Reliability* R-32:366-369

Righter R (1996) Optimal policies for scheduling repairs and allocating heterogeneous servers. *Journal of Applied Probability* 33/2:536-547

Ritchken PH (1985a) Optimal replacement policies for irreparable warranted item. *IEEE Transactions on Reliability,*35/5:621-624

Ritchken PH (1985b) Warranty policies for non-repairable items under risk aversion. *IEEE Transactions on Reliability* 34/2:147-150

Ritchken PH, Tapiero CS (1986) Warranty design under buyer and seller risk aversion. *Naval Research Logistics Quarterly* 33:657-671

Ritchken PH, Wilson JG (1990) (*m, T*) group maintenance policies. *Management Science* 36/5:632-639

Roberts WT Jr, Mann L Jr (1993) Failure predictions in repairable multi-component systems. *International Journal of Production Economics* 29/1:103-110

Rodrigues DJ (1990) Some approximate inspection policies for a system with imperfect inspections. *RAIRO, Recherche Operationnelle* 24/2:191-199

Rolski T, Schmidli H, Schmidt V, Teugels J (1999) *Stochastic Processes for Insurance and Finance*. John Wiley and Sons, Chichester

Romeu JL (1989) Small sample Monte Carlo study for four system reliability bounds. *Computers & Industrial Engineering* 16/1:117-126

Ross SM (1970) *Applied Probability Models with Optimization Applications*. Holden-Day

Ross SM (1983) *Stochastic Processes*. John Wiley and Sons

Rubinstein RY (1981) *Simulation and the Monte Carlo Method*. . John Wiley and Sons

Rustagi JS (1994) *Optimization Techniques in Statistics*. Academic Press

Sahin I (1993) Conformance quality and replacement costs under warranty. *Production and Operational Management* 2:242-261

Sahin I, Polatoglu H (1995) Distributions of manufacturer's and user's replacement costs under warranty. *Naval Research Logistics* 42:1233-1250

Sahin I, Polatoglu H (1996) Maintenance strategies following the expiration of warranty. *IEEE Transactions on Reliability* 45/2:221-228

Sahin I, Polatoglu H (1998) *Quality, Warranty and Preventive Maintenance.* Kluwer Academic Publishers, Boston

Sandve K, Aven T (1999) Cost optimal replacement of monotone, repairable systems. *European Journal of Operational Research* 116:235-248

Savits TH (1988) Some multivariate distributions derived from a nonfatal shock model. *Journal of Applied Probability* 25/2:383-390

Scarf PA (1997) On the application of mathematical models in maintenance. *European Journal of Operational Research* 99/4:493-506

Schneeweiss WG (2005) Toward a Deeper Understanding of the Availability of Series-Systems Without Aging During Repairs. *IEEE Transactions on Reliability 54/1:98-101*

Sengupta B (1980) Maintenance policies under imperfect information. *European Journal of Operational Research* 5/3:198-204

Shaked M, Shanthikumar JG (1986) Multivariate imperfect repair. *Operations Research* 34:437-448

Shaked M, Shanthikumar JG (1988) Multivariate conditional hazard rates and the MIFRA and MIFR properties. *Journal of Applied Probability* 25/1:150-168

Sherif YS, Smith ML (1981) Optimal maintenance models for systems subject to failure - A review. *Naval Research Logistics Quarterly* 28/1:47-74

Sheu SH (1991) A general age replacement model with minimal repair and general random repair cost. *Microelectronics and Reliability* 31/5:1009-1017

Sheu SH (1999) Extended optimal replacement model for deteriorating systems. *European Journal of Operational Research* 112(3):503-516

Sheu SH (2005) Optimal policies with decreasing probability of imperfect maintenance. *IEEE Transactions on Reliability* 54/2:347-357

Sheu SH, Griffith WS (1992) Multivariate imperfect repair. *Journal of Applied Probability* 29/4:947-956

Sheu SH, Jhang J (1997) A generalized group maintenance policy. *European Journal of Operational Research* 96(2):232-247

Sheu SH, Kuo CM, Nakagawa T (1993) Extended optimal age replacement policy with minimal repair. *RAIRO Recherche Operationnelle* 27/3:337-351

Sheu SH, Griffith WS, Nakagawa T (1995) Extended optimal replacement model with random minimal repair costs. *European Journal of Operational Research* 85:636-649

Shiraki W (1989) Extension of iterative fast Monte-Carlo (IFM) procedure and its applications to time-variant structural reliability analysis. *Proceedings of*

ICOSSAR '89, the 5th International Conference on Structural Safety and Reliability, Part II, San Francisco, CA, USA, 7-11 Aug 1989

Shreider IA (ed) (1960) *The Monte Carlo Method: the Method of Statistical Trials*. Pergamon Press, Oxford

Sim SH, Endrenyi J (1993) A failure-repair model with minimal and major maintenance. *IEEE Transactions on Reliability* 42/1:134-139

Singpurwalla ND, Wilson S (1993) The warranty problem: Its statistical and game theoretic aspects. *SIAM Review* 35/1:17-42

Smith MAJ, Dekker R (1997) Preventive maintenance in a 1 out of *n* system: the uptime, downtime and costs. *European Journal of Operational Research* 99/3:565-583

Soboll IM (1974) *The Monte Carlo Method*. University of Chicago Press, Chicago

Srivastava MS, Wu Y (1993) Estimation & testing in an imperfect-inspection model. *IEEE Transactions on Reliability* 42/2:280-286

Stadje W, Zuckerman D (1990) Optimal strategies for some repair replacement models. *Adv. in Appl. Probab.* 22/3, 641-656

Stadje W, Zuckerman D (1996) Generalized maintenance model for stochastically deteriorating equipment. *European Journal of Operational Research* 89/2:285-301

Stadje W, Zuckerman D (1999) Optimal surveillance of a failure system. *Annals of Operations Research* 91:281-288

Su CT, Wu T, Lee T, Huwang C (1986) Capacity planning with flow and reliability evaluation using Monte Carlo simulation. *IEEE Trans. Reliability* R-35:519-522

Subramanian R, Natarajan R (1990) Two-unit redundant system with different types of failure and 'imperfect' repair'. *Microelectronics and Reliability* 30/4:697-699

Sumita U, Shanthikumar JG (1988) An age-dependent counting process generated from a renewal process. *Advances in Applied Probability* 20/4:739-755

Sundt B (2002) Recursive evaluation of aggregate claims distributions. *Insurance: Mathematics and Economics* 30:297-322

Sundt B, Jewell WS (1981) Further results on recursive evaluation of compound distributions. *ASTIN Bulletin* 12:27-39

Suresh PV, Chaudhuri D (1994) Preventive maintenance scheduling for a system with assured reliability using fuzzy set theory. *International Journal of Reliability, Quality and Safety Engineering* 1/4:497-513

Tahara A, Nishida T (1975) Optimal replacement policy for minimal repair model. *Journal of Operations Research Society of Japan* 18/3-4:113-124

Tanaka T, Kumamoto H, Inoue K (1989a) Evaluation of a dynamic reliability problem based on order of component failure. *IEEE Transactions on Reliability* 38/5:573-576

Tanaka T, Kumamoto H, Inoue K (1989b) Monte Carlo evaluation of a dynamic reliability problem with an application to a case of partial cuts. *Annual Reliability and Maintainability Symposium - 1989 Proceedings*, Atlanta, GA, USA,1989 Jan 24-26 (Available from IEEE Service Cent (cat n 89CH2580-9), Piscataway, NJ, USA), pp 108-113

Tango T (1978) Extended block replacement policy with used items. *Journal of Applied Probability* 15:560-572

Tatsuno K, Ohi F, Nishida T (1983) Opportunistic maintenance policy with minimal repair. *Mathematica Japonica* 28/3:327-335

Telcordia, GR-1339-CORE, Generic Reliability Requirements for Digital Cross Connect Systems, Issue Number 01, 1997. http://www.telcordia.com/

Thomas MU (1983a) A prediction model of manufacturer warranty reserves. *Management Science* 35/12:1515-1519

Thomas MU (1983b) Optimum warranty policies for nonreparable items. *IEEE Transactions on Reliability* 32/3:283-288

Thomas MU, Rao SS (1999) Warranty economic decision models: A summary and some suggested directions for future research. *Operation Research* 47/6:807-820

Tian J, Lu P, Palma J (1995) Test-execution-based reliability measurement and modeling for large commercial software. *IEEE Trans. Software Engineering* 21/5: 405-414

Tilquin C, Cleroux R (1975a) Periodic replacement with minimal repair at failure and adjustment costs. *Naval Res. Logist. Quart.* 22/2:243-254

Tilquin C, Cleroux R (1975b) Periodic replacement with minimal repair at failure and general cost function. *J. Statist. Comput. and Simulation* 4/1:63-77

Uematsu K, Nishida T (1987a) One-unit system with a failure rate depending upon the degree of repair. *Mathematica Japonica* 32/1:139-147

Uematsu K, Nishida T (1987b) Branching nonhomogeneous Poisson process and its application to a replacement model. *Microelectronics and Reliability* 27/4:685-691

Valdez-Flores C, Feldman RM (1989) A survey of preventive maintenance models for stochastically deteriorating single-unit systems. *Naval Research Logistics* 36:419-446

Van Der Duyn Schouten F (1995) Maintenance policies for multicomponent systems. In: Ozekici S (ed) *Reliability and maintenance of complex systems.* (NATO ASI series, vol. 154) Springer-Verlag, Berlin, pp 117-136

van-Pul MCJ (1993) *Statistical analysis of software reliability models.* Stichting Mathematisch Centrum, Centrum voor Wiskunde en Informatica, Amsterdam

Venkatakrishnan KS, Venmathi S (1989) Optimal replacement time of an equipment via simulation for truncated failure distributions. *Microelectronics and Reliability* 29/1:49-52

Vergin RC, Scriabin M (1977) Maintenance scheduling for multi-component equipment. AIIE Transactions 9:297-305

Wang C, Sheu S (2005) Optimal lot sizing for products sold under free-repair warranty. *European Journal of Operational Research*

Wang HZ (1997) *Reliability and maintenance modeling for systems with imperfect maintenance and dependence.* PhD Dissertation, Rutgers University, USA

Wang HZ (2002) A survey of maintenance policies of deteriorating systems. *European Journal of Operational Research* 139:469-489

Wang HZ (co-inventors: Kher S, Choudhury N, Rubin H, Franklin P, Remick P, Scarff P, Chien YC) (2004) Method and apparatus for warranty cost calculation. U.S. Patent (pending).

Wang HZ, Pham H (1996a) Optimal age-dependent preventive maintenance policies with imperfect maintenance. *International Journal of Reliability, Quality and Safety Engineering* 3/2:119-135

Wang HZ, Pham H (1996b) A quasi renewal process and its application in the imperfect maintenance. *International Journal of Systems Science* 27/10:1055-1062 and 28/12:1329

Wang HZ, Pham H (1996c) Optimal maintenance policies for several imperfect maintenance models. *International Journal of Systems Science* 27/6:543-549

Wang HZ, Pham H (1996d) Some new software reliability and cost models. *Conference on Performability in Computing Systems*, East Brunswick, New Jersey, 25-26 July 1996

Wang HZ, Pham H (1996e) Estimation Methods for Acceleration Factors, *International Journal of Modelling & Simulation*16/3:166-172.

Wang HZ, Pham H (1997a) Availability and optimal maintenance of series system subject to imperfect repair. *International J. of Plant Engineering and Management* 2

Wang HZ, Pham H (1997b) Optimal opportunistic maintenance of a k-out-of-n:G System. *International Journal of Reliability, Quality and Safety Engineering* 4/4:369-386

Wang HZ, Pham H (1997c) Survey of reliability, availability and MTTF evaluations of complex networks using Monte Carlo techniques. *Microelectronics and Reliability* 37/2:187-209

Wang HZ, Pham H (1999) Some maintenance models and availability with imperfect maintenance in production systems. *Annals of Operations Research* 91:305-318.

Wang HZ, Pham H (2003) Optimal imperfect maintenance models. In: Pham (ed) *Reliability Engineering Handbook*. Springer-Verlag, London

Wang HZ, Pham H (2006) Availability and maintenance of series systems subject to imperfect repair and correlated failure and repair. *European Journal of Operational Research*

Wang HZ, Pham H, Izundu AE (2001) Optimal preparedness maintenance of multi-unit systems with imperfect maintenance and economic dependence. In: Pham H (ed) *Recent Advances in Reliability and Quality Engineering.* World Scientific, New Jersey, pp 75-92

Whitaker LP, Samaniego FJ (1989) Estimating the reliability of systems subject to imperfect repair. *Journal of American statistical Association* 84/405:301-309

Wijnmalen DJD, Hontelez JAM (1997) Coordinated condition-based repair strategies for components of a multi-component maintenance system with discounts. *European Journal of Operational Research* 98/1:52-63

Wildeman RE, Dekker R, Smit ACJM (1997) A dynamic policy for grouping maintenance activities. *European Journal of Operational Research* 99:530-551

Wilson JR (1983) Variance Reduction: the Current State, Mathematics and Computers in Simulation XXV. North Holland Publishing Co.

Wu S, Clements-Croome D (2005) Optimal Maintenance Policies Under Different Operational Schedules. *IEEE Transactions on Reliability* 54/2:338-346

Wu YF, Lewins JD (1991) System reliability perturbation studies by a Monte Carlo method. *Annals of Nuclear Energy* 18/3:141-146

Wu YF, Lewins JD (1992) Monte Carlo studies of engineering system reliability. *Annals of Nuclear Energy* 19/10-12:825-859

Xie M (1991) *Software reliability modeling.* World Scientific, UK.

Xue J, Yang K (1995) Dynamic reliability analysis of coherent multistate systems. *IEEE Trans. Reliability* 44/4:683-688

Yak YW, Dillon TS, Forward KE (1984) The effect of imperfect periodic maintenance on fault tolerant computer systems. *14th Int. Symp. Fault-Tolerant Computing*, pp 67-70

Yang SC, Lin TW (2005) On the application of quasi renewal theory in optimization of imperfect maintenance policies. *Proceedings of 2005 Annual Reliability and Maintainability Symposium*, pp410-415

Yasui K, Nakagawa T, Osaki S (1988) A summary of optimal replacement policies for a parallel redundant system. *Microelectronics and Reliability* 28:635-641

Yeh RH, Ho WT (2000) Optimal preventive-maintenance warranty policy for repairable products. *European Journal of Operational Research* 134:59-69

Yeh RH, Ho WT, Tseng ST (2000) Optimal production run length for products sold with warranty. *European Journal of Operational Research* 120:575-582

Young HC, Chang SL (1992) Optimal replacement policy for a warranted system with imperfect preventive maintenance operations. *Microelectronics and Reliability* 32/6:839-843

Yue D, Cao JH (2001) Some results on successive failure times of a system with minimal instantaneous repairs. *Operations Research Letters* 29:193-197

Yun WY (1989) An age replacement policy with increasing minimal repair cost. *Microelectronics and Reliability* 29:153-157

Yun WY (1997) Expected value and variance of warranty cost of repairable product with two types of warranty. *The international Journal of Quality and Reliability Management* 14/7:661-668

Yun WY, Bai DS (1987) Cost limit replacement policy under imperfect repair. *Reliability Engineering* 19/1:23-28

Yun WY, Bai DS (1988) Repair cost limit replacement policy under imperfect inspection. *Reliability Engineering & System Safety* 23/1:59-64

Zelen M, Severo NC (1964) Probability functions. In: M. Abramowitz M, Stegun IA (eds) *Handbook of Mathematical Functions*. Applied Mathematics Series 55, U.S. Department of Commerce, pp 925-995

Zhao M (1994) Availability for repairable components and series system. *IEEE Trans. on Reliability* 43/2:329-334

Zheng X (1995) All opportunity-triggered replacement policy for multiple-unit system. *IEEE Transactions on Reliability* 44/4:648-652

Zio E (1995) Biasing the transition probabilities in direct Monte Carlo. *Reliability Engineering & System Safety* 47/1:59-63

Zio E, Cammi A., Cioncolini A (2004) Dagger-sampling variance reduction in Monte Carlo reliability analysis. *Monte Carlo Methods and Applications* 10/3-4: 641-52

Zuo MJ, Murthy DNP (2000) Replacement-repair policy for multi-state deteriorating products under warranty. *European Journal of Operational Research* 123:519-530

Zuo MJ, Jiang R, Yam RCM (1999) Approaches for reliability modeling of continuous-state devices. *IEEE Trans. Reliability* 48/1:9-18

.

Index

Printed in the United Kingdom
by Lightning Source UK Ltd.
134677UK00002B/41/A